T0293323

Integral Inequalities and Generalized Convexity

The book covers several new research findings in the area of generalized convexity and integral inequalities. Integral inequalities using various type of generalized convex functions are applicable in many branches of mathematics such as mathematical analysis, fractional calculus and discrete fractional calculus.

The book contains integral inequalities of Hermite–Hadamard type, Hermite–Hadamard–Fejér type and majorization type for the generalized strongly convex functions. It presents Hermite–Hadamard type inequalities for functions defined on Time scales. Further, it provides the generalization and extensions of the concept of preinvexity for interval-valued functions and stochastic processes and gives Hermite–Hadamard type and Ostrowski type inequalities for these functions. These integral inequalities are utilized in numerous areas for the boundedness of generalized convex functions.

Features:
- Covers Interval-valued calculus, Time scale calculus, Stochastic processes – all in one single book.
- Numerous examples to validate results.
- Provides an overview of the current state of integral inequalities and convexity for a much wider audience, including practitioners.
- Applications of some special means of real numbers are also discussed.

The book is ideal for anyone teaching or attending courses in integral inequalities along with researchers in this area.

Integral Inequalities and Generalized Convexity

Shashi Kant Mishra
Nidhi Sharma
Jaya Bisht

CRC Press
Taylor & Francis Group
Boca Raton London New York

CRC Press is an imprint of the
Taylor & Francis Group, an **informa** business

A CHAPMAN & HALL BOOK

First edition published 2024
by CRC Press
6000 Broken Sound Parkway NW, Suite 300, Boca Raton, FL 33487-2742

and by CRC Press
4 Park Square, Milton Park, Abingdon, Oxon, OX14 4RN

CRC Press is an imprint of Taylor & Francis Group, LLC

© 2024 Shashi Kant Mishra, Nidhi Sharma, Jaya Bisht

Reasonable efforts have been made to publish reliable data and information, but the author and publisher cannot assume responsibility for the validity of all materials or the consequences of their use. The authors and publishers have attempted to trace the copyright holders of all material reproduced in this publication and apologize to copyright holders if permission to publish in this form has not been obtained. If any copyright material has not been acknowledged please write and let us know so we may rectify in any future reprint.

Except as permitted under U.S. Copyright Law, no part of this book may be reprinted, reproduced, transmitted, or utilized in any form by any electronic, mechanical, or other means, now known or hereafter invented, including photocopying, microfilming, and recording, or in any information storage or retrieval system, without written permission from the publishers.

For permission to photocopy or use material electronically from this work, access www.copyright. com or contact the Copyright Clearance Center, Inc. (CCC), 222 Rosewood Drive, Danvers, MA 01923, 978-750-8400. For works that are not available on CCC please contact mpkbookspermissions@tandf.co.uk

Trademark notice: Product or corporate names may be trademarks or registered trademarks and are used only for identification and explanation without intent to infringe.

ISBN: 978-1-032-52632-4 (hbk)
ISBN: 978-1-032-52765-9 (pbk)
ISBN: 978-1-003-40828-4 (ebk)
ISBN: 978-1-032-52767-3 (eBook+)

DOI: 10.1201/9781003408284

Typeset in CMR10 font
by KnowledgeWorks Global Ltd.

Publisher's note: This book has been prepared from camera-ready copy provided by the authors.

Contents

Foreword

Integral inequalities constitute a very important topic in mathematics, in view of their applications in the calculus of variations, differential geometry, mathematical analysis, number theory, differential equations, probability theory and statistics, among other fields. On the other hand, generalized convexity is a set of concepts and techniques that provide useful tools in many branches of applied sciences, including economics, engineering, finance, mechanics, variational analysis and, very specially, optimization theory. Even though some articles have explored connections between these two topics, the book "Integral Inequalities and Generalized Convexity", by Shashi Kant Mishra, Nidhi Sharma and Jaya Bisht, is the first monograph devoted to a systematic exploration of the relationship existing between them. It shows the usefulness of generalized convexity in obtaining both classical and new integral inequalities. The fundamental generalized convexity notion considered in the book is that of η-invexity, a very natural generalization of convexity that consists in replacing $x-y$ in the classical sub-gradient inequality with a suitable function $\eta(x,y)$. The relevance of this notion stems from the fact that a differentiable function is η-invex for some η if and only if every stationary point is a global minimum. This book makes an extensive use of this notion as well as of the many generalizations thereof that have been proposed in the literature with the aim of treating non-convex problems with convexity type techniques.

One interesting feature of "Integral Inequalities and Generalized Convexity" is its use of fractional calculus, a branch of differential calculus that has been widely used in applied mathematics and has given rise to the field of fractional differential equations, an extension of the theory of classical differential equations.

Another distinctive feature of "Integral Inequalities and Generalized Convexity" is the consideration of integral inclusions for interval-valued functions. Interval analysis provides a suitable framework for dealing with rounding errors. Since such errors are inevitably present in practical computations, they necessarily have to be considered in every application of integral inequalities to real-world problems, hence the importance of integral inclusions for interval-valued functions.

The last chapter of the book provides an excellent motivation to its contents, as it describes several important applications, including an interesting mass transportation interpretation of the Hermite–Hadamard inequality. It contains the most recent results on the applications of generalized convexity to integral inequalities.

The book "Integral Inequalities and Generalized Convexity" is a valuable addition to the existing literature on integral inequalities. It will be useful for graduate students in mathematics and its applications will undoubtedly become a helpful reference for researchers working in this field.

Juan Enrique Martínez-Legaz
Emeritus Professor
Universitat Autònoma de Barcelona, Spain

Author Biographies

Shashi Kant Mishra PhD, DSc is a Professor at the Department of Mathematics, Institute of Science, Banaras Hindu University, Varanasi, India, with over 24 years of teaching experience. He has authored eight books, including textbooks and monographs, and has been on the editorial boards of several important international journals. He has guest-edited special issues of the *Journal of Global Optimization; Optimization Letters* (both Springer Nature) and *Optimization* (Taylor & Francis). He  has received INSA Teacher Award 2020 from Indian National Science Academy, New Delhi and DST Fast Track Fellow 2001 from Ministry of Science and Technology, Government of India. Prof. Mishra has published over 203 research articles in reputed international journals and supervised 21 PhD students. He has visited around 15 institutes/ universities in countries such as France, Canada, Italy, Spain, Japan, Taiwan, China, Singapore, Vietnam and Kuwait. His current research interest includes mathematical programming with equilibrium, vanishing and switching constraints, invexity, multiobjective optimization, non-linear programming, linear programming, variational inequalities, generalized convexity, integral inequalities, global optimization, non-smooth analysis, convex optimization, non-linear optimization and numerical optimization.

Nidhi Sharma is a Fellow of the Council of Scientific Industrial Research (CSIR) at the Department of Mathematics, Institute of Science, Varanasi, India. She received MSc degree in Mathematics from Banaras Hindu University, Varanasi, India. She is working on generalized convexity and integral inequalities under the supervision of Prof. S. K. Mishra. Her current research interests include integral inequalities, generalized convexities, set-valued functions and stochastic processes.

Jaya Bisht is a DST-INSPIRE Fellow (Senior Re-
search Fellow) at the Department of Mathemat-
ics, Institute of Science, Banaras Hindu University,
Varanasi, India. She received BSc and MSc degrees
in Mathematics from Hemwati Nandan Bahuguna
Garhwal University (Central University), Srinagar,
India. She is awarded Junior Research Fellowship
(JRF) from Human Resource Development Group
of Council of Scientific Industrial Research (HRD-
CSIR), Government of India. Her current research
interest includes integral inequalities, generalized
convexities, mathematical analysis, interval-valued functions and convex
stochastic processes.

Preface

The generalization of convex functions is considered as an original icon in the theoretical study of mathematical inequalities. Integral inequalities involving generalized convexity play an important role in many branches of mathematics and have recently drawn the attention of a large number of researchers interested in both theory and applications. The subject has received tremendous impetus from outside of mathematics from such diverse fields as mathematical economics, game theory, mathematical programming, control theory, variational methods, information theory, probability theory and statistics. It is recognized that some specific inequalities, such as Hermite–Hadamard inequality, Jensen inequality, Ostrowski inequality and Majorization inequality provide a useful and important device in the development of various branches of mathematics. Because of the abundance of applications, the theory of these inequalities is rapidly developing, and it is currently one of the most rapidly developing areas of mathematics. Researchers in various branches of mathematics have discovered generalizations, extensions, refinements, improvements, discretizations, and new applications of these inequalities using generalized convexity over the years.

These developments have inspired the authors to write a monograph devoted to the most recent results in mathematics related to these most important integral inequalities. The monograph covers several new research findings in the area of generalized convexity and integral inequalities. The material of the book concentrated on generalization and extension of the classical convexity in different directions and integral inequalities for these generalized and extended convexity with basic knowledge of calculus, measure theory and real analysis. The book will be useful for graduates and researchers who are working in the field of generalized convexity and integral inequalities. Furthermore, the book can serve as a valuable reference text for anyone including experts working in this research field.

The book is organized as follows: it has eight chapters; Chapter 1 is introductory and contains basic definitions and concepts needed in the book.

Chapter 2 presents integral inequalities for strongly generalized convex functions. We establish some Hermite–Hadamard and Fejér-type inequalities for strongly η-convex functions. We discuss some applications to special means of real numbers with the help of these results. Further, we establish some new weighted Hermite–Hadamard inequalities for strongly GA-convex functions by using geometric symmetry of a continuous positive mapping and a differentiable mapping whose derivatives in absolute value are strongly GA-convex. Furthermore, we introduce the notion of a strongly convex function on time scales and derive some new dynamic inequalities for these strongly convex functions.

Chapter 3 contains integral inequalities for strongly generalized convex functions of higher order. We introduce the concept of strongly η-convex functions of higher order and investigate the Hermite–Hadamard and Hermite–Hadamard-Fejér type inequalities for these functions. Further, we derive Hermite–Hadamard and related integral inequalities for higher order strongly exponentially convex functions. We also discuss Reimann-Liouville fractional estimates via strongly exponentially convex functions of higher order.

Chapter 4 contains integral inequalities for generalized preinvex functions. We prove a new form of Hermite–Hadamard inequality using left and right-sided ψ-Riemann–Liouville fractional integrals for preinvexity. We present two essential results of ψ-Riemann–Liouville fractional integral identities using the first-order derivative of a preinvex function. Further, we propose the concept of generalized (m, h)-preunivex functions and establish some new bounds on Hermite–Hadamard and Simpson's inequalities for mappings whose absolute values of second derivatives are generalized (m, h)-preunivex.

Chapter 5 deals with majorization integral inequalities for strongly convex functions defined on rectangles. We extend several integral majorization type and generalized Favard's inequalities from functions defined on intervals to functions defined on rectangles via strong convexity and apply the results to establish some new integral inequalities for functions defined on rectangles.

Chapter 6 presents Hermite–Hadamard type inclusions for generalized preinvex interval-valued functions. We introduce the concept of (h_1, h_2)-preinvex interval-valued functions, coordinated preinvex interval-valued functions, and harmonically h-preinvex interval-valued functions. Further, we establish inclusions of Hermite–Hadamard type for preinvex and coordinated preinvex interval-valued functions. Furthermore, We prove Hermite–Hadamard type inclusions for harmonically h-preinvex interval-valued functions by using interval-valued Riemann–Liouville fractional integrals.

Chapter 7 presents integral inequalities for generalized convex stochastic processes. We define general h−harmonic preinvex stochastic processes and the multidimensional general h−harmonic preinvex stochastic processes. We prove the Hermite–Hadamard inequality and obtain some important results for these stochastic processes. Further, we introduce the concept of strongly η-convex stochastic processes and obtain the Hermite–Hadamard inequality, Ostrowski inequality and some other interesting inequalities for strongly η-convex stochastic processes.

Chapter 8 contains several applications of Hermite–Hadamard inequality, Jensen's inequality, time-scale analysis and Ostrowski integral inequality for interval-valued functions.

The list of applications related to integral inequalities and generalized convexity is nearly endless, and we are confident that many new and beautiful applications will be developed in the future. A detailed and comprehensive account of typical applications may be found in the various references given at the end.

An undertaking of this kind cannot be completed without some mistakes. Although we have made every effort to ensure that there are no inaccuracies, since we provide the exact references, we sincerely hope that the reader will forgive the occasional mistake and feel free to study the original reference. The authors are thankful to Ms Isha Singh from CRC Press for her patience, support and effort in handling the book.

Symbol Description

$\mathbb{N}$	set of all natural numbers	$\mathbb{T}$	Time scale
$\mathbb{N}_0$	set of all natural numbers included zero	$[c,d]_{\mathbb{T}}$	Time-scaled interval
		$A(c,d)$	Arithmetic mean
$\mathbb{R}$	set of all real numbers	$G(c,d)$	Geometric mean
$\mathbb{R}\backslash\{0\}$	set of all real numbers excluded zero	$H(c,d)$	Harmonic mean
		$L_n(c,d)$	Generalized logarithmic mean
$\overline{\mathbb{R}}$	extended real line		
$\mathbb{R}_+$	set of all nonnegative real numbers	$I(c,d)$	Identric mean
		$H_{w,m}(c,d)$	Heronian mean
$\mathbb{R}_-$	set of all nonpositive real numbers	$\mathbb{R}_{\mathbb{I}}$	set of all closed intervals of $\mathbb{R}$
$\mathbb{R}^n$	Euclidean n-space	$\mathbb{R}_I^+$	set of all positive closed intervals of $\mathbb{R}$
(c,d)	open line segment joining c and d		
$[c,d]$	closed line segment joining c and d	$\mathbb{R}_I^-$	set of all negative closed intervals of $\mathbb{R}$
		$\mathbb{R}_I^-$	set of all negative closed intervals of $\mathbb{R}$
ϕ	empty set		
$\forall$	for all	$\mathbb{R}_I^-$	set of all negative closed intervals of $\mathbb{R}$
$int(K)/K^0$	interior of a set K		
$epi(\xi)$	epigraph of a function ξ	$\mathcal{R}_{([c,d])}$	collection of all Riemann integrable functions on $[c,d]$
L_α	lower level set		
$|.|$	absolute value		
$\|\cdot\|$	Euclidean norm	$\mathcal{IR}_{([c,d])}$	collection of all interval-Riemann integrable functions on $[c,d]$
$\langle\cdot,\cdot\rangle$	Euclidean inner product		
$\xi'(x)$	derivative of ξ at x		
$\xi''(x)$	second derivative of ξ at x	$ID_{(\Delta)}$	collection of all interval double integrable functions on Δ
$\beta(c,d)$	Beta function		
$\Gamma(.)$	Gamma function	K_c	family of all non-empty compact convex subsets of $\mathbb{R}$
$L^1[c,d]$	set of all Lebesgue measurable functions on $[c,d]$		

Chapter 1

Introduction

1.1 Generalized Convexity

Let $\mathbb{R}^n$ be the n-dimensional Euclidean space. We denote the usual inner product by $\langle \cdot, \cdot \rangle$ and for $x \in \mathbb{R}^n$, $\|\cdot\|$ denote the norm defined by

$$\|x\| = \left(\sum_{i=1}^{n} x_i{}^2 \right)^{1/2}.$$

It is basic knowledge in mathematical analysis that a nonempty subset K of $\mathbb{R}^n$ is said to be convex, if and only if for any $x, y \in K$ and $\delta \in [0, 1]$, one has

$$x + \delta(y - x) \in K.$$

Definition 1.1.1. *Let $K \subseteq \mathbb{R}^n$ be a nonempty convex set and let $\xi : K \to \mathbb{R}$.*

(a) The function ξ is said to be convex on K if and only if for any $x, y \in K$ and $\delta \in [0, 1]$, one has

$$\xi(\delta x + (1 - \delta)y) \leq \delta\xi(x) + (1 - \delta)\xi(y).$$

(b) The function ξ is said to be strictly convex on K if and only if for any $x, y \in K$ and $\delta \in [0, 1]$, one has

$$\xi(\delta x + (1 - \delta)y) < \delta\xi(x) + (1 - \delta)\xi(y).$$

A real-valued function ξ defined on a convex set $K \subseteq \mathbb{R}^n$ is concave if and only if $-\xi$ is convex on K.

Definition 1.1.2. *Let ξ be defined on a convex set $K \subseteq \mathbb{R}^n$. Then*

(a) The function ξ is said to be quasiconvex on K if and only if for any $x, y \in K$ and $\delta \in [0, 1]$, one has

$$\xi(\delta x + (1 - \delta)y) \leq max\{\xi(x), \xi(y)\}.$$

DOI: 10.1201/9781003408284-1

1

(b) *The function ξ is said to be strictly quasiconvex on K if and only if for any $x, y \in K$ and $\delta \in [0, 1]$, one has*

$$\xi(\delta x + (1 - \delta)y) < max\{\xi(x), \xi(y)\}.$$

Definition 1.1.3. *Let $K \subseteq \mathbb{R}_+ = (0, \infty)$. The set K is said to be GA-convex set, if*

$$x^\delta y^{1-\delta} \in K, \ \forall x, y \in K, \ \delta \in [0, 1].$$

Definition 1.1.4. *A function $\xi : K \subseteq \mathbb{R}_+ = (0, \infty) \to \mathbb{R}$ is said to be GA-convex on K, if*

$$\xi(x^\delta y^{1-\delta}) \leq \delta\xi(x) + (1 - \delta)\xi(y), \quad \forall x, y \in K, \delta \in [0, 1],$$

where $x^\delta y^{1-\delta}$ and $\delta\xi(x) + (1 - \delta)\xi(y)$ are the weighted geometric mean of two positive numbers x and y and the weighted arithmetic mean of $\xi(x)$ and $\xi(y)$, respectively.

Definition 1.1.5. *A function $\xi : K \subseteq \mathbb{R}_+ = (0, \infty) \to \mathbb{R}$ is said to be geometrically quasiconvex on K if*

$$\xi(x^\delta y^{1-\delta}) \leq sup\{\xi(x), \xi(y)\},$$

$\forall x, y \in K$ and $\delta \in [0, 1]$.

The following concepts of s-convex, tgs-convex and MT-convex are given by Hudzik and Maligranda [62] and Tunc *et al.* [167, 168], respectively.

Definition 1.1.6. *A function $\xi : [0, \infty) \to \mathbb{R}$ is named s-convex in the second sense along with $s \in (0, 1]$ if*

$$\xi(\alpha x + \beta y) \leq \alpha^s \xi(x) + \beta^s \xi(y)$$

holds for all $x, y \in [0, \infty)$ and $\alpha, \beta \geq 0$ along with $\alpha + \beta = 1$.

Definition 1.1.7. *A function $\xi : K \subseteq \mathbb{R} \to \mathbb{R}$ is named tgs-convex on X if ξ is non negative and*

$$\xi(\delta x + (1 - \delta)y) \leq \delta(1 - \delta)[\xi(x) + \xi(y)]$$

holds for all $x, y \in K$ and $\delta \in (0, 1)$.

Definition 1.1.8. *A function $\xi : K \subseteq \mathbb{R} \to \mathbb{R}$ is called MT-convex if ξ is non-negative and*

$$\xi(\delta x + (1 - \delta)y) \leq \frac{\sqrt{\delta}}{2\sqrt{1 - \delta}}\xi(x) + \frac{\sqrt{1 - \delta}}{2\sqrt{\delta}}\xi(y)$$

holds for all $x, y \in K$ and $\delta \in (0, 1)$.

Karamardian [72] gave the definition of strongly convex function.

Definition 1.1.9. *A function $\xi : K \subseteq \mathbb{R}^n \longrightarrow \mathbb{R}$ is said to be strongly convex on a convex set $K \subseteq \mathbb{R}^n$ if there exists a constant $\mu > 0$ such that*

$$\xi(\delta x + (1-\delta)y) \leq \delta\xi(x) + (1-\delta)\xi(y) - \mu\delta(1-\delta)\|y-x\|^2 \qquad (1.1)$$

for any $x, y \in K$ and $\delta \in [0,1]$.

The following theorem shows that the definition of the convex functions may be extended to any weighted average of its values at a finite number of points.

Theorem 1.1.1. *(Jensen Inequality) Let $K \subseteq \mathbb{R}^n$ be a nonempty convex set and let $\xi : K \to \mathbb{R}$.*

(a) *The function ξ is convex on K if and only if for any $x_1, x_2, ..., x_n \in K$ and $\delta_i \geq 0$, $i = 1, 2, ..., n$, one has*

$$\xi\left(\sum_{i=1}^{n} \delta_i x_i\right) \leq \sum_{i=1}^{n} \delta_i\xi(x_i), \sum_{i=1}^{n} \delta_i = 1.$$

(b) *The function ξ is strictly convex on K if and only if for any $x_1, x_2, ..., x_n \in K$ and $\delta_i \geq 0$, $i = 1, 2, ..., n$, one has*

$$\xi\left(\sum_{i=1}^{n} \delta_i x_i\right) < \sum_{i=1}^{n} \delta_i\xi(x_i), \sum_{i=1}^{n} \delta_i = 1.$$

Definition 1.1.10. *The epigraph of any function ξ is given by*

$$epi\ \xi = \{(x, \alpha) : x \in K, \xi(x) \leq \alpha, \alpha \in \mathbb{R}\}.$$

In the following theorem the convex theorem is characterized by the convexity of its epigraph.

Theorem 1.1.2. *Let $K \subseteq \mathbb{R}^n$ be a nonempty convex set and let $\xi : K \to \mathbb{R}$. Then ξ is convex on K if and only if epi ξ is convex set.*

The lower-level set of any function $\xi : K \to \mathbb{R}$ is given by

$$L_\alpha = \{x \in K : \xi(x) \leq \alpha\}.$$

The convexity of the lower-level set is a necessary condition for a function to be convex.

Theorem 1.1.3. *Let $K \subseteq \mathbb{R}^n$ be a nonempty convex set and let $\xi : K \to \mathbb{R}$. If ξ is convex on K, then L_α is convex for any $\alpha \in \mathbb{R}$.*

The following theorem gives the algebraic structure of the convex functions.

Theorem 1.1.4. *Let $\xi_1, \xi_2, ..., \xi_m$ be real-valued functions defined on a nonempty convex set $K \subseteq \mathbb{R}^n$ and let $\xi(x) = \sum_{i=1}^{m} \alpha_i \xi_i(x)$, $\alpha_i \geq 0$. Then*

(a) If $\xi_i, i = 1, 2, ..., m$ are convex on K, then ξ is convex on K.

(b) If $\xi_i, i = 1, 2, ..., m$ are strictly convex on K, then ξ is strictly convex on K.

The following theorem related to the composition of functions is important in the construction of convex functions.

Theorem 1.1.5. *Let $\xi : K \to \mathbb{R}$ be a convex function defined on a convex set $K \subseteq \mathbb{R}^n$ and let $\psi : A \to \mathbb{R}$ be a nondecreasing convex function, with $\xi(K) \subseteq A$. Then, the composite function $\phi(x) = \psi(\xi(x))$ is convex on K. Furthermore, if ξ is strictly convex and ψ is an increasing convex function, then ϕ is strictly convex.*

The following theorems characterize differentiable convex and strongly convex functions.

Theorem 1.1.6. *Let ξ be a differentiable function defined on a nonempty open convex set $K \subseteq \mathbb{R}^n$. Then ξ is convex on K if and only if for every $x, y \in K$, one has*

$$\xi(x) - \xi(y) \geq \langle \nabla \xi(y), x - y \rangle.$$

Theorem 1.1.7. *Let ξ be a differentiable function defined on a nonempty open convex set $K \subseteq \mathbb{R}^n$. Then ξ is strongly convex on K with modulus $\mu > 0$ if and only if for every $x, y \in K$, one has*

$$\xi(x) - \xi(y) \geq \langle \nabla \xi(y), x - y \rangle + \mu \|x - y\|^2.$$

The following concepts of coordinate convex functions are given by Dragomir [45].

Definition 1.1.11. *A function $\xi : \Delta = [a, b] \times [c, d] \to \mathbb{R}$ is said to be coordinate convex if the partial mappings $\xi_y : [a, b] \to \mathbb{R}$ defined as $\xi_y(u) = \xi(u, y)$ for all $y \in [c, d]$ and $\xi_x : [c, d] \to \mathbb{R}$ defined as $\xi_x(v) = \xi(x, v)$ for all $x \in [a, b]$ are convex.*

Lemma 1.1.1. *Every convex function defined on a rectangle is coordinate convex, but the converse is not true, in general.*

Example 1.1.1. *The mapping $\xi : [0, 1]^2 \to [0, \infty)$ given by $\xi(x, y) = xy$. Then ξ is convex on the coordinates but not convex on $[0, 1]^2$.*

The following concepts of coordinate strongly convex functions are given by Khan et al. [84].

Definition 1.1.12. *A function $\xi : \Delta = [a, b] \times [c, d] \to \mathbb{R}$ is said to be coordinate strongly convex if the partial mappings $\xi_y : [a, b] \to \mathbb{R}$ defined as $\xi_y(u) = \xi(u, y)$ for all $y \in [c, d]$ and $\xi_x : [c, d] \to \mathbb{R}$ defined as $\xi_x(v) = \xi(x, v)$ for all $x \in [a, b]$ are strongly convex.*

Lemma 1.1.2. *Every strongly convex function $\psi : \Delta = [a, b] \times [c, d] \to \mathbb{R}$ is coordinate strongly convex, but the converse is not true in general.*

1.2 Invexity

Hanson [57] introduced a new class of functions which was later termed as the class of invex functions by Craven [33].

Definition 1.2.1. *Given a nonempty open subset K of $\mathbb{R}^n$, a mapping $\eta : K \times K \to \mathbb{R}^n$ and a differentiable scalar function $\xi : K \to \mathbb{R}$. The function ξ is said to be invex at $y \in K$, if and only if for all $x \in K$, one has*

$$\xi(x) - \xi(y) \geq \langle \nabla \xi(y), \eta(x, y) \rangle.$$

Example 1.2.1. *Every real-valued convex function is invex for $\eta(x, y) = x - y$.*

The following important characterization of invexity was first given by Craven and Glover [34] and later Ben-Israel and Mond [16] provided a simple proof of the result.

Theorem 1.2.1. *ξ is invex if and only if every stationary point is a global minimum.*

Remark 1.2.1. *If ξ has no stationary points, then ξ is invex.*

Ben-Israel and Mond [16] gave the concept of invex sets as follows:

Definition 1.2.2. *The set $K \subseteq \mathbb{R}^n$ is said to be invex with respect to vector function $\eta : \mathbb{R}^n \times \mathbb{R}^n \to \mathbb{R}^n$, if*

$$y + \delta \eta(x, y) \in K, \quad \forall x, y \in K, \quad \delta \in [0, 1].$$

It is well known that every convex set is invex with respect to $\eta(x, y) = x - y$, but not conversely.

Remark 1.2.2. *(a) The concept of invex set is a generalization of the notion of convex set.*

(b) Every set in $\mathbb{R}^n$ is an invex set with respect to $\eta(x, y) = 0, \forall x, y \in \mathbb{R}^n$.

(c) The only function $\xi : \mathbb{R}^n \to \mathbb{R}$ invex with respect to $\eta(x, y) = 0$ is the constant function $\xi(x) = c, c \in \mathbb{R}$.

Ben-Israel and Mond [16] and Weir and Jeyakumar [172] studied the class of preinvex functions to relax the differentiability requirements in invexity given as follows:

Definition 1.2.3. *Let $K \subseteq \mathbb{R}^n$ be invex with respect to $\eta : K \times K \to \mathbb{R}^n$. A function $\xi : K \to \mathbb{R}$ is said to be preinvex if and only if for any $x, y \in K$ and $\delta \in [0, 1]$, one has*

$$\xi(y + \delta \eta(x, y)) \leq \delta \xi(x) + (1 - \delta)\xi(y).$$

It is well known that every convex function is a preinvex with respect to $\eta(x, y) = x - y$, but not conversely.

Ben-Israel and Mond [16] obtained that a differentiable preinvex function is also invex as follows:

Proposition 1.2.1. *Let $K \subseteq \mathbb{R}^n$ be an open invex set with respect to η : $K \times K \to \mathbb{R}^n$ and let $\xi : K \to \mathbb{R}$ be differentiable on K. If ξ is preinvex with respect to η, then ξ is invex with respect to η.*

The converse of the above result does not hold in general.

Example 1.2.2. *The function $\xi(x) = e^x, x \in \mathbb{R}$ is invex with respect to $\eta(x, y) = -1$, but not preinvex with respect to the same η.*

Mohan and Neogy [108] gave the following Condition C imposed on η, a differentiable function which is invex on K, with respect to η, is also preinvex.

Definition 1.2.4. *Let $K \subseteq \mathbb{R}^n$ be an open invex subset with respect to η : $K \times K \to \mathbb{R}$. The function η satisfies the Condition C if for any $x, y \in K$ and any $\delta \in [0, 1]$,*

$$\eta(y, y + \delta\eta(x, y)) = -\delta\eta(x, y),$$
$$\eta(x, y + \delta\eta(x, y)) = (1 - \delta)\eta(x, y).$$

Note that $\forall x, y \in K$ and $\delta \in [0, 1]$, then from Condition C, we have

$$\eta(y + \delta_2\eta(x, y), y + \delta\eta(x, y)) = (\delta_2 - \delta)\eta(x, y).$$

Example of a set which is not convex but invex is given as follows:

Example 1.2.3. *The set $\mathbb{R} \backslash \{0\}$ is invex with respect to η satisfying Condition C given by*

$$\eta(x, y) = \begin{cases} x - y, & x \geq 0, y \geq 0 \\ y - x, & x \leq 0, y \leq 0 \\ -y, & otherwise. \end{cases}$$

Definition 1.2.5. *At $y \in K$, where K is a convex set, a function ξ is said to be B-vex with respect to b, if for every $x \in K$ and $0 \leq \delta \leq 1$,*

$$\xi(\delta x + (1 - \delta)y) \leq \delta b(x, y, \delta)\xi(x) + (1 - \delta b(x, y, \delta))\xi(y)$$
$$= \xi(y) + \delta b(x, y, \delta)(\xi(x) - \xi(y)).$$

Definition 1.2.6. *At $y \in K$, where K is a invex set, the function ξ is said to be pre-quasiunivex with respect to η, ϕ and b, if for every $x \in K$ and $0 \leq \delta \leq 1$,*

$$\phi(\xi(x) - \xi(y)) \leq 0 \implies b\xi(y + \delta\eta(x, y)) \leq b\xi(y).$$

Bector *et al.* [15] introduced preunivex and univex functions with respect to a vector function η, scalar function ϕ and b as generalizations of preinvex, invex, b-preinvex, b-invex and b-vex functions.

Definition 1.2.7. *Given $K \subseteq \mathbb{R}^n \times \mathbb{R}$. K is said to be univex set with respect to η, ϕ and b, if every*

$$(x,\alpha),(y,\beta) \in K \text{ and } 0 \leq \delta \leq 1 \implies (y+\delta\eta(x,y),\beta+\delta b\phi(\alpha-\beta)) \in K.$$

Definition 1.2.8. *At $y \in K$, the function ξ is said to be univex with respect to η, ϕ and b, if for every $x \in K$ there exists a function $b(x,y)$ such that*

$$b(x,y)\phi[\xi(x) - \xi(y)] \geq \eta(x,y)^T \nabla\xi(y).$$

Example 1.2.4. *If $\xi : \mathbb{R} \to \mathbb{R}$ be defined by $\xi(x) = x^3$, where*

$$b(x,y) = \begin{cases} \frac{y^2}{x-y}, & x > y, \\ 0, & x \leq y, \end{cases}$$

and

$$\eta(x,y) = \begin{cases} x^2 + y^2 + xy, & x > y, \\ x - y, & x \leq y. \end{cases}$$

Let $\phi : \mathbb{R} \to \mathbb{R}$ be defined by $\phi(c) = 3c$. Then function ξ is univex.

Remark 1.2.3. *Every invex function is univex, where $\phi : \mathbb{R} \to \mathbb{R}$ can be defined by $\phi(c) = c$, $b(x,y) \equiv 1$, but not conversely.*

Example 1.2.5. *The function considered in Example 1.2.4 is univex but not invex because for $x = -3$, $y = 1$, $(\xi(x) - \xi(y)) < \eta(x,y)^T \nabla\xi(y)$.*

Remark 1.2.4. *Every convex function is univex, where $\phi : \mathbb{R} \to \mathbb{R}$ can be defined by $\phi(c) = c$, $b(x,y) \equiv 1$ and $\eta(x,y) = x - y$, but not conversely.*

Example 1.2.6. *The function considered in Example 1.2.4 is univex but not convex because for $x = -2$, $y = 1$, $(\xi(x) - \xi(y)) < \eta(x,y)^T \nabla\xi(y)$.*

Remark 1.2.5. *Every B-vex function is univex, where $\phi : \mathbb{R} \to \mathbb{R}$ can be defined by $\phi(c) = c$, and $\eta(x,y) = x - y$, but not conversely.*

Example 1.2.7. *The function considered in Example 1.2.4 is univex but not B-vex because for $x = \frac{1}{10}$, $y = \frac{1}{100}$, $b(x,y)(\xi(x) - \xi(y)) < \eta(x,y)^T \nabla\xi(y)$.*

Definition 1.2.9. *if $\xi : K \to \mathbb{R}$ is differentiable and prequasiunivex with respect to η, ϕ and b then ξ is quasiunivex with respect to η, ϕ and b, where $b(x,u) = \lim_{\delta \to 0+} b(x,u,\delta)$.*

Remark 1.2.6. *Every univex function with respect to η, ϕ and b is quasiunivex with respect to same η, ϕ and b. However, the converse does not hold, as it shown by the following example.*

Example 1.2.8. $\xi : \mathbb{R} \to \mathbb{R}$ *be defined by* $\xi(x) = -x^2$,

$$b(x, y) = \begin{cases} 1, & x = -y, \\ 0, & otherwise, \end{cases}$$

and

$$\eta(x, y) = \begin{cases} y - x, & x = -y, \\ x - y, & otherwise, \end{cases}$$

and $\phi : \mathbb{R} \to \mathbb{R}$ *is defined by* $\phi(c) = 2c$. *The function* ξ *is quasiunivex but not univex, because for* $x = 1, y = 2$,

$$b(x, y)\phi(\xi(x) - \xi(y)) < \eta(x, y)^T \nabla \xi(y).$$

Matłoka [104] gave the following concept of invex set and preinvex functions on the coordinates. Let K_1 and K_2 be two nonempty subsets of $\mathbb{R}^n$, let $\eta_1 : K_1 \times K_1 \to \mathbb{R}^n$ and $\eta_2 : K_2 \times K_2 \to \mathbb{R}^n$.

Definition 1.2.10. *Let* $(u, v) \in K_1 \times K_2$. *The set* $K_1 \times K_2$ *is said to be invex at* (u, v) *with respect to* η_1 *and* η_2, *if for each* $(x, y) \in K_1 \times K_2$ *and* $\lambda_1, \lambda_2 \in [0, 1]$,

$$(u + \lambda_1 \eta_1(x, u), v + \lambda_2 \eta_2(y, v)) \in K_1 \times K_2.$$

$K_1 \times K_2$ is said to be invex set with respect to η_1 and η_2 if $K_1 \times K_2$ is invex at each $(u, v) \in K_1 \times K_2$.

Definition 1.2.11. *A non-negative function* ξ *on the invex set* $K_1 \times K_2$ *is said to be coordinated preinvex with respect to* η_1 *and* η_2 *if the partial mappings* $\xi_y : K_1 \to \mathbb{R}$ *defined as* $\xi_y(u) = \xi(u, y)$ *for all* $y \in K_2$ *and* $\xi_x : K_2 \to \mathbb{R}$ *defined as* $\xi_x(v) = \xi(x, v)$ *for all* $x \in K_1$ *are preinvex with respect to* η_1 *and* η_2, *respectively.*

Remark 1.2.7. *If* $\eta_1(x, u) = x - u$ *and* $\eta_2(y, v) = y - v$, *then the function* ξ *is called convex on the coordinates.*

The following concept of harmonic convex functions are given by Anderson et al. [5] and Işcan [66].

Definition 1.2.12. *A set* $K = [c, d] \subseteq \mathbb{R} \backslash \{0\}$ *is called a harmonic convex set, if*

$$\frac{xy}{\delta x + (1 - \delta)y} \in K, \quad \forall x, y \in K, \quad \delta \in [0, 1].$$

Definition 1.2.13. *A function* $\xi : K = [c, d] \subseteq \mathbb{R} \backslash \{0\} \to \mathbb{R}$ *is called harmonic convex, if*

$$\xi\left(\frac{xy}{\delta x + (1 - \delta)y}\right) \le (1 - \delta)\xi(x) + \delta\xi(y), \quad \forall x, y \in K, \quad \delta \in [0, 1].$$

The concept of harmonic invex set and harmonic preinvex function was first given by Noor *et al.* [126].

Definition 1.2.14. *A set $K = [c, c + \eta(d, c)] \subseteq \mathbb{R} \backslash \{0\}$ is said to be harmonic invex set with respect to the bifunction $\eta(., .)$, if*

$$\frac{x(x + \eta(y, x))}{x + (1 - \delta)\eta(y, x)} \in K, \forall \ x, y \in K, \delta \in [0, 1]. \tag{1.2}$$

It is well known that every harmonic convex set is harmonic invex with respect to $\eta(y, x) = y - x$ but not conversely.

Definition 1.2.15. *A function $\xi : K = [c, c + \eta(d, c))] \subseteq \mathbb{R} \backslash \{0\} \to \mathbb{R}$ is said to be harmonic preinvex with respect to the bifunction $\eta(., .)$, if*

$$\xi\left(\frac{x(x + \eta(y, x))}{x + (1 - \delta)\eta(y, x)}\right) \leq (1 - \delta)\xi(x) + \delta\xi(y), \quad \forall x, y \in K, \quad \delta \in [0, 1].$$

Noor *et al.* [121] defined the relative harmonic preinvex functions.

Definition 1.2.16. *Let $h : [0, 1] \subseteq J \to \mathbb{R}$ be a nonnegative function. A function $\xi : K = [c, c + \eta(d, c)] \subseteq \mathbb{R} \backslash \{0\} \to \mathbb{R}$ is a relative harmonic preinvex function with respect to an arbitrary nonnegative function h and an arbitrary bifunction $\eta(., .)$, if*

$$\xi\left(\frac{x(x + \eta(y, x))}{x + (1 - \delta)\eta(y, x)}\right) \leq h(1 - \delta)\xi(x) + h(\delta)\xi(y), \forall x, y \in K, \ \delta \in [0, 1].$$
$$\tag{1.3}$$

1.3 Integral Inequalities

The classical Hermite–Hadamard type inequality provides lower and upper estimates for the integral average of any convex function defined on a compact interval, involving the midpoint and the endpoints of the domain. This interesting inequality was first discovered by Hermite in 1883 in the journal Mathesis (see [56, 107]). However, this beautiful result was nowhere mentioned in the mathematical literature and was not widely known as Hermite's result (see [133]).

Theorem 1.3.1. *(Hermite–Hadamard inequality) Let $\xi : [c, d] \to \mathbb{R}$ be a convex function with $c < d$. Then,*

$$\xi\left(\frac{c + d}{2}\right) \leq \frac{1}{d - c} \int_c^d \xi(x)dx \leq \frac{\xi(c) + \xi(d)}{2}.$$

The following double inequality is known as Hermite–Hadamard Fejér inequality [50] in the literature:

Theorem 1.3.2. *(Hermite–Hadamard Fejér inequality) Let $\xi : [c, d] \to \mathbb{R}$ be a convex function with $c < d$. Then*

$$\xi\left(\frac{c+d}{2}\right)\int_c^d \psi(x)dx \leq \frac{1}{d-c}\int_c^d \xi(x)\psi(x)dx \leq \frac{\xi(c)+\xi(d)}{2}\int_c^d \xi(x)dx,$$

where $\psi : [c, d] \to \mathbb{R}$ is non-negative, integrable, and symmetric about $x = \frac{c+d}{2}$.

Theorem 1.3.3. *(Simpson's inequality) Let $\xi : [c, d] \to \mathbb{R}$ be a four times continuously differentiable mapping on (c, d) and $\|\xi^{(4)}\|_\infty = \sup_{x \in (c,d)}|\xi^{(4)}(x)| < \infty$. Then the following inequality holds:*

$$\left|\frac{1}{3}\left[\frac{\xi(c)+\xi(d)}{2}+2\xi\left(\frac{c+d}{2}\right)\right] - \frac{1}{d-c}\int_c^d f(x)dx\right| \leq \frac{1}{2880}\|\xi^{(4)}\|_\infty (d-c)^4.$$

Theorem 1.3.4. *(Ostrowski inequality) Let $\xi : I \subset [0, \infty) \to \mathbb{R}$ be a differentiable mapping on the interior of the interval I, such that $\xi' \in L[c, d]$, where $c, d \in I$ with $c < d$. If $|\xi'(x)| \leq M$, then the following inequality holds:*

$$\left|\xi(x) - \frac{1}{d-c}\int_c^d \xi(x)dx\right| \leq \frac{M}{d-c}\left[\frac{(x-c)^2+(d-x)^2}{2}\right].$$

Theorem 1.3.5. *(Trapezoid type inequality) Let $\xi : int(I) \subseteq \mathbb{R} \to \mathbb{R}$ be a differentiable mapping on $int(I)$, $c, d \in int(I)$ with $c < d$. If $|\xi'|$ is convex on $[c, d]$, then the following inequality holds:*

$$\left|\int_c^d \xi(x)dx - (d-c)\frac{\xi(c)+\xi(d)}{2}\right| \leq \frac{1}{8}(d-c)^2(|\xi'(c)|+|\xi'(d)|). \qquad (1.4)$$

Theorem 1.3.6. *(Mid-point type inequality) Let $\xi : int(I) \subseteq \mathbb{R} \to \mathbb{R}$ be a differentiable mapping on $int(I)$, $c, d \in int(I)$ with $c < d$. If $|\xi'|$ is convex on $[c, d]$, then the following inequality holds:*

$$\left|\int_c^d \xi(x)dx - (d-c)\xi\left(\frac{c+d}{2}\right)\right| \leq \frac{1}{8}(d-c)^2(|\xi'(c)|+|\xi'(d)|). \qquad (1.5)$$

1.4   Fractional Calculus

In recent years, fractional calculus has been proven a powerful tool for the study of dynamical properties of many interesting systems in physics, chemistry and engineering. It serves as a powerful application in nonlinear oscillations of earthquakes, many physical phenomena such as seepage flow

in porous media and in fluid dynamical traffic modelling. For more recent developments on fractional calculus and application to fractional differential equations, refer to [14, 171] and references therein. Now, we shall discuss the following special functions:
The Gamma function

$$\Gamma(\alpha) = \int_0^\infty e^{-\delta}\delta^{(\alpha-1)}d\delta.$$

The Beta function:

$$\beta(x,y) = \frac{\Gamma(x)\Gamma(y)}{\Gamma(x+y)} = \int_0^1 \delta^{(x-1)}(1-\delta)^{(y-1)}d\delta, \quad x,y > 0.$$

The incomplete Beta function:

$$\beta(a;x,y) = \int_0^a \delta^{(x-1)}(1-\delta)^{(y-1)}d\delta, \quad 0 < a < 1, \; x,y > 0.$$

The Hypergeometric function:

$$_2F_1(a,c;d;z) = \frac{1}{\beta(c,d-c)} \int_0^1 \delta^{(c-1)}(1-\delta)^{(d-c-1)}(1-z\delta)^{-a}d\delta,$$

$d > c > 0, \; |z| < 1.$

Definition 1.4.1. *Let* $\xi \in L^1[c,d]$. *The symbol* $J_{c+}^\alpha \xi$ *and* $J_{d-}^\alpha \xi$ *denote the left-sided and right-sided Riemann–Liouville fractional integrals of the order* $\alpha \in \mathbb{R}_+$ *are defined by*

$$J_{c+}^\alpha \xi(x) = \frac{1}{\Gamma(\alpha)} \int_c^x (x-\delta)^{\alpha-1}\xi(\delta)d\delta, \quad (0 \le c < x < d)$$

and

$$J_{d-}^\alpha \xi(x) = \frac{1}{\Gamma(\alpha)} \int_x^d (\delta-x)^{\alpha-1}\xi(\delta)d\delta, \quad (0 \le c < x < d)$$

respectively, where $\Gamma(.)$ *is the Gamma function.*

Mubeen and Habibullah [112] introduced the following Riemann–Liouville $k-$fractional integrals.

Definition 1.4.2. *Let* $\xi \in L^1[c,d]$ *then the Riemann–Liouville k-fractional integrals* $_kJ_{c+}^\alpha \xi(x)$ *and* $_kJ_{d-}^\alpha \xi(x)$ *of order* $\alpha > 0$ *are given as*

$$_kJ_{c+}^\alpha \xi(x) = \frac{1}{k\Gamma_k(\alpha)} \int_c^x (x-\delta)^{\frac{\alpha}{k}-1}\xi(\delta)d\delta, \quad (0 \le c < x < d)$$

and

$$_kJ^\alpha_{d-}\xi(x) = \frac{1}{k\Gamma_k(\alpha)}\int_x^d (\delta - x)^{\frac{\alpha}{k}-1}\xi(\delta)d\delta, \quad (0 \le c < x < d)$$

respectively, where $k > 0$ and $\Gamma_k(\alpha)$ is the k-Gamma function defined by $\Gamma_k(\alpha) = \int_0^\infty e^{-\frac{\delta^k}{k}}\delta^{(\alpha-1)}d\delta$. Furthermore, $\Gamma_k(\alpha + k) = \alpha\Gamma_k(\alpha)$ and $_kJ^0_{c+}\xi(x) =_k J^0_{d-}\xi(x) = \xi(x)$.

Sarikaya *et al.* [149] studied inequalities of Hermite–Hadamard type involving Riemann–Liouville fractional integrals and produced a new fractional Hermite–Hadamard type inequality as follows:

Theorem 1.4.1. *Let $\xi : [c, d] \to \mathbb{R}$ be a positive function along with $0 \le c < d$, and let $\xi \in L^1[c, d]$. Suppose that ξ is a convex function on $[c, d]$, then the following inequalities for fractional integrals hold:*

$$\xi\left(\frac{c+d}{2}\right) \le \frac{\Gamma(\alpha+1)}{2(d-c)^\alpha}[J^\alpha_{c+}\xi(d) + J^\alpha_{d-}\xi(c)] \le \frac{\xi(c) + \xi(d)}{2},$$

where the symbols $J^\alpha_{c+}\xi$ and $J^\alpha_{d-}\xi$ denote, respectively, the left-sided and right-sided Riemann–Liouville fractional integrals of order $\alpha > 0$ defined by

$$J^\alpha_{c+}\xi(x) = \frac{1}{\Gamma(\alpha)}\int_c^x (x-\delta)^{(\alpha-1)}\xi(\delta)d\delta, \quad c < x,$$

and

$$J^\alpha_{d-}\xi(x) = \frac{1}{\Gamma(\alpha)}\int_x^d (\delta-x)^{(\alpha-1)}\xi(\delta)d\delta, \quad x < d,$$

where $\Gamma(\alpha)$ is the gamma function and defined as $\Gamma(\alpha) = \int_0^\infty e^{-\delta}\delta^{(\alpha-1)}d\delta$.

1.5 Majorization Inequalities

The theory of majorization is a very significant topic in mathematics; a remarkable and complete reference on the majorization subject is the book by Marshall *et al.* [103]. For example, the theory of majorization is an essential tool that permits us to transform nonconvex complicated constrained optimization problems that involve matrix-valued variables into simple problems with scalar variables that can be easily solved [32, 61].

A vector $\mathbf{x} = (x_1, x_2, ..., x_n)$ is said to be majorized by a vector $\mathbf{y} = (y_1, y_2, ..., y_n)$, in symbol $y \succ x$, if $x_1 \ge ... \ge x_n$, $y_1 \ge ... \ge y_n$, $\sum_{i=1}^m y_i \ge \sum_{i=1}^m x_i$, $m = 1, 2, ..., n - 1$ and $\sum_{i=1}^n y_i = \sum_{i=1}^n x_i$. It means that the sum of m largest entries of x does not exceed the sum of m largest entries of y; for all $m = 1, 2, ..., n$.

The following theorem is well-known in the literature as the majorization theorem, and for its proof, we refer to [103]. This result is due to Hardy *et al.* [58] and it can also be found in [73].

Theorem 1.5.1. *Let K be an interval in $\mathbb{R}$, and $x = (x_1, x_2, ..., x_n)$ and $y = (y_1, y_2, ..., y_n)$ be two n-tuples such that $x_i, y_i \in K (i = 1, 2, ..., n)$. Then the inequality*

$$\sum_{i=1}^{n} \xi(c_i) \geq \sum_{i=1}^{n} \xi(d_i)$$

holds for every continuous convex function $\xi : K \to \mathbb{R}$ if and only if $c \succ d$.

The following theorem is a weighted version of above theorem and is given by Fuchs [52].

Theorem 1.5.2. *Let $c = (c_1, c_2, ..., c_n)$ and $d = (d_1, d_2, ..., d_n)$ be two decreasing n-tuples such that $c_i, d_i \in K (i = 1, 2, ..., n)$ and $p = (p_1, p_2, ..., p_n)$ be a real n-tuple with*

$$\sum_{i=1}^{k} p_i c_i \geq \sum_{i=1}^{k} p_i d_i, \quad k = 1, 2, ..., n - 1,$$

$$\sum_{i=1}^{n} p_i c_i \geq \sum_{i=1}^{n} p_i d_i.$$

Then the inequality

$$\sum_{i=1}^{n} p_i \xi(c_i) \geq \sum_{i=1}^{n} p_i \xi(d_i)$$

holds for every continuous convex function $\xi : K \to \mathbb{R}$.

The definition of majorization for integrable functions can be stated as follows (see, [133]).

Definition 1.5.1. *Let f and g be two decreasing real-valued integrable functions on the interval $[a, b]$. Then f is said to majorize g, in symbol, $f \succ g$, if the inequality*

$$\int_a^x g(u) du \leq \int_a^x f(u) du \quad \text{holds for all } x \in [a, b)$$

and

$$\int_a^b g(u) du = \int_a^b f(u) du.$$

1.6   Time Scale Calculus

In 1988, Hilger [60] introduced the theory of time scale which is a unification of the discrete theory with the continuous theory. Recently, much attention has been given to the time scales calculus by many researchers, see for instance [22, 39, 161, 165]. Consequently, the concept of time scale theory has been extended and generalized. Time scales calculus has applications in various fields such as Economics, Engineering, Physics, Signal processing, Aerospace, Dynamic programming, Recurrent neural networks and Control theory, see references [9, 23, 151, 152, 176, 178].

Time scale is a nonempty closed subset of the set of real numbers $\mathbb{R}$. The set of integers $\mathbb{Z}$, the set of real numbers $\mathbb{R}$, finite unions of disjoint intervals, limit sets such as $\{0\} \cup \{\frac{1}{n}\} : n = 1, 2, ...,$ Cantor sets etc are the examples of time scales. We denote time scale by $\mathbb{T}$, time-scaled interval by $[c, d]_{\mathbb{T}}$, and the interior of K by K^0. There are two types of the operator: The forward jump operator $\diamond(\delta) = inf\{x \in \mathbb{T} : x > \delta\}$ and the backward jump operator $\varsigma(\delta) = sup\{x \in \mathbb{T} : x < \delta\}$ for all $\delta \in \mathbb{T}$. The forward jump operator represents the next element and the backward jump operator represents the previous element in the domain. If $\mathbb{T}$ has a maximum δ, then $\diamond(\delta) = \delta$, and if $\mathbb{T}$ has a minimum δ, then $\varsigma(\delta) = \delta$.

If $\diamond(\delta) > \delta$, then δ is called right-scattered and if $\varsigma(\delta) < \delta$, then δ is called left-scattered. The point δ is said to be isolated if it is both right-scattered and left-scattered: $\diamond(\delta) > \delta > \varsigma(\delta)$ for $\delta \in \mathbb{T}$. It is a characteristic of discrete domains that all points within them are isolated. δ is said to be right-dense if $\diamond(\delta) = \delta$ and δ is said to be left-dense if $\varsigma(\delta) = \delta$. The point δ is said to be dense if it is both left-dense and right-dense: $\diamond(\delta) = \delta = \varsigma(\delta)$ for $\delta \in \mathbb{T}$.

The mappings $\eta, \zeta : \mathbb{T} \to [0, \infty)$ defined by

$$\eta(\delta) := \diamond(\delta) - \delta, \quad \zeta(\delta) := \delta - \varsigma(\delta)$$

are said to be forward and backward graininess functions, respectively. The graininess function measures the step size between two consecutive points in $\mathbb{T}$. The set $\mathbb{T}^k$ which is derived from time scale $\mathbb{T}$ is defined as follows: If $\mathbb{T}$ has a left-scattered maximum m, then $\mathbb{T}^k = \mathbb{T} - m$; otherwise $\mathbb{T}^k = \mathbb{T}$.

The delta derivative is a basic time scale derivative and is denoted by $\xi^\Delta(\delta)$. Let $\xi : \mathbb{T} \to \mathbb{R}$ be a function. Then the delta derivative $\xi^\Delta(\delta)$ of ξ at a point $\delta \in \mathbb{T}^k$ is defined to be the number such that given $\epsilon > 0$, there exists a neighbourhood N of δ, such that

$$|\xi(\diamond(\delta)) - \xi(s) - \xi^\Delta(\delta)(\diamond(\delta) - s)| \leq \epsilon|\diamond(\delta) - s|,$$

for all $s \in \mathbb{N}$.

If $\mathbb{T} = \mathbb{R}$, then the delta derivative $\xi^\Delta = \xi'$ where ξ' is the derivative from continuous calculus.

If $\mathbb{T} = \mathbb{Z}$, then the delta derivative $\xi^\Delta = \Delta\xi$ where $\Delta\xi$ is the forward difference operator from discrete calculus.

The following definitions and results are given by Bohner and Peterson [22].

Definition 1.6.1. *A function $\xi : \mathbb{T} \to \mathbb{R}$ is called rd-continuous if it is continuous at every right-dense point of $\mathbb{T}$ and if its left-sided limit is finite at any left-dense point of $\mathbb{T}$. All rd-continuous functions are denoted by C_{rd}.*

Definition 1.6.2. *A function $F : \mathbb{T} \to \mathbb{R}$ is called an antiderivative of $\xi : \mathbb{T} \to \mathbb{R}$ if $F^\Delta(\delta) = \xi(\delta)$, for all $\delta \in \mathbb{T}^k$. Then, delta integral is defined by*

$$\int_c^s \xi(\delta)\Delta\delta = F(s) - F(c),$$

where $s, c \in \mathbb{T}$.

Theorem 1.6.1. *If $\xi \in C_{rd}$ and $\delta \in \mathbb{T}^k$ then*

$$\int_\delta^{\diamond(\delta)} \xi(x)\Delta(x) = \eta(\delta)\xi(\delta).$$

Theorem 1.6.2. *Let $\xi_1, \xi_2 \in \mathbb{C}_{rd}, \lambda \in \mathbb{R}$ and $c, d, \alpha \in \mathbb{T}$ then*

(i) $\int_c^d (\xi_1(x) + \xi_2(x))\Delta x = \int_c^d \xi_1(x)\Delta x + \int_c^d \xi_2(x)\Delta x;$

(ii) $\int_c^d \lambda\xi(x)\Delta x = \lambda \int_c^d \xi(x)\Delta x;$

(iii) $\int_c^d \xi(x)\Delta x = -\int_d^c \xi(x)\Delta x;$

(iv) $\int_c^d \xi(x)\Delta x = \int_c^\alpha \xi(x)\Delta x + \int_\alpha^d \xi(x)\Delta x;$

(v) $\int_c^d \xi_1^\diamond(x)\xi_2^\theta(x)\Delta x = (\xi_1\xi_2)(d) - (\xi_1\xi_2)(c) - \int_c^d \xi_1^\theta(x)\xi_2(x)\Delta x;$

(vi) $\int_c^d \xi_1(x)\xi_2^\theta(x)\Delta x = (\xi_1\xi_2)(d) - (\xi_1\xi_2)(c) - \int_c^d \xi_1^\theta(x)\xi_2^\diamond(x)\Delta x;$

(vii) $\int_c^c \xi(x)\Delta x = 0;$

(viii) *If $\xi(x) \geq 0$ for all x, then $\int_c^d \xi(x)\Delta x \geq 0;$*

(ix) *If $|\xi_1(x)| \leq \xi_2(x)$ on $[c, d]$, then $|\int_c^d \xi_1(x)\Delta x| \leq \int_c^d \xi_2(x)\Delta x.$*

From assertion (ix) of Theorem 1.6.2 for $\xi_2(x) = |\xi_1(x)|$ on $[c, d]$, we have

$$\left| \int_c^d \xi_1(x)\Delta x \right| \leq \int_c^d |\xi_1(x)|\Delta x.$$

1.7   Interval Analysis

The theory of interval analysis was initiated by Moore [109]. We can compute arbitrarily sharp upper and lower bounds on exact solutions of many problems in applied mathematics by using interval arithmetic, interval-valued functions, integrals of interval-valued functions, etc. Moore [109] supplied techniques for keeping track of errors, developed and applied, for the machine computation of rigorous error bounds on approximate solutions. Computational tests for machine convergence of iterative methods existence and nonexistence of solutions for a variety of equations are obtained via interval analysis. Several researchers focused on studying the literature and applications of interval analysis in automatic error analysis, computer graphics, neural network output optimization, robotics, computational physics, etc.

1.7.1   Interval arithmetic

The rules for interval addition, subtraction, product and quotient [109] are

$$[\underline{X}, \overline{X}] + [\underline{Y}, \overline{Y}] = [\underline{X} + \underline{Y}, \overline{X} + \overline{Y}].$$

$$[\underline{X}, \overline{X}] - [\underline{Y}, \overline{Y}] = [\underline{X} - \overline{Y}, \overline{X} - \underline{Y}].$$

$$X.Y = \{xy : x \in X, y \in Y\}.$$

It is easy to see that X.Y is again an interval, whose ends points can be computed from

$$\underline{X}.\underline{Y} = min\{\underline{X}\ \underline{Y}, \underline{X}\ \overline{Y}, \overline{X}\ \underline{Y}, \overline{X}\ \overline{Y}\}$$

and

$$\overline{X}.\overline{Y} = max\{\underline{X}\ \underline{Y}, \underline{X}\ \overline{Y}, \overline{X}\ \underline{Y}, \overline{X}\ \overline{Y}\}.$$

The reciprocal of an interval as follows

$$1/X = \{1/x : x \in X\}. \tag{1.6}$$

If X is an interval not containing the number 0, then

$$1/X = [1/\overline{X}, 1/\underline{X}].$$

$$X/Y = X.(1/Y) = \{x/y : x \in X, y \in Y\},$$

where $1/y$ is defined by (1.6).

Scalar multiplication of the interval X is defined by

$$\lambda X = \lambda[\underline{X}, \overline{X}] = \begin{cases} [\lambda\underline{X}, \lambda\overline{X}], & \text{if } \lambda > 0, \\ \{0\}, & \text{if } \lambda = 0, \\ [\lambda\overline{X}, \lambda\underline{X}], & \text{if } \lambda < 0, \end{cases}$$

where $\lambda \in \mathbb{R}$.

The Hausdorff distance between $X = [\underline{X}, \overline{X}]$ and $Y = [\underline{Y}, \overline{Y}]$ is defined as

$$d(X, Y) = d([\underline{X}, \overline{X}], [\underline{Y}, \overline{Y}]) = max\{|\underline{X} - \underline{Y}|, |\overline{X} - \overline{Y}|\}.$$

Let $\mathbb{R}_I$, $\mathbb{R}_I^+$ and $\mathbb{R}_I^-$ be the sets of all closed intervals of $\mathbb{R}$, sets of all positive closed intervals of $\mathbb{R}$ and sets of all negative closed intervals of $\mathbb{R}$, respectively. Now, we discuss some algebraic properties of interval arithmetic [109].

(1) (Associativity of addition) $(X+Y)+Z = X+(Y+Z)$, $\forall\, X, Y, Z \in \mathbb{R}_I$.

(2) (Additive element) $X + 0 = 0 + X = X$, $\forall\, X \in \mathbb{R}_I$.

(3) (Commutativity of addition) $X + Y = Y + X$, $\forall\, X, Y \in \mathbb{R}_I$.

(4) (Cancellation law) $X + Z = Y + Z \implies X = Y$, $\forall\, X, Y, Z \in \mathbb{R}_I$.

(5) (Associativity of multiplication) $(X.Y).Z = X.(Y.Z)$, $\forall\, X, Y, Z \in \mathbb{R}_I$.

(6) (Commutativity of multiplication) $X.Y = Y.X$, $\forall\, X, Y \in \mathbb{R}_I$.

(7) (Unit element) $X.1 = 1.X = X$, $\forall\, X \in \mathbb{R}_I$.

(8) (Associate law) $\lambda(\mu X) = (\lambda\mu)X$, $\forall\, X \in \mathbb{R}_I$ and $\forall\, \lambda, \mu \in \mathbb{R}$.

(9) (First distributive law) $\lambda(X+Y) = \lambda X + \lambda Y$, $\forall\, X, Y \in \mathbb{R}_I$ and $\forall\, \lambda \in \mathbb{R}$.

(10) (Second distributive law) $(\lambda+\mu)X = \lambda X + \mu X$, $\forall\, X \in \mathbb{R}_I$ and $\forall\, \lambda, \mu \in \mathbb{R}$.

However, the distributive law does not always hold.

Example 1.7.1. $X = [-2, -1], Y = [-1, 0]$ *and* $Z = [1, 3]$.
$X.(Y + Z) = [-2, -1].([-1, 0] + [1, 3]) = [-6, 0]$
whereas
$X.Y + X.Z = [-2, -1].[-1, 0] + [-2, -1].[1, 3] = [-6, 1]$.

1.7.2 Integral of interval-valued functions

A function ξ is said to be an interval-valued function of δ on $[c, d]$ if it assigns a nonempty interval to each $\delta \in [c, d]$

$$\xi(\delta) = [\underline{\xi}(\delta), \overline{\xi}(\delta)],$$

where $\underline{\xi}$ and $\overline{\xi}$ are real-valued functions.
A tagged partition P of $[a, b]$ is a set of numbers $\{t_{i-1}, u_i, t_i\}_{i=1}^m$ such that

$$P : a = t_0 < t_1 < \ldots < t_m = b.$$

with $t_{i-1} \leq u_i \leq t_i$ for all $i = 1, 2, 3 \ldots m$. Partition P is said to be ρ-fine if $\Delta t_i < \rho$ for all i, where $\Delta t_i = t_i - t_{i-1}$. Let $\mathfrak{P}(\rho, [a, b])$ be the family of all

ρ-fine partitions of $[a, b]$. Then, we define the sum

$$S(\xi, P, \rho) = \sum_{i=1}^{m} \xi(u_i)[t_i - t_{i-1}],$$

where $\xi : [c, d] \to \mathbb{R}_I$. $S(\xi, P, \rho)$ denotes the Riemann sum of ξ corresponding to the $P \in \mathfrak{P}(\rho, [a, b])$.

The following definition is given by Piatek [134].

Definition 1.7.1. *A function $\xi : [a, b] \to \mathbb{R}_I$ is called interval Riemann integrable (IR-integrable) on $[a, b]$ if there exist $K \in \mathbb{R}_I$ such that, for each $\epsilon > 0$, there exist $\rho > 0$ such that*

$$d(S(\xi, P, \rho), K) < \epsilon$$

for every Riemann sum S of ξ corresponding to each $P \in \mathfrak{P}(\rho, [a, b])$ and independent of choice of $u_i \in [t_{i-1}, t_i]$ for $1 \leq i \leq m$.
$\mathfrak{R}$ is called the $IR-$integral of ξ on $[a, b]$ and is denoted by

$$\mathfrak{R} = (IR) \int_a^b \xi(\delta) d\delta.$$

The collection of all (IR)–integrable functions on $[a, b]$ denoted by $IR_{([a,b])}$.

Zhao et al. [184] gave the concept of interval double integral for interval-valued functions as follows:

If $P_1 = \{t_{i-1}, u_i, t_i\}_{i=1}^m$ such that $P_1 \in \mathfrak{P}(\rho, [a, b])$ and $P_2 = \{s_{j-1}, v_j, s_j\}_{j=1}^n$ such that $P_2 \in \mathfrak{P}(\rho, [c, d])$, then the rectangles

$$\Delta_{i,j} = [t_{i-1}, t_i] \times [s_{j-1}, s_j]$$

partition rectangle $\Delta = [a, b] \times [c, d]$ with the points (u_i, v_j) are inside the rectangles $[t_{i-1}, t_i] \times [s_{j-1}, s_j]$. Let $\mathfrak{P}(\rho, \Delta)$ be the family of all ρ-fine partitions P of Δ such that $P = P_1 \times P_2$. Let $\Delta A_{i,j}$ be the area of the rectangle $\Delta_{i,j}$. Choose an arbitrary (u_i, v_j) from each rectangle $\Delta_{i,j}$, where $1 \leq i \leq m, 1 \leq j \leq n$, and we get

$$S(\xi, P, \rho, \Delta) = \sum_{i=1}^{m} \sum_{j=1}^{n} \xi(u_i, v_j) \Delta A_{i,j},$$

where $\xi : \Delta \to \mathbb{R}_I$. $S(\xi, P, \rho, \Delta)$ denotes integral sum of ξ corresponding to the $P \in \mathfrak{P}(\rho, \Delta)$.

Definition 1.7.2. *A function $\xi : \Delta = [a, b] \times [c, d] \to \mathbb{R}_I$ is called interval double integrable (ID-integrable) on Δ with (ID)-integral $I = (ID) \int \int_{\Delta} \xi(\delta_1, \delta_2) dA$ if for each $\epsilon > 0$, there exists $\rho > 0$ such that*

$$d(S(\xi, P, \rho, \Delta), I) < \epsilon$$

for each $P \in \mathfrak{P}(\rho, \Delta)$.

The collection of all (ID)-integrable functions on Δ denoted by $ID_{(\Delta)}$.

Theorem 1.7.1. *If $\xi : \Delta = [a,b] \times [c,d] \to \mathbb{R}_I$ be an interval-valued function such that $\xi = [\underline{\xi}, \overline{\xi}]$ and $\xi \in ID_{(\Delta)}$, then we have*

$$(ID) \int \int_{\Delta} \xi(\delta_1, \delta_2) dA = (IR) \int_a^b (IR) \int_c^d \xi(\delta_1, \delta_2) d\delta_1 d\delta_2.$$

1.8   Stochastic Processes

The notion of stochastic processes for convexity is of great importance in optimization and also useful for numerical approximations when there exist probabilistic quantities in the literature [28]. A stochastic process $\xi(t)$ is a function which maps the index set T into the space S of random variable defined on (Ω, A, P). Sobczyk [164] introduced the concepts of continuity, mean square continuity, monotonicity and differentiability for stochastic processes.

Definition 1.8.1. *The stochastic process $\xi : K \times \Omega \to \mathbb{R}$ is called*

- *continuous in probability in interval K, if for all $u_0 \in K$*

$$P - \lim_{u \to u_0} \xi(u,.) = \xi(u_0,.),$$

  where $P - \lim$ denotes the limit in probability;

- *mean square continuous in the interval K, if for all $u_0 \in K$*

$$\lim_{u \to u_0} E[(\xi(u,.) - \xi(u_0,.))^2] = 0,$$

  where $E[\xi(u,.)]$ denotes the expectation value of the random variable $\xi(u,.)$;

- *increasing(decreasing), if for all $u,v \in K$ such that $u < v$,*

$$\xi(u,.) \leq \xi(v,.), \quad (\xi(u,.) \geq \xi(v,.));$$

- *monotonic if it is increasing or decreasing;*

- *differentiable at a point $u_0 \in K$ if there is a random variable $\xi'(u_0,.) : \Omega \to \mathbb{R}$*

$$\xi'(u_0,.) = P - \lim_{u \to u_0} \frac{\xi(u,.) - \xi(u_0,.)}{u - u_0}.$$

The definition of mean square integrability for stochasic processes is given by Sobczyk [164].

Definition 1.8.2. *Suppose that* $\xi : K \times \Omega \to \mathbb{R}$ *is a stochastic process with* $E[\xi(u)^2] < \infty$ *for all* $u \in K$ *and* $[c, d] \in K$, $c = u_0 < u_1 < u_2 < ... < u_n = d$ *is a partition of* $[c, d]$ *and* $\Theta_k \in [u_{k-1}, u_k]$ *for all* $k = 1, 2, ...n$. *Further, suppose that* $\phi : K \times \Omega \to \mathbb{R}$ *be a random variable. Then, it is said to be mean square integrable of the process* ξ *on* $[c, d]$, *if for each normal sequence of partitions of the interval* $[c, d]$ *and for each* $\Theta_k \in [u_{k-1}, u_k]$, $k = 1, 2, ...n$, *we have*

$$\lim_{n \to \infty} E\left[\left(\sum_{k=1}^{n} \xi(\Theta_k)\Delta(u_k - u_{k-1}) - \phi \right)^2 \right] = 0.$$

Then, we can write

$$\phi(.) = \int_c^d \xi(v,.)dv \quad almost \; everywhere.$$

Also, mean square integral operator is increasing, that is,

$$\int_c^d \xi(u,.)du \leq \int_c^d \phi(u,.)du \quad almost \; everywhere,$$

where $\xi(u,.) \leq \phi(u,.) in \; [c, d]$.

Chapter 2

Integral Inequalities for Strongly Generalized Convex Functions

2.1 Introduction

Karamardian [72] introduced strongly convex function. However, there are references citing Polyak [136] has introduced strongly convex functions as a generalization of convex functions, see [105, 117]. It is well known that every differentiable function is strongly convex if and only if its gradient map is strongly monotone [72]. Karamardian [72] showed that every bidifferentiable function is strongly convex if and only if its Hessian matrix is strongly positive definite.

Azócar *et al.* [13] derived an appropriate counterpart of the Fejér inequalities and presented refinement of Hermite–Hadamard inequalities for strongly convex functions. Işcan [65] established Hermite–Hadamard-Fejér inequality for fractional integrals. Further, Park [132] obtained new estimates on the generalization of Hermite–Hadamard-Fejér type inequalities for differentiable functions whose derivatives in absolute value at certain powers are convex. Gordji *et al.* [54] introduced the concept of η-convex/(φ-convex [55]) functions as a generalization of convex functions and obtained the Hermite–Hadamard, Fejér, Jensen and Slater type inequalities for η-convex functions. Further, Rostamian Delavar and De La Sen [143] gave some applications for Hermite–Hadamard-Fejér type integral inequalities for differentiable η-convex functions. Further, Awan *et al.* [12] introduced the notion of strongly η-convex functions and formulated some new integral inequalities of Hermite–Hadamard type for strongly η-convex functions. For more details, one can refer to [35, 43].

The fractional inequalities play an important role in calculating different means for generalized convexity, so researchers are attracting to develop fractional integral inequalities for generalized convexity. Kwun *et al.* [90] established Hermite–Hadamard and Fejér-type inequalities and derived fractional integral inequalities for η-convex functions. Further, Yang *et al.* [177] investigated some Hermite–Hadamard-type fractional integral inequalities for generalized h-convex functions.

Niculescu [113] investigated the class of multiplicatively convex functions by replacing the arithmetic mean to the geometric mean. It is well known that every polynomial $p(x)$ with non-negative coefficients is a multiplicatively

DOI: 10.1201/9781003408284-2

convex function on $[0, \infty)$. More generally, every real analytic function $\xi(x) = \sum_{n=0}^{\infty} c_n x^n$ with non-negative coefficients is a multiplicatively convex function on $(0, R)$, where R denotes the radius of convergence [113]. Niculescu [113] showed that a continuous function $\xi : K \subset (0, \infty) \to [0, \infty)$ is multiplicatively convex if and only if $\xi(\sqrt{xy}) \leq \sqrt{\xi(x)\xi(y)}, \forall x, y \in K$.

Qi and Xi [139] introduced a new concept of geometrically quasi-convex functions and established some integral inequalities of Hermite–Hadamard type for the function whose derivatives are of geometric quasi-convexity [139]. Noor *et al.* [127] introduced generalized geometrically convex functions and derived some basic inequalities related to generalized geometrically convex functions. Noor *et al.* [127] also established new Hermite–Hadamard type inequalities for generalized geometrically convex functions. For more details, one can refer to [94, 114, 124, 125, 163, 181].

Recently, Obeidat and Latif [128] established some new weighted Hermite–Hadamard type inequalities for geometrically quasi-convex functions and showed how we can use inequalities of Hermite–Hadamard type to obtain the inequalities for special means. For more details on Hermite–Hadamard inequalities, we refer to [43, 94, 138, 163, 181] and references therein.

Dinu [39] obtained Hermite–Hadamard inequality for convex functions on time scales. In 2019, Tahir *et al.* [165] established some new Hermite–Hadamard type integral inequalities using the concept of time scales. Recently, Rashid *et al.* [140] investigated the time scales version of two non-negative auxiliary functions for the class of convex functions and obtained several dynamical variants that are essentially based on Hermite–Hadamard inequality.

The organization of this chapter is as follows: In Section 2.2, we recall some basic results that are necessary for our main results. In Section 2.3, we establish some Hermite–Hadamard-and Fejér-type inequalities for strongly η-convex functions. We derive some integral inequalities for strongly η-convex functions. Further, in Section 2.3.1 we discuss some applications to special means of real numbers with the help of these results. In Section 2.4, we establish some new weighted Hermite–Hadamard inequalities for strongly GA-convex functions by using geometric symmetry of a continuous positive mapping and a differentiable mapping whose derivatives in absolute value are strongly GA-convex. In Section 2.5, we introduce the notion of a strongly convex function with respect to two auxiliary functions ψ_1 and ψ_2 on Time scales $\mathbb{T}$. Also, we derive some new dynamic inequalities for these strongly convex functions. Some examples are also mentioned in the support of our theory.

2.2   Preliminaries

In this section, we mention some definitions and related results required for this chapter.

Gordji *et al.* [55] introduced the concept of η-convex functions.

Definition 2.2.1. *A function* $\xi : K \subseteq \mathbb{R} \longrightarrow \mathbb{R}$ *is said to be* η-*convex function with respect to* $\eta : \mathbb{R} \times \mathbb{R} \longrightarrow \mathbb{R}$, *if*

$$\xi(\delta x + (1-\delta)y) \leq \xi(y) + \delta \eta(\xi(x), \xi(y)), \quad \forall x, y \in K, \quad \delta \in [0,1].$$

Awan *et al.* [12] gave the following concept of strongly η-convex functions.

Definition 2.2.2. *A function* $\xi : K \subseteq \mathbb{R} \to \mathbb{R}$ *is said to be strongly* η-*convex function with respect to* $\eta : \mathbb{R} \times \mathbb{R} \to \mathbb{R}$ *and modulus* $\mu > 0$ *if*

$$\xi(\delta x + (1-\delta)y) \leq \xi(y) + \delta \eta(\xi(x), \xi(y)) - \mu \delta(1-\delta)(x-y)^2, \quad \forall \, x, y \in K, \, \delta \in [0,1].$$
$$(2.1)$$

The concepts of geometrically symmetric and strongly GA-convex functions established by Obeidat and Latif [128] and Maden *et al.* [66], respectively.

Definition 2.2.3. *A function* $\xi : K \subseteq \mathbb{R}_+ = (0, \infty) \to \mathbb{R}$ *is said to be geometrically symmetric with respect to* $\sqrt{cd}$ *if* $\xi\left(\frac{cd}{x}\right) = \xi(x)$ *for every* $x \in K$.

Definition 2.2.4. *A function* $\xi : K \subseteq \mathbb{R}_+ = (0, \infty) \to \mathbb{R}$ *is said to be strongly GA-convex with modulus* $\mu > 0$, *if*

$$\xi(x^\delta y^{1-\delta}) \leq \delta \xi(x) + (1-\delta)\xi(y) - \mu \delta(1-\delta)\|\ln y - \ln x\|^2, \quad \forall x, y \in K, \delta \in [0,1].$$

Rashid *et. al* [140] gave the following concept of convex functions on time scales.

Definition 2.2.5. *Consider a time scale* $\mathbb{T}$ *and let* $\psi_1, \psi_2 : (0,1) \to \mathbb{R}$ *be two nonnegative funcions. A function* $\xi : K = [c,d]_\mathbb{T} \to \mathbb{R}$ *is said to be a* (ψ_1, ψ_2)-*convex function with respect to two nonnegative functions* ψ_1 *and* ψ_2 *if*

$$\xi((1-\delta)x + \delta y) \leq \psi_1(1-\delta)\psi_2(\delta)\xi(x) + \psi_2(1-\delta)\psi_1(\delta)\xi(y),$$
$$\forall \, x, y \in K, \; \delta \in [0,1].$$

The following result is given by Martin and Thomas [21].

Definition 2.2.6. *Let* $\gamma_k : \mathbb{T}^2 \to \mathbb{R}$, $k \in N_0$ *be defined by*

$$\gamma_0(\delta, \delta_1) = 1, \; \forall \, \delta, \delta_1 \in \mathbb{T}$$

and then recursively by

$$\gamma_{k+1}(\delta, \delta_1) = \int_{\delta_1}^{\delta} \gamma_k(\vartheta, \delta_1)\Delta\vartheta, \; \forall \, \delta, \delta_1 \in \mathbb{T}.$$

Gordji *et al.* [54] established the Hermite–Hadamard and Hermite–Hadamard-Fejér inequalities for η-convex functions.

Theorem 2.2.1. *Suppose that $\xi : [c,d] \to \mathbb{R}$ is a η-convex function such that η is bounded from above on $\xi([c,d]) \times \xi([c,d])$. Then,*

$$\xi\left(\frac{c+d}{2}\right) - \frac{M_\eta}{2} \leq \frac{1}{d-c}\int_c^d \xi(x)dx$$

$$\leq \frac{\xi(c)+\xi(d)}{2} + \frac{\eta(\xi(c),\xi(d)) + \eta(\xi(d),\xi(c))}{4}$$

$$\leq \frac{\xi(c)+\xi(d)}{2} + \frac{M_\eta}{2},$$

where M_η is upper bound of η.

Theorem 2.2.2. *(Hermite–Hadamard-Fejér Left Inequality). Suppose that $\xi : [c,d] \to \mathbb{R}$ is a η-convex function, such that η is bounded from above on $\xi([c,d]) \times \xi([c,d])$. Also suppose that $\psi : [c,d] \to \mathbb{R}^+$ is integrable and symmetric with respect to $\frac{c+d}{2}$. Then,*

$$\xi\left(\frac{c+d}{2}\right)\int_c^d \psi(x)dx - \frac{1}{2}\int_c^d \eta(\xi(c+d-x),\xi(x))\psi(x)dx \leq \int_c^d \xi(x)\psi(x)dx.$$

Theorem 2.2.3. *(Hermite–Hadamard-Fejér Right Inequality). Suppose that $\xi : [c,d] \to \mathbb{R}$ is a η-convex function, such that η is bounded from above on $\xi([c,d]) \times \xi([c,d])$. Also suppose that $\psi : [c,d] \to \mathbb{R}^+$ is integrable and symmetric with respect to $\frac{c+d}{2}$. Then,*

$$\int_c^d \xi(x)\psi(x)dx \leq \frac{\xi(c)+\xi(d)}{2}\int_c^d \psi(x)dx$$

$$+ \frac{\eta(\xi(c),\xi(d)) + \eta(\xi(d),\xi(c))}{2(d-c)}\int_c^d (d-x)\psi(x)dx.$$

The following lemmas proposed by Yan Xi and Qi [175].

Lemma 2.2.1. *Let $\xi : K \subseteq \mathbb{R} \to \mathbb{R}$ be a differentiable function on K^0 such that $\xi' \in L^1[c,d]$, where $c,d \in K$ with $c < d$. If $\beta, \gamma \in \mathbb{R}$, then*

$$\frac{\beta\xi(c)+\gamma\xi(d)}{2} + \frac{2-\beta-\gamma}{2}\xi\left(\frac{c+d}{2}\right) - \frac{1}{d-c}\int_c^d \xi(x)dx$$

$$= \frac{d-c}{4}\int_0^1 \left[(1-\beta-\delta)\xi'\left(\delta c + (1-\delta)\frac{c+d}{2}\right)\right.$$

$$\left. + (\gamma-\delta)\xi'\left(\delta\frac{c+d}{2} + (1-\delta)d\right)\right]d\delta.$$

Lemma 2.2.2. *For $m > 0$ and $0 \leq \rho \leq 1$, we have*

$$\int_0^1 |\rho-\delta|^m d\delta = \frac{\rho^{m+1} + (1-\rho)^{m+1}}{m+1}.$$

and

$$\int_0^1 \delta |\rho - \delta|^m d\delta = \frac{\rho^{m+2} + (m + \rho + 1)(1 - \rho)^{m+1}}{(m+1)(m+2)}.$$

We have proposed the following lemma.

Lemma 2.2.3. *For $m > 0$ and $0 \le \rho \le 1$, we have*

$$\int_0^1 \delta^2 |\rho - \delta|^m d\delta = \frac{2\rho^{m+3} + (1-\rho)^{m+3}}{m+3} + \frac{2\rho(1-\rho)^{m+2}}{m+2} + \frac{\rho^2(1-\rho)^{m+1}}{m+1}.$$

The following results are taken by Kwun *et. al* [90].

Lemma 2.2.4. *Let $\xi : K \subseteq \mathbb{R} \to \mathbb{R}$ be a differentiable mapping on K^0 with $\xi'' \in L^1[c, d]$, where $c, d \in K$ and $c < d$. Then*

$$\frac{1}{d-c} \int_c^d \xi(x) dx - \xi\left(\frac{c+d}{2}\right) = \frac{(d-c)^2}{16} \left[\int_0^1 \delta^2 \xi'' \left(\delta \frac{c+d}{2} + (1-\delta)c\right) d\delta \right.$$
$$\left. + \int_0^1 (\delta - 1)^2 \xi'' \left(\delta d + (1-\delta)\frac{c+d}{2}\right) d\delta \right].$$

Theorem 2.2.4. *Let $\xi : K \subseteq \mathbb{R} \to \mathbb{R}$ be an η-convex function with $\xi \in L^1[c, d]$, where $c, d \in K$ with $c < d$. Then*

$$\xi\left(\frac{c+d}{2}\right) - \frac{1}{2(d-c)} \int_c^d \eta(\xi(c+d-x), \xi(x)) dx$$
$$\le \frac{1}{d-c} \int_c^d \xi(x) dx \le \xi(d) + \frac{1}{2}\eta(\xi(c), \xi(d)).$$

The following lemmas proposed by Jiang *et al.* [68] and Obeidat and Latif [128], respectively.

Lemma 2.2.5. *If $\xi^{(n)}$ for $n \in \mathbb{N}$ exists and is integrable on $[c, d]$, then*

$$\frac{\xi(c) + \xi(d)}{2} - \frac{1}{d-c} \int_c^d \xi(x) dx - \sum_{k=2}^{n-1} \frac{(k-1)(d-c)^k}{2(k+1)!} \xi^{(k)}(c)$$
$$= \frac{(d-c)^n}{2n!} \int_0^1 \delta^{n-1}(n - 2\delta)\xi^{(n)}(\delta c + (1-\delta)d) d\delta.$$

Lemma 2.2.6. *For $0 < c < d$, we have*
(1) $\triangle_1(c, d) = \int_0^1 |\ln(c^{\frac{1-\delta}{2}} d^{\frac{1+\delta}{2}})| d\delta$

$$= \begin{cases} \frac{(\ln c)^2 - 3(\ln d)^2 + 2(\ln c)(\ln d)}{4(\ln d - \ln c)}, & \text{if } d \le 1, \\ \frac{-(\ln c)^2 + 3(\ln d)^2 - 2(\ln c)(\ln d)}{4(\ln d - \ln c)}, & \text{if } \sqrt{cd} \ge 1, \\ \frac{(\ln c)^2 + 5(\ln d)^2 + 2(\ln c)(\ln d)}{4(\ln d - \ln c)}, & \text{if } \sqrt{cd} < 1 < d. \end{cases}$$

(2) $\Delta_2(c,d) = \int_0^1 c^{\frac{1-\delta}{2}} d^{\frac{1+\delta}{2}} |\ln(c^{\frac{1-\delta}{2}} d^{\frac{1+\delta}{2}})| d\delta$

$$= \begin{cases} \frac{2[d - d\ln d - \sqrt{cd} + \sqrt{cd}\ln(\sqrt{cd})]}{\ln d - \ln c}, & \text{if } d \le 1, \\ \frac{2[-d + d\ln d + \sqrt{cd} - \sqrt{cd}\ln(\sqrt{cd})]}{\ln d - \ln c}, & \text{if } \sqrt{cd} \ge 1, \\ \frac{2[2 - d + d\ln d - \sqrt{cd} + \sqrt{cd}\ln(\sqrt{cd})]}{\ln d - \ln c}, & \text{if } \sqrt{cd} < 1 < d. \end{cases}$$

We have proposed the following lemma which are useful for our main results.

Lemma 2.2.7. *For $0 < c < d$, we have*

(1) $\Delta_3(c,d) = \int_0^1 \delta c^{\frac{1-\delta}{2}} d^{\frac{1+\delta}{2}} |\ln(c^{\frac{1-\delta}{2}} d^{\frac{1+\delta}{2}})| d\delta$

$$= \begin{cases} \frac{4[d(-(\ln d)^2 + (\ln d)(\ln\sqrt{cd}) + 2\ln d - \ln\sqrt{cd} - 2) - \sqrt{cd}(\ln\sqrt{cd} - 2)]}{(\ln d - \ln c)^2}, & \text{if } d \le 1, \\ \frac{4[d((\ln d)^2 - (\ln d)(\ln\sqrt{cd}) - 2\ln d + \ln\sqrt{cd} + 2) + \sqrt{cd}(\ln\sqrt{cd} - 2)]}{(\ln d - \ln c)^2}, & \text{if } \sqrt{cd} \ge 1, \\ \frac{4[d((\ln d)^2 - (\ln d)(\ln\sqrt{cd}) - 2\ln d + \ln\sqrt{cd} + 2) + \sqrt{cd}(2 - \ln\sqrt{cd}) - 2\ln\sqrt{cd} - 4]}{(\ln d - \ln c)^2}, & \text{if } \sqrt{cd} < 1 < d. \end{cases}$$

(2) $\Delta_4(c,d) = \int_0^1 \delta^2 c^{\frac{1-\delta}{2}} d^{\frac{1+\delta}{2}} |\ln(c^{\frac{1-\delta}{2}} d^{\frac{1+\delta}{2}})| d\delta$

$$= \begin{cases} \frac{8}{(\ln d - \ln c)^3}[d((\ln d)^2(-\ln d + 2\ln\sqrt{cd} + 3) - \ln d((\ln\sqrt{cd})^2 \\ + 4\ln\sqrt{cd} + 6) + (\ln\sqrt{cd})^2 + 4\ln\sqrt{cd} + 6) - 2\sqrt{cd}(3 - \ln\sqrt{cd})], & \text{if } d \le 1, \\ \frac{8}{(\ln d - \ln c)^3}[d((\ln d)^2(\ln d - 2\ln\sqrt{cd} - 3) + \ln d((\ln\sqrt{cd})^2 \\ + 4\ln\sqrt{cd} + 6) - (\ln\sqrt{cd})^2 - 4\ln\sqrt{cd} - 6) + 2\sqrt{cd}(3 - \ln\sqrt{cd})], & \text{if } \sqrt{cd} \ge 1, \\ \frac{8}{(\ln d - \ln c)^3}[d((\ln d)^3 - (\ln\sqrt{cd})^2 - 3(\ln d)^2 + 6\ln d + \ln d(\ln\sqrt{cd})^2 \\ - 2(\ln d)^2(\ln\sqrt{cd}) + 4\ln d\ln\sqrt{cd} - 4\ln\sqrt{cd} - 6) \\ + 2(\ln\sqrt{cd})^2 + 2\sqrt{cd}\ln\sqrt{cd} + 8\ln\sqrt{cd} - 6\sqrt{cd} + 12], & \text{if } \sqrt{cd} < 1 < d. \end{cases}$$

(3) $\Delta_5(c,d) = \int_0^1 \delta |\ln(c^{\frac{1-\delta}{2}} d^{\frac{1+\delta}{2}})| d\delta$

$$= \begin{cases} \frac{2[-(\ln\sqrt{cd})^3 - 2(\ln d)^3 + 3(\ln\sqrt{cd})(\ln d)^2]}{3(\ln d - \ln c)^2}, & \text{if } d \le 1, \\ \frac{2[(\ln\sqrt{cd})^3 + 2(\ln d)^3 - 3(\ln\sqrt{cd})(\ln d)^2]}{3(\ln d - \ln c)^2}, & \text{if } \sqrt{cd} \ge 1, \\ \frac{2[-(\ln\sqrt{cd})^3 + 2(\ln d)^3 - 3(\ln\sqrt{cd})(\ln d)^2]}{3(\ln d - \ln c)^2}, & \text{if } \sqrt{cd} < 1 < d. \end{cases}$$

(4) $\Delta_6(c,d) = \int_0^1 \delta^2 |\ln(c^{\frac{1-\delta}{2}} d^{\frac{1+\delta}{2}})| d\delta$

$$= \begin{cases} \frac{2[(\ln\sqrt{cd})^4 + 8(\ln d)^3(\ln\sqrt{cd}) - 6(\ln d)^2(\ln\sqrt{cd})^2 - 3(\ln d)^4]}{3(\ln d - \ln c)^3}, & \text{if } d \le 1, \\ \frac{2[-(\ln\sqrt{cd})^4 - 8(\ln d)^3(\ln\sqrt{cd}) + 6(\ln d)^2(\ln\sqrt{cd})^2 + 3(\ln d)^4]}{3(\ln d - \ln c)^3}, & \text{if } \sqrt{cd} \ge 1, \\ \frac{2[(\ln\sqrt{cd})^4 - 8(\ln d)^3(\ln\sqrt{cd}) + 6(\ln d)^2(\ln\sqrt{cd})^2 + 3(\ln d)^4]}{3(\ln d - \ln c)^3}, & \text{if } \sqrt{cd} < 1 < d. \end{cases}$$

For the simplicity, we will use the following notations throughout the chapter:

$$\varrho_1(\delta) = c^{\frac{1-\delta}{2}} d^{\frac{1+\delta}{2}} \quad \text{and} \quad \varrho_2(\delta) = c^{\frac{1+\delta}{2}} d^{\frac{1-\delta}{2}}.$$

Obeidat and Latif [128] established the following lemma.

Lemma 2.2.8. *Let $\xi : K \subseteq \mathbb{R}_+ = (0,\infty) \to \mathbb{R}$ be a differentiable function on K^0 and $c,d \in K^0$ with $c < d$, and let $\lambda : [c,d] \to [0,\infty)$ be a continuous positive mapping and geometrically symmetric to $\sqrt{cd}$. If $\xi' \in L^1[c,d]$ and $\xi : K \subseteq \mathbb{R}_+ = (0,\infty) \to \mathbb{R}$ is geometrically symmetric with respect to $\sqrt{cd}$, then*

$$\frac{(\ln d)\xi(d) + (\ln c)\xi(c)}{\ln d + \ln c} \int_c^d \frac{(\ln x)\lambda(x)}{x}dx - \int_c^d \frac{(\ln x)\lambda(x)\xi(x)}{x}dx$$

$$= \frac{(\ln d - \ln c)}{2(\ln d + \ln c)} \left[\int_0^1 \left(\int_{\varrho_2(\delta)}^{\varrho_1(\delta)} \frac{(\ln x)\lambda(x)}{x}dx \right) \varrho_1(\delta)\ln(\varrho_1(\delta))\xi'(\varrho_1(\delta))d\delta \right.$$

$$\left. - \int_0^1 \left(\int_{\varrho_2(\delta)}^{\varrho_1(\delta)} \frac{(\ln x)\lambda(x)}{x}dx \right) \varrho_2(\delta)\ln(\varrho_2(\delta))\xi'(\varrho_2(\delta))d\delta \right].$$

Followings lemmas are proposed by Tahir *et al.* [165]

Lemma 2.2.9. *Let $\xi : \mathbb{T} \to \mathbb{R}$ be a delta differentiable mapping and $c,d \in \mathbb{T}$ with $c < d$. If $\xi^\Delta \in C_{rd}$, then the following equality holds:*

$$\xi(c)\{1 - \gamma_2(1,0)\} + \xi(d)\gamma_2(1,0) - \frac{1}{d-c}\int_c^d \xi^\diamond(x)\Delta x$$

$$= \frac{d-c}{2}\int_0^1\int_0^1 [\xi^\Delta(\delta c + (1-\delta)d) - \xi^\Delta(\delta_1 c + (1-\delta_1)d)](\delta - \delta_1)\Delta\delta\Delta\delta_1.$$

Lemma 2.2.10. *Let $\xi : [c,d]_\mathbb{T} \to \mathbb{R}$ be a delta differentiable mapping on $\mathbb{T}^0$ and $c,d \in \mathbb{T}$ with $c < d$. If $\xi^\Delta \in C_{rd}$, then the following equality holds :*

$$\xi\left(\frac{c+d}{2}\right) - \frac{1}{d-c}\int_c^d \xi^\diamond(x)\Delta x$$

$$= \frac{d-c}{2}\int_0^1\int_0^1 [\xi^\Delta(\delta c + (1-\delta)d) - \xi^\Delta(\delta_1 c + (1-\delta_1)d)](\psi(\delta_1)$$

$$- \psi(\delta))\Delta\delta\Delta\delta_1,$$

where

$$\psi(x) = \begin{cases} x, & x \in [0,\frac{1}{2}] \\ x-1, & x \in (\frac{1}{2},1]. \end{cases}$$

Rashid *et al.* [140] gave the following corollary for delta differentiable function.

Corollary 2.2.1. *Consider a time scale $\mathbb{T}$ and $K = [c,d]_\mathbb{T}$ such that $c < d$ and $c,d \in \mathbb{T}$. Suppose that there is a delta differentiable function $\xi : K \to \mathbb{R}$*

on K^0. If $\xi^\Delta \in C_{rd}$, then

$$\frac{\xi(c) + \xi(d)}{2} - \frac{1}{d-c} \int_c^d \xi^\diamond(x)\Delta x$$

$$= \frac{d-c}{2}\left[\int_0^1 \delta\xi^\Delta(\delta d + (1-\delta)c)\Delta\delta - \int_0^1 \delta\xi^\Delta(\delta c + (1-\delta)d)\Delta\delta\right].$$

Corollary 2.2.2. *Consider a time scale $\mathbb{T}$ and $K = [c,d]_\mathbb{T}$ such that $c < d$ and $c, d \in \mathbb{T}$. Suppose that there is a delta differentiable function $\xi : K \to \mathbb{R}$ on K^0. If $\xi^\Delta \in C_{rd}$, then*

$$\xi\left(\frac{c+d}{2}\right) - \frac{1}{d-c}\int_c^d \xi^\diamond(x)\Delta x$$

$$= (d-c)\left[\int_0^{1/2} \delta\xi^\Delta(\delta d + (1-\delta)c)\Delta\delta + \int_{1/2}^1 (\delta - 1)\xi^\Delta(\delta d + (1-\delta)c)\Delta\delta\right].$$

2.3   Hermite–Hadamard Type Inequalities for Functions Whose Derivatives are Strongly η-Convex

In this section, we prove some Hermite–Hadamard-and Fejér-type inequalities for strongly η-convex functions [153].

Theorem 2.3.1. *Let ξ and ψ be nonnegative strongly η-convex functions with modulus μ_1 and μ_2, respectively, and $\xi\psi \in L^1[c,d]$, where $c, d \in K$, $c < d$. Then*

$$\frac{1}{(d-c)}\int_c^d \xi(x)\psi(x)dx \le P(c,d),$$

where

$$P(c,d) = \xi(d)\psi(d) + \frac{1}{2}[\xi(d)\eta(\psi(c),\psi(d)) + \psi(d)\eta(\xi(c),\xi(d))]$$

$$- \frac{(c-d)^2}{6}(\mu_1\psi(d) + \mu_2\xi(d)) + \frac{1}{3}\eta(\xi(c),\xi(d))\eta(\psi(c),\psi(d))$$

$$- \frac{(c-d)^2}{12}(\mu_1\eta(\psi(c),\psi(d)) + \mu_2\eta(\xi(c),\xi(d))) + \frac{\mu_1\mu_2}{30}(c-d)^4.$$

Proof Since ξ and ψ are strongly η-convex functions with modulus μ_1 and μ_2, respectively, therefore

$$\xi(\delta c + (1-\delta)d) \le \xi(d) + \delta\eta(\xi(c),\xi(d)) - \mu_1\delta(1-\delta)(c-d)^2, \ \forall \ \delta \in [0,1] \quad (2.2)$$

and

$$\psi(\delta c + (1-\delta)d) \le \psi(d) + \delta\eta(\psi(c),\psi(d)) - \mu_2\delta(1-\delta)(c-d)^2, \ \forall \ \delta \in [0,1]. \quad (2.3)$$

From (2.2) and (2.3), we obtain

$$\xi(\delta c + (1-\delta)d)\psi(\delta c + (1-\delta)d) \leq \xi(d)\psi(d) + \delta(\xi(d)\eta(\psi(c),\psi(d))$$
$$+ \psi(d)\eta(\xi(c),\xi(d))) + \delta^2 \eta(\xi(c),\xi(d))\eta(\psi(c),\psi(d))$$
$$- \delta^2(1-\delta)(c-d)^2(\mu_1\eta(\psi(c),\psi(d)) + \mu_2\eta(\xi(c),\xi(d)))$$
$$- \delta(1-\delta)(c-d)^2(\mu_1\psi(d) + \mu_2\xi(d)) + \delta^2(1-\delta)^2\mu_1\mu_2(c-d)^4.$$

Integrating above inequality from 0 to 1 on both sides with respect to δ, we have

$$\int_0^1 \xi(\delta c + (1-\delta)d)\psi(\delta c + (1-\delta)d)d\delta \leq \xi(d)\psi(d)$$
$$+ \frac{1}{2}[\xi(d)\eta(\psi(c),\psi(d)) + \psi(d)\eta(\xi(c),\xi(d))] + \frac{1}{3}\eta(\xi(c),\xi(d))\eta(\psi(c),\psi(d))$$
$$- \frac{(c-d)^2}{12}(\mu_1\eta(\psi(c),\psi(d)) + \mu_2\eta(\xi(c),\xi(d))) - \frac{(c-d)^2}{6}(\mu_1\psi(d)$$
$$+ \mu_2\xi(d)) + \frac{\mu_1\mu_2}{30}(c-d)^4.$$

This implies,

$$\frac{1}{(d-c)}\int_c^d \xi(x)\psi(x)dx \leq P(c,d).$$

This completes the proof.

Remark 2.3.1. *When* $\mu_1 = 0$ *and* $\mu_2 = 0$, *then above theorem reduces to Theorem 2.2 of [90]: i.e.*

$$\frac{1}{(d-c)}\int_c^d \xi(x)\psi(x)dx \leq P_1(c,d),$$

where

$$P_1(c,d) = \xi(d)\psi(d) + \frac{1}{2}[\xi(d)\eta(\psi(c),\psi(d)) + \psi(d)\eta(\xi(c),\xi(d))]$$
$$+ \frac{1}{3}\eta(\xi(c),\xi(d))\eta(\psi(c),\psi(d)).$$

Remark 2.3.2. *If* $\mu = 0$ *and* $\eta(x,y) = x - y$, *then above theorem reduces to Theorem 1 of [130]: i.e.*

$$\frac{1}{(d-c)}\int_c^d \xi(x)\psi(x)dx \leq P_2(c,d),$$

where

$$P_2(c,d) = \xi(d)\psi(d) + \frac{1}{2}[\xi(d)(\psi(c)-\psi(d)) + \psi(d)(\xi(c)-\xi(d))]$$
$$+ \frac{1}{3}(\xi(c)-\xi(d))(\psi(c)-\psi(d)).$$

Theorem 2.3.2. *Let ξ be an strongly η-convex function with modulus μ and $\xi \in L^1[c, d]$, where $c, d \in K$, $c < d$ and $\psi : [c, d] \to \mathbb{R}$ be nonnegative, integrable and symmetric about $\left(\frac{c+d}{2}\right)$. Then*

$$\int_c^d \xi(x)\psi(x)dx \leq \left(\xi(d) + \frac{1}{2}\eta(\xi(c), \xi(d))\right) \int_c^d \psi(x)dx$$

$$- \mu \int_c^d (x - c)(d - x)\psi(x)dx.$$

Proof Since, ξ be an strongly η-convex function with modulus μ, and ψ nonnegative, integrable and symmetric about $\left(\frac{c+d}{2}\right)$, therefore we have

$$\int_c^d \xi(x)\psi(x)dx$$

$$= \frac{1}{2}\left[\int_c^d \xi(x)\psi(x)dx + \int_c^d \xi(c + d - x)\psi(c + d - x)dx\right]$$

$$= \frac{1}{2}\left[\int_c^d \xi(x)\psi(x)dx + \int_c^d \xi(c + d - x)\psi(x)dx\right]$$

$$= \frac{1}{2}\int_c^d \left[\xi\left(\frac{d - x}{d - c}c + \frac{x - c}{d - c}d\right) + \xi\left(\frac{x - c}{d - c}c + \frac{d - x}{d - c}d\right)\right]\psi(x)dx$$

$$\leq \frac{1}{2}\int_c^d \left[\left(\xi(d) + \left(\frac{d - x}{d - c}\right)\eta(\xi(c), \xi(d))\right.\right.$$

$$- \mu\left(\frac{x - c}{d - c}\right)\left(\frac{d - x}{d - c}\right)(c - d)^2\right) + \left(\xi(d) + \left(\frac{x - c}{d - c}\right)\eta(\xi(c), \xi(d))\right.$$

$$\left.\left.- \mu\left(\frac{x - c}{d - c}\right)\left(\frac{d - x}{d - c}\right)(c - d)^2\right)\right]\psi(x)dx$$

$$= \left(\xi(d) + \frac{1}{2}\eta(\xi(c), \xi(d))\right)\int_c^d \psi(x)dx - \mu\int_c^d (x - c)(d - x)\psi(x)dx.$$

This completes the proof.

Remark 2.3.3. *When $\mu = 0$, then above theorem reduces to Theorem 2.3 of [90]: i.e*

$$\int_c^d \xi(x)\psi(x)dx \leq \left(\xi(d) + \frac{1}{2}\eta(\xi(c), \xi(d))\right)\int_c^d \psi(x)dx.$$

Remark 2.3.4. *If $\mu = 0, \eta(x, y) = x - y$ and $\psi(x) = 1$, then above theorem reduces to the second inequality of Theorem 1.3.1, i.e*

$$\frac{1}{(d - c)}\int_c^d \xi(x)dx \leq \frac{1}{2}(\xi(c) + \xi(d)).$$

Now we establish the results on integral inequalities for strongly η-convex functions.

Theorem 2.3.3. *Let $\xi : K \subseteq \mathbb{R} \to \mathbb{R}$, be a differentiable mapping on K^0 with $\xi' \in L^1[c,d]$, where $c, d \in K, c < d$. If $|\xi'(x)|^q$ for $q \geq 1$ is strongly η-convex with modulus μ on $[c,d]$ and $0 \leq \beta, \gamma \leq 1$, then*

$$\left| \frac{\beta\xi(c) + \gamma\xi(d)}{2} + \frac{2 - \beta - \gamma}{2}\xi\left(\frac{c+d}{2}\right) - \frac{1}{d-c}\int_c^d \xi(x)dx \right|$$

$$\leq \left(\frac{d-c}{8}\right)\left(\frac{1}{24}\right)^{1/q}\left[(2\beta^2 - 2\beta + 1)^{1-\frac{1}{q}}(24(2\beta^2 - 2\beta + 1)|\xi'(d)|^q \right.$$
$$+ 4(-2\beta^3 + 12\beta^2 - 9\beta + 4)\eta(|\xi'(c)|^q, |\xi'(d)|^q)$$
$$- \mu(-7\beta^4 + 28\beta^3 - 30\beta^2 + 12\beta)(c-d)^2)^{1/q}$$
$$+ (2\gamma^2 - 2\gamma + 1)^{1-\frac{1}{q}}(24(2\gamma^2 - 2\gamma + 1)|\xi'(d)|^q$$
$$+ 4(2\gamma^3 - 3\gamma + 2)\eta(|\xi'(c)|^q, |\xi'(d)|^q)$$
$$\left. - \mu(-7\gamma^4 + 8\gamma^3 - 8\gamma + 5)(c-d)^2)^{1/q} \right].$$

Proof Recall Lemma 2.2.1;

$$\frac{\beta\xi(c) + \gamma\xi(d)}{2} + \frac{2 - \beta - \gamma}{2}\xi\left(\frac{c+d}{2}\right) - \frac{1}{d-c}\int_c^d \xi(x)dx$$
$$= \frac{d-c}{4}\int_0^1 \left[(1 - \beta - \delta)\xi'\left(\delta c + (1-\delta)\frac{c+d}{2}\right) \right.$$
$$\left. + (\gamma - \delta)\xi'\left(\delta\frac{c+d}{2} + (1-\delta)d\right) \right] d\delta.$$

This implies,

$$\left| \frac{\beta\xi(c) + \gamma\xi(d)}{2} + \frac{2 - \beta - \gamma}{2}\xi\left(\frac{c+d}{2}\right) - \frac{1}{d-c}\int_c^d \xi(x)dx \right|$$
$$\leq \frac{d-c}{4}\left[\int_0^1 |1 - \beta - \delta|\left|\xi'\left(\delta c + (1-\delta)\frac{c+d}{2}\right)\right| d\delta \right.$$
$$\left. + \int_0^1 |\gamma - \delta|\left|\xi'\left(\delta\frac{c+d}{2} + (1-\delta)d\right)\right| d\delta \right]. \tag{2.4}$$

Applying Hölder's inequality and the definition of strong η-convexity in (2.4), we have

$$
\left| \frac{\beta\xi(c) + \gamma\xi(d)}{2} + \frac{2 - \beta - \gamma}{2}\xi\left(\frac{c+d}{2}\right) - \frac{1}{d-c}\int_c^d \xi(x)dx \right|
$$

$$
\leq \frac{d-c}{4}\left[\left(\int_0^1 |1 - \beta - \delta|d\delta\right)^{1-\frac{1}{q}} \left(\int_0^1 |1 - \beta - \delta|\left(|\xi'(d)|^q\right.\right.\right.
$$

$$
\left. + \left(\frac{1+\delta}{2}\right)\eta(|\xi'(c)|^q, |\xi'(d)|^q) - \mu\left(\frac{1+\delta}{2}\right)\left(\frac{1-\delta}{2}\right)(c-d)^2\right)d\delta\right)^{1/q}
$$

$$
+ \left(\int_0^1 |\gamma - \delta|d\delta\right)^{1-\frac{1}{q}} \left(\int_0^1 |\gamma - \delta|\left(|\xi'(d)|^q + \left(\frac{\delta}{2}\right)\eta(|\xi'(c)|^q, |\xi'(d)|^q)\right.\right.
$$

$$
\left.\left.\left. - \mu\left(\frac{\delta}{2}\right)\left(1 - \frac{\delta}{2}\right)(c-d)^2\right)d\delta\right)^{1/q}\right].
\tag{2.5}
$$

Using Lemmas 2.2.2 and 2.2.3, we calculate

$$
\int_0^1 |1 - \beta - \delta|\left(|\xi'(d)|^q + \left(\frac{1+\delta}{2}\right)\eta(|\xi'(c)|^q, |\xi'(d)|^q) - \frac{\mu}{4}(1 - \delta^2)(c-d)^2\right)d\delta
$$

$$
= \left(|\xi'(d)|^q + \frac{1}{2}\eta(|\xi'(c)|^q, |\xi'(d)|^q) - \frac{\mu}{4}(c-d)^2\right)\int_0^1 |1 - \beta - \delta|d\delta
$$

$$
+ \frac{1}{2}\eta(|\xi'(c)|^q, |\xi'(d)|^q)\int_0^1 \delta|1 - \beta - \delta|d\delta + \frac{\mu}{4}(c-d)^2\int_0^1 \delta^2|1 - \beta - \delta|d\delta
$$

$$
= \frac{1}{2}(2\beta^2 - 2\beta + 1)|\xi'(d)|^q + \frac{1}{12}(-2\beta^3 + 12\beta^2 - 9\beta + 4)\eta(|\xi'(c)|^q, |\xi'(d)|^q)
$$

$$
- \frac{\mu}{48}(-7\beta^4 + 28\beta^3 - 30\beta^2 + 12\beta)(c-d)^2
\tag{2.6}
$$

and

$$
\int_0^1 |\gamma - \delta|\left(|\xi'(d)|^q + \left(\frac{\delta}{2}\right)\eta(|\xi'(c)|^q, |\xi'(d)|^q) - \mu\left(\frac{\delta}{2}\right)\left(1 - \frac{\delta}{2}\right)(c-d)^2\right)d\delta
$$

$$
= |\xi'(d)|^q\int_0^1 |\gamma - \delta|d\delta + \left(\frac{1}{2}\eta(|\xi'(c)|^q, |\xi'(d)|^q) - \frac{\mu}{2}(c-d)^2\right)\int_0^1 \delta|\gamma - \delta|d\delta
$$

$$
+ \frac{\mu}{4}(c-d)^2\int_0^1 \delta^2|\gamma - \delta|d\delta
$$

$$
= \frac{1}{2}(2\gamma^2 - 2\gamma + 1)|\xi'(d)|^q + \frac{1}{12}(2\gamma^3 - 3\gamma + 2)\eta(|\xi'(c)|^q, |\xi'(d)|^q)
$$

$$
- \frac{\mu}{48}(-7\gamma^4 + 8\gamma^3 - 8\gamma + 5)(c-d)^2.
\tag{2.7}
$$

From (2.5)–(2.7) and Lemma 2.2.2, we have

$$\left| \frac{\beta\xi(c) + \gamma\xi(d)}{2} + \frac{2 - \beta - \gamma}{2}\xi\left(\frac{c+d}{2}\right) - \frac{1}{d-c}\int_c^d \xi(x)dx \right|$$

$$\leq \left(\frac{d-c}{8}\right)\left(\frac{1}{24}\right)^{1/q}\left[(2\beta^2 - 2\beta + 1)^{1-\frac{1}{q}}(24(2\beta^2 - 2\beta + 1)|\xi'(d)|^q\right.$$

$$+ 4(-2\beta^3 + 12\beta^2 - 9\beta + 4)\eta(|\xi'(c)|^q, |\xi'(d)|^q)$$

$$- \mu(-7\beta^4 + 28\beta^3 - 30\beta^2 + 12\beta)(c-d)^2)^{1/q}$$

$$+ (2\gamma^2 - 2\gamma + 1)^{1-\frac{1}{q}}(24(2\gamma^2 - 2\gamma + 1)|\xi'(d)|^q$$

$$+ 4(2\gamma^3 - 3\gamma + 2)\eta(|\xi'(c)|^q, |\xi'(d)|^q)$$

$$\left. -\mu(-7\gamma^4 + 8\gamma^3 - 8\gamma + 5)(c-d)^2)^{1/q}\right].$$

This completes the proof.

Corollary 2.3.1. *If $\beta = \gamma$ in above theorem, then*

$$\left| \frac{\beta}{2}(\xi(c) + \xi(d)) + (1 - \beta)\xi\left(\frac{c+d}{2}\right) - \frac{1}{d-c}\int_c^d \xi(x)dx \right|$$

$$\leq \left(\frac{d-c}{8}\right)\left(\frac{1}{24}\right)^{1/q}(2\beta^2 - 2\beta + 1)^{1-\frac{1}{q}}$$

$$\times [(24(2\beta^2 - 2\beta + 1)|\xi'(d)|^q + 4(-2\beta^3 + 12\beta^2 - 9\beta + 4)\eta(|\xi'(c)|^q, |\xi'(d)|^q)$$

$$- \mu(-7\beta^4 + 28\beta^3 - 30\beta^2 + 12\beta)(c-d)^2)^{1/q} + (24(2\beta^2 - 2\beta + 1)|\xi'(d)|^q$$

$$+ 4(2\beta^3 - 3\beta + 2)\eta(|\xi'(c)|^q, |\xi'(d)|^q) - \mu(-7\beta^4 + 8\beta^3 - 8\beta + 5)(c-d)^2)^{1/q}].$$

Corollary 2.3.2. *If $\beta = \gamma = \frac{1}{2}$ in Corollary 2.3.1, then*

$$\left| \frac{1}{2}\left[\frac{\xi(c) + \xi(d)}{2} + \xi\left(\frac{c+d}{2}\right)\right] - \frac{1}{d-c}\int_c^d \xi(x)dx \right|$$

$$\leq \left(\frac{d-c}{16}\right)\left(\frac{1}{192}\right)^{1/q}[(192|\xi'(d)|^q + 144\eta(|\xi'(c)|^q, |\xi'(d)|^q) - 25\mu(c-d)^2)^{1/q}$$

$$+ (192|\xi'(d)|^q + 48\eta(|\xi'(c)|^q, |\xi'(d)|^q) - 25\mu(c-d)^2)^{1/q}].$$

Corollary 2.3.3. *If $q = 1$ in Corollary 2.3.2, then*

$$\left| \frac{1}{2} \left[\frac{\xi(c) + \xi(d)}{2} + \xi\left(\frac{c+d}{2}\right) \right] - \frac{1}{d-c} \int_c^d \xi(x)dx \right|$$

$$\leq \left(\frac{d-c}{1536}\right) \left[192|\xi'(d)| + 96\eta(|\xi'(c)|, |\xi'(d)|) - 25\mu(c-d)^2 \right].$$

Remark 2.3.5. *When $\mu = 0$, then above theorem reduces to Theorem 3.1 of [90]: i.e.*

$$\left| \frac{\beta\xi(c) + \gamma\xi(d)}{2} + \frac{2 - \beta - \gamma}{2}\xi\left(\frac{c+d}{2}\right) - \frac{1}{d-c}\int_c^d \xi(x)dx \right|$$

$$\leq \left(\frac{d-c}{8}\right)\left(\frac{1}{24}\right)^{1/q} \left[(2\beta^2 - 2\beta + 1)^{1-\frac{1}{q}}(24(2\beta^2 - 2\beta + 1)|\xi'(d)|^q \right.$$
$$+ 4(-2\beta^3 + 12\beta^2 - 9\beta + 4)\eta(|\xi'(c)|^q, |\xi'(d)|^q)$$
$$+ (2\gamma^2 - 2\gamma + 1)^{1-\frac{1}{q}}(24(2\gamma^2 - 2\gamma + 1)|\xi'(d)|^q$$
$$\left. + 4(2\gamma^3 - 3\gamma + 2)\eta(|\xi'(c)|^q, |\xi'(d)|^q)) \right].$$

Theorem 2.3.4. *Let $\xi : K \subseteq \mathbb{R} \to \mathbb{R}$, be a differentiable mapping on K^0 with $\xi' \in L^1[c,d]$, where $c, d \in K$, $c < d$. If $|\xi'(x)|^q$ for $q \geq 1$ is strongly η-convex with modulus μ on $[c,d]$ and $0 \leq \beta, \gamma \leq 1$, then*

$$\left| \frac{\beta\xi(c) + \gamma\xi(d)}{2} + \frac{2 - \beta - \gamma}{2}\xi\left(\frac{c+d}{2}\right) - \frac{1}{d-c}\int_c^d \xi(x)dx \right|$$

$$\leq \frac{d-c}{4}\left[\left(\left(\frac{(1-\beta)^{q+1} + \beta^{q+1}}{q+1}\right)|\xi'(d)|^q \right.\right.$$

$$+ \left(\frac{(q+2)((1-\beta)^{q+1} + 2\beta^{q+1}) + (1-\beta)^{q+2} - \beta^{q+2}}{2(q+1)(q+2)}\right)\eta(|\xi'(c)|^q, |\xi'(d)|^q)$$

$$- \frac{\mu}{4}(c-d)^2\left(\frac{(1-\beta)^{q+1} - \beta^{q+3} + 2\beta^{q+2}}{q+1} - \frac{2(1-\beta)\beta^{q+2}}{q+2}\right.$$

$$\left.\left.- \frac{2(1-\beta)^{q+3}\beta^{q+3}}{q+3}\right)\right)^{1/q} + \left(\left(\frac{(1-\gamma)^{q+1} + \gamma^{q+1}}{q+1}\right)|\xi'(d)|^q\right.$$

$$+ \left(\frac{(q+\gamma+1)(1-\gamma)^{q+1} + \gamma^{q+2}}{2(q+1)(q+2)}\right)\eta(|\xi'(c)|^q, |\xi'(d)|^q)$$

$$- \frac{\mu}{4}(c-d)^2\left(\frac{-2(q+1)\gamma(1-\gamma)^{q+2} - (q+2)\gamma^2(1-\gamma)^{q+1} + 2\gamma^{q+2}}{(q+1)(q+2)}\right.$$

$$\left.\left.+ \frac{2(q+\gamma+1)(1-\gamma)^{q+1}}{(q+1)(q+2)} - \frac{2\gamma^{q+3} + (1-\gamma)^{q+3}}{q+3}\right)\right)^{1/q}\right].$$

Proof From Lemma 2.2.1, we have

$$\left| \frac{\beta\xi(c) + \gamma\xi(d)}{2} + \frac{2 - \beta - \gamma}{2}\xi\left(\frac{c+d}{2}\right) - \frac{1}{d-c}\int_c^d \xi(x)dx \right|$$

$$\leq \frac{d-c}{4}\left[\int_0^1 |1 - \beta - \delta| \left| \xi'\left(\delta c + (1-\delta)\frac{c+d}{2}\right) \right| d\delta \right.$$

$$\left. + \int_0^1 |\gamma - \delta| \left| \xi'\left(\delta\frac{c+d}{2} + (1-\delta)d\right) \right| d\delta \right].$$

Using Hölder's inequality and the definition of strong η-convexity, we have

$$\left| \frac{\beta\xi(c) + \gamma\xi(d)}{2} + \frac{2 - \beta - \gamma}{2}\xi\left(\frac{c+d}{2}\right) - \frac{1}{d-c}\int_c^d \xi(x)dx \right|$$

$$\leq \frac{d-c}{4}\left[\left(\int_0^1 d\delta\right)^{1-\frac{1}{q}} \left(\int_0^1 |1 - \beta - \delta|^q \left| \xi'\left(\delta c + (1-\delta)\frac{c+d}{2}\right) \right|^q d\delta\right)^{1/q} \right.$$

$$\left. + \left(\int_0^1 d\delta\right)^{1-\frac{1}{q}} \left(\int_0^1 |\gamma - \delta|^q \left| \xi'\left(\delta\frac{c+d}{2} + (1-\delta)d\right) \right|^q d\delta\right)^{1/q} \right]$$

$$\leq \frac{d-c}{4}\left[\left(\int_0^1 |1 - \beta - \delta|^q \left(|\xi'(d)|^q + \left(\frac{1+\delta}{2}\right)\eta(|\xi'(c)|^q, |\xi'(d)|^q) \right. \right. \right.$$

$$\left. \left. - \frac{\mu}{4}(1 - \delta^2)(c-d)^2 \right) d\delta\right)^{1/q} + \left(\int_0^1 |\gamma - \delta|^q \left(|\xi'(d)|^q \right. \right.$$

$$\left. \left. + \left(\frac{\delta}{2}\right)\eta(|\xi'(c)|^q, |\xi'(d)|^q) - \mu\left(\frac{\delta}{2}\right)\left(1 - \frac{\delta}{2}\right)(c-d)^2 \right) d\delta\right)^{1/q} \right]$$

$$= \frac{d-c}{4}\left[\left(\left(|\xi'(d)|^q + \frac{1}{2}\eta(|\xi'(c)|^q, |\xi'(d)|^q) - \frac{\mu}{4}(c-d)^2 \right) \right. \right.$$

$$\times \int_0^1 |1 - \beta - \delta|^q d\delta + \frac{1}{2}\eta(|\xi'(c)|^q, |\xi'(d)|^q)\int_0^1 \delta|1 - \beta - \delta|^q d\delta$$

$$\left. + \frac{\mu}{4}(c-d)^2 \int_0^1 \delta^2|1 - \beta - \delta|^q d\delta\right)^{1/q}$$

$$+ \left(|\xi'(d)|^q \int_0^1 |\gamma - \delta|^q d\delta + \left(\frac{1}{2}\eta(|\xi'(c)|^q, |\xi'(d)|^q) - \frac{\mu}{2}(c-d)^2 \right) \right.$$

$$\left. \left. \times \int_0^1 \delta|\gamma - \delta|^q d\delta + \frac{\mu}{4}(c-d)^2 \int_0^1 \delta^2|\gamma - \delta|^q d\delta\right)^{1/q} \right].$$

Applying Lemmas 2.2.2 and 2.2.3, we obtain

$$\left| \frac{\beta\xi(c) + \gamma\xi(d)}{2} + \frac{2 - \beta - \gamma}{2}\xi\left(\frac{c+d}{2}\right) - \frac{1}{d-c}\int_c^d \xi(x)dx \right|$$

$$\leq \frac{d-c}{4}\left[\left(\left(\frac{(1-\beta)^{q+1} + \beta^{q+1}}{q+1}\right)|\xi'(d)|^q \right.\right.$$

$$+ \left(\frac{(q+2)((1-\beta)^{q+1} + 2\beta^{q+1}) + (1-\beta)^{q+2} - \beta^{q+2}}{2(q+1)(q+2)}\right)$$

$$\times \ \eta(|\xi'(c)|^q, |\xi'(d)|^q) - \frac{\mu}{4}(c-d)^2\left(\frac{(1-\beta)^{q+1} - \beta^{q+3} + 2\beta^{q+2}}{q+1}\right.$$

$$\left.\left.- \frac{2(1-\beta)\beta^{q+2}}{q+2} - \frac{2(1-\beta)^{q+3} + \beta^{q+3}}{q+3}\right)\right)^{1/q}\right]$$

$$+ \frac{d-c}{4}\left[\left(\left(\frac{(1-\gamma)^{q+1} + \gamma^{q+1}}{q+1}\right)|\xi'(d)|^q \right.\right.$$

$$+ \left(\frac{(q+\gamma+1)(1-\gamma)^{q+1} + \gamma^{q+2}}{2(q+1)(q+2)}\right)\eta(|\xi'(c)|^q, |\xi'(d)|^q) - \frac{\mu}{4}(c-d)^2$$

$$\times \left(\frac{-2(q+1)\gamma(1-\gamma)^{q+2} - (q+2)\gamma^2(1-\gamma)^{q+1} + 2\gamma^{q+2}}{(q+1)(q+2)}\right.$$

$$\left.\left.+ \frac{2(q+\gamma+1)(1-\gamma)^{q+1}}{(q+1)(q+2)} - \frac{2\gamma^{q+3} + (1-\gamma)^{q+3}}{q+3}\right)\right)^{1/q}\right].$$

This completes the proof.

Corollary 2.3.4. *If $\beta = \gamma$ in above theorem, then*

$$\left| \frac{\beta}{2}(\xi(c) + \xi(d)) + (1-\beta)\xi\left(\frac{c+d}{2}\right) - \frac{1}{d-c}\int_c^d \xi(x)dx \right|$$

$$\leq \frac{d-c}{4}\left[\left(\left(\frac{(1-\beta)^{q+1} + \beta^{q+1}}{q+1}\right)|\xi'(d)|^q \right.\right.$$

$$+ \left(\frac{(q+2)((1-\beta)^{q+1} + 2\beta^{q+1}) + (1-\beta)^{q+2} - \beta^{q+2}}{2(q+1)(q+2)}\right)\eta(|\xi'(c)|^q, |\xi'(d)|^q)$$

$$- \frac{\mu}{4}(c-d)^2\left(\frac{(1-\beta)^{q+1} - \beta^{q+3} + 2\beta^{q+2}}{q+1} - \frac{2(1-\beta)\beta^{q+2}}{q+2}\right.$$

$$\left.\left.- \frac{2(1-\beta)^{q+3}\beta^{q+3}}{q+3}\right)\right)^{1/q} + \left(\left(\frac{(1-\beta)^{q+1} + \beta^{q+1}}{q+1}\right)|\xi'(d)|^q \right.$$

$$+ \left(\frac{(q+\beta+1)(1-\beta)^{q+1} + \beta^{q+2}}{2(q+1)(q+2)}\right)\eta(|\xi'(c)|^q, |\xi'(d)|^q)$$

$$- \frac{\mu}{4}(c-d)^2\left(\frac{-2(q+1)\beta(1-\beta)^{q+2} - (q+2)\beta^2(1-\beta)^{q+1} + 2\beta^{q+2}}{(q+1)(q+2)}\right.$$

$$\left.\left.+ \frac{2(q+\beta+1)(1-\beta)^{q+1}}{(q+1)(q+2)} - \frac{2\beta^{q+3} + (1-\beta)^{q+3}}{q+3}\right)\right)^{1/q}\right].$$

Remark 2.3.6. *When $\mu = 0$, then above theorem reduces to Theorem 3.2 of [90], i.e.*

$$\left| \frac{\beta\xi(c) + \gamma\xi(d)}{2} + \frac{2 - \beta - \gamma}{2}\xi\left(\frac{c+d}{2}\right) - \frac{1}{d-c}\int_c^d \xi(x)dx \right|$$

$$\leq \frac{d-c}{4}\left[\left(\left(\frac{(1-\beta)^{q+1} + \beta^{q+1}}{q+1} \right) |\xi'(d)|^q \right. \right.$$

$$+ \left. \left(\frac{(q+2)((1-\beta)^{q+1} + 2\beta^{q+1}) + (1-\beta)^{q+2} - \beta^{q+2}}{2(q+1)(q+2)} \right) \eta(|\xi'(c)|^q, |\xi'(d)|^q) \right)^{1/q}$$

$$+ \left(\left(\frac{(1-\gamma)^{q+1} + \gamma^{q+1}}{q+1} \right) |\xi'(d)|^q + \left(\frac{(q+\gamma+1)(1-\gamma)^{q+1} + \gamma^{q+2}}{2(q+1)(q+2)} \right) \right.$$

$$\left. \times \; \eta(|\xi'(c)|^q, |\xi'(d)|^q))^{1/q} \right].$$

Theorem 2.3.5. *Let $\xi : K \subset [0,\infty) \to \mathbb{R}$ be a differentiable mapping on K^0 with $\xi'' \in L^1[c,d]$, where $c, d \in K$ and $c < d$. If $|\xi''|$ is strongly η-convex with modulus μ on $[c,d]$, then*

$$\left| \xi\left(\frac{c+d}{2}\right) - \frac{1}{d-c}\int_c^d \xi(x)dx \right|$$

$$\leq \frac{(d-c)^2}{16}\left[\frac{1}{3}\left(|\xi''(c)| + \left|\xi''\left(\frac{c+d}{2}\right)\right| \right) + \frac{1}{4}\left(\eta\left(\left|\xi''\left(\frac{c+d}{2}\right)\right|, |\xi''(c)| \right) \right. \right.$$

$$+ \frac{1}{3}\eta\left(|\xi''(d)|, \left|\xi''\left(\frac{c+d}{2}\right)\right| \right) \right) - \frac{\mu}{40}(d-c)^2 \Bigg].$$

Proof Recall Lemma 2.2.4, we have

$$\left| \xi\left(\frac{c+d}{2}\right) - \frac{1}{d-c}\int_c^d \xi(x)dx \right|$$

$$\leq \frac{(d-c)^2}{16}\left[\int_0^1 \delta^2 \left|\xi''\left(\delta\frac{c+d}{2} + (1-\delta)c \right)\right| d\delta \right.$$

$$+ \left. \int_0^1 (\delta-1)^2 \left|\xi''\left(\delta d + (1-\delta)\frac{c+d}{2} \right)\right| d\delta \right].$$

Using the definition of strong η-convexity, we obtain

$$
\left| \xi\left(\frac{c+d}{2}\right) - \frac{1}{d-c}\int_c^d \xi(x)dx \right|
$$

$$
\leq \frac{(d-c)^2}{16} \left[\int_0^1 \delta^2 \left(|\xi''(c)| + \delta\eta\left(\left|\xi''\left(\frac{c+d}{2}\right)\right|, |\xi''(c)| \right) \right) \right.
$$

$$
- \mu\delta(1-\delta)\left(\frac{d-c}{2}\right)^2 \right) d\delta
$$

$$
+ \int_0^1 (\delta-1)^2 \left(\left|\xi''\left(\frac{c+d}{2}\right)\right| + \delta\eta\left(|\xi''(d)|, \left|\xi''\left(\frac{c+d}{2}\right)\right| \right) \right)
$$

$$
\left. - \mu\delta(1-\delta)\left(\frac{d-c}{2}\right)^2 \right) d\delta \right]
$$

$$
= \frac{(d-c)^2}{16} \left[\left(\frac{1}{3}|\xi''(c)| + \frac{1}{4}\eta\left(\left|\xi''\left(\frac{c+d}{2}\right)\right|, |\xi''(c)| \right) - \frac{\mu}{20}\left(\frac{d-c}{2}\right)^2 \right) \right.
$$

$$
\left. + \left(\frac{1}{3}\left|\xi''\left(\frac{c+d}{2}\right)\right| + \frac{1}{12}\eta\left(|\xi''(d)|, \left|\xi''\left(\frac{c+d}{2}\right)\right| \right) - \frac{\mu}{20}\left(\frac{d-c}{2}\right)^2 \right) \right]
$$

$$
= \frac{(d-c)^2}{16} \left[\frac{1}{3}\left(|\xi''(c)| + \left|\xi''\left(\frac{c+d}{2}\right)\right| \right) + \frac{1}{4}\left(\eta\left(\left|\xi''\left(\frac{c+d}{2}\right)\right|, |\xi''(c)| \right) \right) \right.
$$

$$
\left. + \frac{1}{3}\eta\left(|\xi''(d)|, \left|\xi''\left(\frac{c+d}{2}\right)\right| \right) \right) - \frac{\mu}{40}(d-c)^2 \right].
$$

This completes the proof.

Remark 2.3.7. *When $\mu = 0$, then above theorem reduces to Theorem 3.3 of [90]: i.e.*

$$
\left| \xi\left(\frac{c+d}{2}\right) - \frac{1}{d-c}\int_c^d \xi(x)dx \right|
$$

$$
\leq \frac{(d-c)^2}{16} \left[\frac{1}{3}\left(|\xi''(c)| + \left|\xi''\left(\frac{c+d}{2}\right)\right| \right) + \frac{1}{4}\left(\eta\left(\left|\xi''\left(\frac{c+d}{2}\right)\right|, |\xi''(c)| \right) \right.\right.
$$

$$
\left.\left. + \frac{1}{3}\eta\left(|\xi''(d)|, \left|\xi''\left(\frac{c+d}{2}\right)\right| \right) \right) \right].
$$

Theorem 2.3.6. *Let $\xi : K \subset [0,\infty) \to \mathbb{R}$ be a differentiable mapping on K^0 with $\xi'' \in L^1[c,d]$, where $c, d \in K$ and $c < d$. If $|\xi''|^q$ for $q \geq 1$ with $\frac{1}{p} + \frac{1}{q} = 1$*

is strongly η-convex with modulus μ on $[c, d]$, then

$$\left| \xi \left(\frac{c+d}{2} \right) - \frac{1}{d-c} \int_c^d \xi(x) dx \right|$$

$$\leq \frac{(d-c)^2}{16} \left(\frac{1}{3} \right)^{\frac{1}{p}} \left[\left(\frac{1}{3} |\xi''(c)|^q + \frac{1}{4} \eta \left(\left| \xi'' \left(\frac{c+d}{2} \right) \right|^q , |\xi''(c)|^q \right) \right. \right.$$

$$\left. - \frac{\mu}{80} (d-c)^2 \right)^{\frac{1}{q}} + \left(\frac{1}{3} \left| \xi'' \left(\frac{c+d}{2} \right) \right|^q \right.$$

$$\left. \left. + \frac{1}{12} \eta \left(|\xi''(d)|^q , \left| \xi'' \left(\frac{c+d}{2} \right) \right|^q \right) - \frac{\mu}{80} (d-c)^2 \right)^{\frac{1}{q}} \right].$$

Proof From Lemma 2.2.4, we have

$$\left| \xi \left(\frac{c+d}{2} \right) - \frac{1}{d-c} \int_c^d \xi(x) dx \right|$$

$$\leq \frac{(d-c)^2}{16} \left[\int_0^1 \delta^2 \left| \xi'' \left(\delta \frac{c+d}{2} + (1-\delta)c \right) \right| d\delta \right.$$

$$\left. + \int_0^1 (\delta - 1)^2 \left| \xi'' \left(\delta d + (1-\delta) \frac{c+d}{2} \right) \right| d\delta \right].$$

Using Hölder's inequality, we have

$$\left| \xi \left(\frac{c+d}{2} \right) - \frac{1}{d-c} \int_c^d \xi(x) dx \right|$$

$$\leq \frac{(d-c)^2}{16} \left[\int_0^1 \delta^{\frac{2}{p}} \delta^{\frac{2}{q}} \left| \xi'' \left(\delta \frac{c+d}{2} + (1-\delta)c \right) \right| d\delta \right.$$

$$\left. + \int_0^1 (\delta - 1)^{\frac{2}{p}} (\delta - 1)^{\frac{2}{q}} \left| \xi'' \left(\delta d + (1-\delta) \frac{c+d}{2} \right) \right| d\delta \right]$$

$$\leq \frac{(d-c)^2}{16} \left[\left(\int_0^1 \delta^2 d\delta \right)^{\frac{1}{p}} \left(\int_0^1 \delta^2 \left| \xi'' \left(\delta \frac{c+d}{2} + (1-\delta)c \right) \right|^q d\delta \right)^{\frac{1}{q}} \right.$$

$$\left. + \left(\int_0^1 (\delta - 1)^2 d\delta \right)^{\frac{1}{p}} \left(\int_0^1 (\delta - 1)^2 \left| \xi'' \left(\delta d + (1-\delta) \frac{c+d}{2} \right) \right|^q d\delta \right)^{\frac{1}{q}} \right].$$

Since $|\xi''|$ is strongly η-convex function with modulus $\mu > 0$, therefore

$$\left| \xi\left(\frac{c+d}{2}\right) - \frac{1}{d-c}\int_c^d \xi(x)dx \right|$$

$$\leq \frac{(d-c)^2}{16}\left[\left(\int_0^1 \delta^2 d\delta\right)^{\frac{1}{p}} \left(\int_0^1 \delta^2\left(|\xi''(c)|^q + \delta\eta\left(\left|\xi''\left(\frac{c+d}{2}\right)\right|^q, |\xi''(c)|^q\right) \right. \right.\right.$$

$$\left.\left. - \frac{\mu}{4}\delta(1-\delta)(d-c)^2\right)d\delta\right)^{\frac{1}{q}} + \left(\int_0^1 (\delta-1)^2 d\delta\right)^{\frac{1}{p}} \left(\int_0^1 (\delta-1)^2\left(\left|\xi''\left(\frac{c+d}{2}\right)\right|^q\right.\right.$$

$$\left.\left.\left. + \delta\eta\left(|\xi''(d)|^q, \left|\xi''\left(\frac{c+d}{2}\right)\right|^q\right) - \frac{\mu}{4}\delta(1-\delta)(d-c)^2\right)d\delta\right)^{\frac{1}{q}}\right]$$

$$= \frac{(d-c)^2}{16}\left[\left(\frac{1}{3}\right)^{\frac{1}{p}}\left(\frac{1}{3}|\xi''(c)|^q + \frac{1}{4}\eta\left(\left|\xi''\left(\frac{c+d}{2}\right)\right|^q, |\xi''(c)|^q\right)\right.\right.$$

$$\left. - \frac{\mu}{80}(d-c)^2\right)^{\frac{1}{q}} + \left(\frac{1}{3}\right)^{\frac{1}{p}}\left(\frac{1}{3}\left|\xi''\left(\frac{c+d}{2}\right)\right|^q\right.$$

$$\left.\left. + \frac{1}{12}\eta\left(|\xi''(d)|^q, \left|\xi''\left(\frac{c+d}{2}\right)\right|^q\right) - \frac{\mu}{80}(d-c)^2\right)^{\frac{1}{q}}\right].$$

This completes the proof.

Remark 2.3.8. *When $\mu = 0$, then above theorem reduces to Theorem 3.4 of [90]: i.e.*

$$\left| \xi\left(\frac{c+d}{2}\right) - \frac{1}{d-c}\int_c^d \xi(x)dx \right|$$

$$\leq \frac{(d-c)^2}{16}\left(\frac{1}{3}\right)^{\frac{1}{p}}\left[\left(\frac{1}{3}|\xi''(c)|^q + \frac{1}{4}\eta\left(\left|\xi''\left(\frac{c+d}{2}\right)\right|^q, |\xi''(c)|^q\right)\right)^{\frac{1}{q}}\right.$$

$$\left. + \left(\frac{1}{3}\left|\xi''\left(\frac{c+d}{2}\right)\right|^q + \frac{1}{12}\eta\left(|\xi''(d)|^q, \left|\xi''\left(\frac{c+d}{2}\right)\right|^q\right)\right)^{\frac{1}{q}}\right].$$

2.3.1 Application to special means

Now, we consider the following special means for positive real numbers $c, d > 0$ [153]:

$$\text{Arithmetic mean: } A(c,d) = \frac{c+d}{2}.$$

$$\text{Geometric mean: } G(c,d) = \sqrt{cd}.$$

$$\text{Harmonic mean : } H(c,d) = \frac{2}{\frac{1}{c}+\frac{1}{d}}.$$

Generalized logarithmic mean:

$$L(c,d) = \begin{cases} \left[\frac{d^{m+1}-c^{m+1}}{(m+1)(d-c)}\right]^{1/m}, & \text{if } c \neq d, m \neq -1, 0 \\ c, & \text{if } c = d. \end{cases}$$

Identric mean: $\quad I(c,d) = \begin{cases} \frac{1}{e}\left(\frac{d^d}{c^c}\right)^{1/(d-c)}, & \text{if } d \neq c, \\ c, & \text{if } c = d. \end{cases}$

Heronian mean: $\quad H_{w,m}(c,d) = \begin{cases} \left[\frac{c^m+w(cd)^{\frac{m}{2}}+d^m}{(w+2)}\right]^{1/m}, & \text{if } m \neq 0, \\ \sqrt{cd}, & \text{if } m = 0, \end{cases}$

for $0 \leq w < \infty$.

Now, using the above results in previous theorems, we have some applications to the special means of positive real numbers.

Theorem 2.3.7. *Let $c, d > 0, c \neq d, q \geq 1$, and either $m > 1$ and $(m-1)q \geq 1$ or $m < 0$. Then*

$$\left| A(\beta c^m, \gamma d^m) + \frac{2-\beta-\gamma}{2} A^m(c,d) - L^m(c,d) \right|$$

$$\leq \left(\frac{d-c}{8}\right)\left(\frac{1}{24}\right)^{1/q} \left[(2\beta^2-2\beta+1)^{1-\frac{1}{q}}(24(2\beta^2-2\beta+1)|md^{m-1}|^q \right.$$

$$+ 4(-2\beta^3+12\beta^2-9\beta+4)\eta(|mc^{m-1}|^q, |md^{m-1}|)|^q)$$

$$- \mu(-7\beta^4+28\beta^3-30\beta^2+12\beta)(c-d)^2)^{1/q} + (2\gamma^2-2\gamma+1)^{1-\frac{1}{q}}$$

$$\times (24(2\gamma^2-2\gamma+1)|md^{m-1}|)|^q + 4(2\gamma^3-3\gamma+2)\eta(|mc^{m-1}|^q, |md^{m-1}|^q)$$

$$- \mu(-7\gamma^4+8\gamma^3-8\gamma+5)(c-d)^2)^{1/q}\right].$$

Proof Applying Theorem 2.3.3 with $\xi(x) = x^m$. Then we obtain the result immediately.

Example 2.3.1. *Let $\xi(x) = x^2, \eta(x,y) = x+y+(x-y)^2, \mu = 1, \beta = \gamma = 1, c = 1, d = 2, q = 1$. Then above theorem is verified.*

Theorem 2.3.8. *Let $c, d > 0, c \neq d, q \geq 1$, Then*

$$\left| \frac{\ln G^2(c^\beta, d^\gamma)}{2} + \frac{2 - \beta - \gamma}{2} \ln A(c, d) - \ln I(c, d) \right|$$

$$\leq \left(\frac{d - c}{8} \right) \left(\frac{1}{24} \right)^{1/q} \left[(2\beta^2 - 2\beta + 1)^{1 - \frac{1}{q}} \left(24(2\beta^2 - 2\beta + 1) \left(\frac{1}{d} \right)^q \right. \right.$$

$$+ 4(-2\beta^3 + 12\beta^2 - 9\beta + 4)\eta \left(\left(\frac{1}{c} \right)^q, \left(\frac{1}{d} \right)^q \right)$$

$$- \mu(-7\beta^4 + 28\beta^3 - 30\beta^2 + 12\beta)(c - d)^2 \Big)^{1/q} + (2\gamma^2 - 2\gamma + 1)^{1 - \frac{1}{q}}$$

$$\times \left(24(2\gamma^2 - 2\gamma + 1) \left(\frac{1}{d} \right)^q + 4(2\gamma^3 - 3\gamma + 2)\eta \left(\left(\frac{1}{c} \right)^q, \left(\frac{1}{d} \right)^q \right) \right.$$

$$- \mu(-7\gamma^4 + 8\gamma^3 - 8\gamma + 5)(c - d)^2 \Big)^{1/q} \Big].$$

Proof Applying Theorem 2.3.3 with $\xi(x) = \ln x$. Then we obtain the result immediately.

Theorem 2.3.9. *For $d > c > 0, c \neq dw \geq 0$, and $m \geq 4$ or $0 \neq m < 1$, we have*

$$\left| \frac{1}{2} \left[\frac{H_{w,m}^m(c, d)}{H(c^m, d^m)} + H_{w,m}^m \left(\frac{c}{d} + \frac{d}{c}, 1 \right) \right] - H_{w,m}^m \left(L \left(\frac{c}{d}, \frac{d}{c} \right), 1 \right) \right|$$

$$\leq \frac{(d - c)A(c, d)}{768 \, G^2(c, d)} \left[\frac{192|m|}{w + 2} \left(G^{2(m-1)} \left(d, \frac{1}{c} \right) + \frac{w}{2} G^{2(\frac{m}{2} - 1)} \left(d, \frac{1}{c} \right) \right) \right.$$

$$+ 96\eta \left(\frac{|m|}{w + 2} \left(G^{2(m-1)} \left(c, \frac{1}{d} \right) + \frac{w}{2} G^{2(\frac{m}{2} - 1)} \left(c, \frac{1}{d} \right) \right) \right.,$$

$$\frac{|m|}{w + 2} \left(G^{2(m-1)} \left(d, \frac{1}{c} \right) + \frac{w}{2} G^{2(\frac{m}{2} - 1)} \left(d, \frac{1}{c} \right) \right) \right) - \frac{100\mu(d - c)^2 A^2(c, d)}{G^2(c^2, d^2)} \right].$$

Proof From Corollary 2.3.3, we have

$$\left| \frac{1}{2} \left[\frac{\xi \left(\frac{c}{d} \right) + \xi \left(\frac{d}{c} \right)}{2} + \xi \left(\frac{\frac{c}{d} + \frac{d}{c}}{2} \right) \right] - \frac{1}{\frac{d}{c} - \frac{c}{d}} \int_{\frac{c}{d}}^{\frac{d}{c}} \xi(x) dx \right|$$

$$\leq \left(\frac{\frac{d}{c} - \frac{c}{d}}{1536} \right) \left[192 \left| \xi' \left(\frac{d}{c} \right) \right| + 96\eta \left(\left| \xi' \left(\frac{c}{d} \right) \right|, \left| \xi' \left(\frac{d}{c} \right) \right| \right) - 25\mu \left(\frac{c}{d} - \frac{d}{c} \right)^2 \right].$$

$$(2.8)$$

Applying $\xi(x) = \frac{x^m + wx^{\frac{m}{2}} + 1}{w+2}$ for $x > 0$ and $m \notin (1,4)$ in above inequality, we obtain

$$
\begin{aligned}
&\frac{\xi\left(\frac{c}{d}\right) + \xi\left(\frac{d}{c}\right)}{2} \\
&= \frac{\left(\frac{c}{d}\right)^m + w\left(\frac{c}{d}\right)^{\frac{m}{2}} + 1}{2(w+2)} + \frac{\left(\frac{d}{c}\right)^m + w\left(\frac{d}{c}\right)^{\frac{m}{2}} + 1}{2(w+2)} \qquad (2.9) \\
&= \frac{1}{2(w+2)}\left[\frac{c^{2m} + wc^m(cd)^{\frac{m}{2}} + 2c^m d^m + + wd^m(cd)^{\frac{m}{2}} + d^{2m}}{c^m d^m}\right] \\
&= \frac{1}{2(w+2)}\left[\frac{(c^m + w(cd)^{\frac{m}{2}} + d^m)(c^m + d^m)}{c^m d^m}\right] \\
&= \frac{H_{w,m}^m(c,d)}{H(c^m, d^m)}, \qquad (2.10)
\end{aligned}
$$

$$
\xi\left(\frac{\frac{c}{d} + \frac{d}{c}}{2}\right) = \frac{\left(\frac{\frac{c}{d}+\frac{d}{c}}{2}\right)^m + w\left(\frac{\frac{c}{d}+\frac{d}{c}}{2}\right)^{\frac{m}{2}} + 1}{(w+2)} = H_{w,m}^m\left(\frac{\frac{c}{d}+\frac{d}{c}}{2}, 1\right), \quad (2.11)
$$

$$
\begin{aligned}
&\frac{1}{\frac{d}{c} - \frac{c}{d}} \int_{\frac{c}{d}}^{\frac{d}{c}} \xi(x)\,dx \\
&= \frac{1}{(w+2)}\left[\left\{\frac{\left(\frac{d}{c}\right)^{m+1} - \left(\frac{c}{d}\right)^{m+1}}{(m+1)(\frac{d}{c} - \frac{c}{d})}\right\} + w\left\{\frac{\left(\frac{d}{c}\right)^{\frac{m}{2}+1} - \left(\frac{c}{d}\right)^{\frac{m}{2}+1}}{(\frac{m}{2}+1)(\frac{d}{c} - \frac{c}{d})}\right\} + 1\right] \\
&= H_{w,m}^m\left(L\left(\frac{c}{d}, \frac{d}{c}\right), 1\right), \qquad (2.12)
\end{aligned}
$$

and

$$
\begin{aligned}
&\left(\frac{\frac{d}{c} - \frac{c}{d}}{1536}\right)\left[192\left|\xi'\left(\frac{d}{c}\right)\right| + 96\eta\left(\left|\xi'\left(\frac{c}{d}\right)\right|, \left|\xi'\left(\frac{d}{c}\right)\right|\right) - 25\mu\left(\frac{c}{d} - \frac{d}{c}\right)^2\right] \\
&= \frac{(d-c)A(c,d)}{768\,G^2(c,d)}\left[\frac{192|m|}{w+2}\left(G^{2(m-1)}\left(d, \frac{1}{c}\right) + \frac{w}{2}G^{2(\frac{m}{2}-1)}\left(d, \frac{1}{c}\right)\right)\right. \\
&\quad + 96\eta\left(\frac{|m|}{w+2}\left(G^{2(m-1)}\left(c, \frac{1}{d}\right) + \frac{w}{2}G^{2(\frac{m}{2}-1)}\left(c, \frac{1}{d}\right)\right),\right. \\
&\quad \left.\frac{|m|}{w+2}\left(G^{2(m-1)}\left(d, \frac{1}{c}\right) + \frac{w}{2}G^{2(\frac{m}{2}-1)}\left(d, \frac{1}{c}\right)\right)\right) \\
&\quad \left. - \frac{100\mu(d-c)^2 A^2(c,d)}{G^2(c^2, d^2)}\right]. \qquad (2.13)
\end{aligned}
$$

Applying (2.9) and (2.11)–(2.13) in (2.8), we have

$$\left| \frac{1}{2} \left[\frac{H_{w,m}^m(c,d)}{H(c^m,d^m)} + H_{w,m}^m \left(\frac{c}{d} + \frac{d}{c}, 1 \right) \right] - H_{w,m}^m \left(L \left(\frac{c}{d}, \frac{d}{c} \right), 1 \right) \right|$$

$$\leq \frac{(d-c)A(c,d)}{768\, G^2(c,d)} \left[\frac{192|m|}{w+2} \left(G^{2(m-1)} \left(d, \frac{1}{c} \right) + \frac{w}{2} G^{2\left(\frac{m}{2}-1 \right)} \left(d, \frac{1}{c} \right) \right) \right.$$

$$+ 96\eta \left(\frac{|m|}{w+2} \left(G^{2(m-1)} \left(c, \frac{1}{d} \right) + \frac{w}{2} G^{2\left(\frac{m}{2}-1 \right)} \left(c, \frac{1}{d} \right) \right), \right.$$

$$\frac{|m|}{w+2} \left(G^{2(m-1)} \left(d, \frac{1}{c} \right) + \frac{w}{2} G^{2\left(\frac{m}{2}-1 \right)} \left(d, \frac{1}{c} \right) \right) \right)$$

$$\left. - \frac{100\mu(d-c)^2 A^2(c,d)}{G^2(c^2,d^2)} \right].$$

2.4 Weighted Version of Hermite–Hadamard Type Inequalities for Strongly GA-Convex Functions

We shall establish some weighted Hermite–Hadamard inequalities for strongly GA-convex functions by using geometric symmetry of a continuous positive mapping and a differentiable mapping whose derivatives in absolute value are strongly GA-convex [159].

Theorem 2.4.1. *Let $\xi : K \subseteq \mathbb{R}_+ = (0,\infty) \to \mathbb{R}$ be a differentiable function on K^0 and $c,d \in K^0$ with $c < d$, and let $\lambda : [c,d] \to [0,\infty)$ be a continuous positive mapping and geometrically symmetric to $\sqrt{cd}$. If $\xi' \in L^1[c,d]$, $\xi : K \subseteq \mathbb{R}_+ = (0,\infty) \to \mathbb{R}$ is geometrically symmetric with respect to $\sqrt{cd}$ and $|\xi'|$ is strongly GA-convex on $[c,d]$ with modulus $\mu > 0$, then*

$$\left| \frac{(\ln d)\xi(d) + (\ln c)\xi(c)}{\ln d + \ln c} \int_c^d \frac{(\ln x)\lambda(x)}{x}\, dx - \int_c^d \frac{(\ln x)\lambda(x)\xi(x)}{x}\, dx \right|$$

$$\leq \frac{(\ln d - \ln c)^2}{8} \|\lambda\|_\infty \left[\frac{|\xi'(c)|}{2} (\Delta_2(c,d) + \Delta_2(d,c) - \Delta_3(c,d) + \Delta_3(d,c)) \right.$$

$$+ \frac{|\xi'(d)|}{2} (\Delta_2(c,d) + \Delta_2(d,c) + \Delta_3(c,d) - \Delta_3(d,c))$$

$$\left. - \frac{\mu}{4} \|\ln d - \ln c\|^2 (\Delta_2(c,d) + \Delta_2(d,c) - \Delta_4(c,d) - \Delta_4(d,c)) \right],$$

where $\|\lambda\|_\infty = \sup_{x \in [c,d]} |\lambda(x)|$.

Proof For the proof of this theorem, we will use Lemma 2.2.8.

$$\frac{(\ln d)\xi(d) + (\ln c)\xi(c)}{\ln d + \ln c} \int_c^d \frac{(\ln x)\lambda(x)}{x}dx - \int_c^d \frac{(\ln x)\lambda(x)\xi(x)}{x}dx$$

$$= \frac{(\ln d - \ln c)}{2(\ln d + \ln c)}\left[\int_0^1 \left(\int_{\varrho_2(\delta)}^{\varrho_1(\delta)} \frac{(\ln x)\lambda(x)}{x}dx\right) \varrho_1(\delta)\ln(\varrho_1(\delta))\xi'(\varrho_1(\delta))d\delta\right.$$

$$\left. - \int_0^1 \left(\int_{\varrho_2(\delta)}^{\varrho_1(\delta)} \frac{(\ln x)\lambda(x)}{x}dx\right) \varrho_2(\delta)\ln(\varrho_2(\delta))\xi'(\varrho_2(\delta))d\delta\right].$$

This implies,

$$\left|\frac{(\ln d)\xi(d) + (\ln c)\xi(c)}{\ln d + \ln c} \int_c^d \frac{(\ln x)\lambda(x)}{x}dx - \int_c^d \frac{(\ln x)\lambda(x)\xi(x)}{x}dx\right|$$

$$\leq \frac{(\ln d - \ln c)}{2(\ln d + \ln c)}\|\lambda\|_\infty\left[\int_0^1 \left(\int_{\varrho_2(\delta)}^{\varrho_1(\delta)} \frac{\ln x}{x}dx\right) \varrho_1(\delta)|\ln(\varrho_1(\delta))||\xi'(\varrho_1(\delta))|d\delta\right.$$

$$\left. + \int_0^1 \left(\int_{\varrho_2(\delta)}^{\varrho_1(\delta)} \frac{\ln x}{x}dx\right) \varrho_2(\delta)|\ln(\varrho_2(\delta))||\xi'(\varrho_2(\delta))|d\delta\right]. \qquad (2.14)$$

Since $|\xi'|$ is strongly GA-convex function on $[c,d]$ with modulus $\mu > 0$, we have

$$|\xi'(\varrho_1(\delta))| = |\xi'(c^{\frac{1-\delta}{2}}d^{\frac{1+\delta}{2}})|$$

$$\leq \frac{(1-\delta)}{2}|\xi'(c)| + \frac{(1+\delta)}{2}|\xi'(d)| - \frac{\mu}{4}(1-\delta)(1+\delta)\|\ln d - \ln c\|^2 \qquad (2.15)$$

and

$$|\xi'(\varrho_2(\delta))| = |\xi'(c^{\frac{1+\delta}{2}}d^{\frac{1-\delta}{2}})|$$

$$\leq \frac{(1+\delta)}{2}|\xi'(c)| + \frac{(1-\delta)}{2}|\xi'(d)| - \frac{\mu}{4}(1+\delta)(1-\delta)\|\ln d - \ln c\|^2. \qquad (2.16)$$

Using (2.15) and (2.16) in (2.14), we have

$$\left|\frac{(\ln d)\xi(d) + (\ln c)\xi(c)}{\ln d + \ln c} \int_c^d \frac{(\ln x)\lambda(x)}{x}dx - \int_c^d \frac{(\ln x)\lambda(x)\xi(x)}{x}dx\right|$$

$$\leq \frac{(\ln d - \ln c)^2}{8}\|\lambda\|_\infty\left[\int_0^1 \left(\frac{(1-\delta)}{2}|\xi'(c)| + \frac{(1+\delta)}{2}|\xi'(d)|\right.\right.$$

$$\left. - \frac{\mu}{4}(1-\delta^2)\|\ln d - \ln c\|^2\right) \varrho_1(\delta)|\ln(\varrho_1(\delta))|d\delta + \int_0^1 \left(\frac{(1+\delta)}{2}|\xi'(c)|\right.$$

$$\left.\left. + \frac{(1-\delta)}{2}|\xi'(d)| - \frac{\mu}{4}(1-\delta^2)\|\ln d - \ln c\|^2\right) \varrho_2(\delta)|\ln(\varrho_2(\delta))|d\delta\right].$$

By applying Lemmas 2.2.6 and 2.2.7, we have

$$
\left| \frac{(\ln d)\xi(d) + (\ln c)\xi(c)}{\ln d + \ln c} \int_c^d \frac{(\ln x)\lambda(x)}{x} dx - \int_c^d \frac{(\ln x)\lambda(x)\xi(x)}{x} dx \right|
$$

$$
\leq \frac{(\ln d - \ln c)^2}{8} \|\lambda\|_\infty \left[\frac{|\xi'(c)|}{2}(\Delta_2(c,d) + \Delta_2(d,c) - \Delta_3(c,d) + \Delta_3(d,c)) \right.
$$

$$
+ \frac{|\xi'(d)|}{2}(\Delta_2(c,d) + \Delta_2(d,c) + \Delta_3(c,d) - \Delta_3(d,c))
$$

$$
\left. - \frac{\mu}{4}\|\ln d - \ln c\|^2(\Delta_2(c,d) + \Delta_2(d,c) - \Delta_4(c,d) - \Delta_4(d,c)) \right].
$$

This completes the proof.

Corollary 2.4.1. *If* $\lambda(x) = \frac{1}{(\ln x)(\ln d - \ln c)}$, $\forall\, x \in [c,d]$ *with* $1 < c < d < \infty$ *in Theorem 2.4.1, then*

$$
\left| \frac{(\ln d)\xi(d) + (\ln c)\xi(c)}{\ln d + \ln c} - \frac{1}{\ln d - \ln c} \int_c^d \frac{\xi(x)}{x} dx \right|
$$

$$
\leq \frac{(\ln d - \ln c)}{8(\ln c)} \left[\frac{|\xi'(c)|}{2}(\Delta_2(c,d) + \Delta_2(d,c) - \Delta_3(c,d) + \Delta_3(d,c)) \right.
$$

$$
+ \frac{|\xi'(d)|}{2}(\Delta_2(c,d) + \Delta_2(d,c) + \Delta_3(c,d) - \Delta_3(d,c))
$$

$$
\left. - \frac{\mu}{4}\|\ln d - \ln c\|^2(\Delta_2(c,d) + \Delta_2(d,c) - \Delta_4(c,d) - \Delta_4(d,c)) \right].
$$

Corollary 2.4.2. *If* $\mu = 0$ *in Theorem 2.4.1, then*

$$
\left| \frac{(\ln d)\xi(d) + (\ln c)\xi(c)}{\ln d + \ln c} \int_c^d \frac{(\ln x)\lambda(x)}{x} dx - \int_c^d \frac{(\ln x)\lambda(x)\xi(x)}{x} dx \right|
$$

$$
\leq \frac{(\ln d - \ln c)^2}{8} \|\lambda\|_\infty \left[\frac{|\xi'(c)|}{2}(\Delta_2(c,d) + \Delta_2(d,c) - \Delta_3(c,d) + \Delta_3(d,c)) \right.
$$

$$
\left. + \frac{|\xi'(d)|}{2}(\Delta_2(c,d) + \Delta_2(d,c) + \Delta_3(c,d) - \Delta_3(d,c)) \right],
$$

where $\|\lambda\|_\infty = sup_{x \in [c,d]} |\lambda(x)|$.

Remark 2.4.1. *If* $|\xi'|$ *is geometrically quasi convex, then above theorem reduces to Theorem 1 of [128], i.e.*

$$
\left| \frac{(\ln d)\xi(d) + (\ln c)\xi(c)}{\ln d + \ln c} \int_c^d \frac{(\ln x)\lambda(x)}{x} dx - \int_c^d \frac{(\ln x)\lambda(x)\xi(x)}{x} dx \right|
$$

$$
\leq \frac{(\ln d - \ln c)^2}{8} \|\lambda\|_\infty \left[\Delta_2(c,d)(sup\{|\xi'(\sqrt{cd})|, |\xi'(d)|\}) \right.
$$

$$
\left. + \Delta_2(d,c)(sup\{|\xi'(c)|, |\xi'(\sqrt{cd})|\}) \right].
$$

Theorem 2.4.2. *Let $\xi : K \subseteq \mathbb{R}_+ = (0, \infty) \to \mathbb{R}$ be a differentiable function on K^0 and $c, d \in K^0$ with $c < d$, and let $\lambda : [c, d] \to [0, \infty)$ be a continuous positive mapping and geometrically symmetric to $\sqrt{cd}$. If $\xi' \in L^1[c, d]$, $\xi : K \subseteq \mathbb{R}_+ = (0, \infty) \to \mathbb{R}$ is geometrically symmetric with respect to $\sqrt{cd}$ and $|\xi'|^\alpha$ is strongly GA-convex on $[c, d]$ for $\alpha > 1$ with modulus $\mu > 0$, then*

$$\left| \frac{(\ln d)\xi(d) + (\ln c)\xi(c)}{\ln d + \ln c} \int_c^d \frac{(\ln x)\lambda(x)}{x} dx - \int_c^d \frac{(\ln x)\lambda(x)\xi(x)}{x} dx \right|$$

$$\leq \frac{(\ln d - \ln c)^2}{8} \|\lambda\|_\infty \left(\frac{\alpha - 1}{\alpha} \right)^{1-\frac{1}{\alpha}} \left[\left(\Delta_2(c^{\frac{\alpha}{\alpha-1}}, d^{\frac{\alpha}{\alpha-1}}) \right)^{1-\frac{1}{\alpha}} \right.$$

$$\times \left(\frac{|\xi'(c)|^\alpha}{2} (\Delta_1(c, d) - \Delta_5(c, d)) + \frac{|\xi'(d)|^\alpha}{2} (\Delta_1(c, d) + \Delta_5(c, d)) \right.$$

$$\left. - \frac{\mu}{4} \|\ln d - \ln c\|^2 (\Delta_1(c, d) - \Delta_6(c, d)) \right)^{1/\alpha} + \left(\Delta_2(d^{\frac{\alpha}{\alpha-1}}, c^{\frac{\alpha}{\alpha-1}}) \right)^{1-\frac{1}{\alpha}}$$

$$\times \left(\frac{|\xi'(c)|^\alpha}{2} (\Delta_1(d, c) + \Delta_5(d, c)) + \frac{|\xi'(d)|^\alpha}{2} (\Delta_1(d, c) - \Delta_5(d, c)) \right.$$

$$\left. \left. - \frac{\mu}{4} \|\ln d - \ln c\|^2 (\Delta_1(d, c) - \Delta_6(d, c)) \right)^{1/\alpha} \right],$$

where $\|\lambda\|_\infty = \sup_{x \in [c,d]} |\lambda(x)|$.

Proof From Lemma 2.2.8, we have

$$\left| \frac{(\ln d)\xi(d) + (\ln c)\xi(c)}{\ln d + \ln c} \int_c^d \frac{(\ln x)\lambda(x)}{x} dx - \int_c^d \frac{(\ln x)\lambda(x)\xi(x)}{x} dx \right|$$

$$\leq \frac{(\ln d - \ln c)}{2(\ln d + \ln c)} \|\lambda\|_\infty \left[\int_0^1 \left(\int_{\varrho_2(\delta)}^{\varrho_1(\delta)} \frac{\ln x}{x} dx \right) \varrho_1(\delta) |\ln(\varrho_1(\delta))| |\xi'(\varrho_1(\delta))| d\delta \right.$$

$$\left. + \int_0^1 \left(\int_{\varrho_2(\delta)}^{\varrho_1(\delta)} \frac{\ln x}{x} dx \right) \varrho_2(\delta) |\ln(\varrho_2(\delta))| |\xi'(\varrho_2(\delta))| d\delta \right].$$

Applying Hölder's inequality, we have

$$\left| \frac{(\ln d)\xi(d) + (\ln c)\xi(c)}{\ln d + \ln c} \int_c^d \frac{(\ln x)\lambda(x)}{x} dx - \int_c^d \frac{(\ln x)\lambda(x)\xi(x)}{x} dx \right|$$

$$\leq \frac{(\ln d - \ln c)^2}{8} \|\lambda\|_\infty \left[\left(\frac{\alpha - 1}{\alpha} \int_0^1 \varrho_1^{\frac{\alpha}{\alpha-1}}(\delta) |\ln(\varrho_1^{\frac{\alpha}{\alpha-1}}(\delta))| d\delta \right)^{1-\frac{1}{\alpha}} \right.$$

$$\times \left(\int_0^1 |\ln(\varrho_1(\delta))| |\xi'(\varrho_1(\delta))|^\alpha d\delta \right)^{1/\alpha}$$

$$+ \left(\frac{\alpha - 1}{\alpha} \int_0^1 \varrho_2^{\frac{\alpha}{\alpha-1}}(\delta) |\ln(\varrho_2^{\frac{\alpha}{\alpha-1}}(\delta))| d\delta \right)^{1-\frac{1}{\alpha}}$$

$$\times \left. \left(\int_0^1 |\ln(\varrho_2(\delta))| |\xi'(\varrho_2(\delta))|^\alpha d\delta \right)^{1/\alpha} \right].$$

Using Lemmas 2.2.6 and 2.2.7, and strong GA-convexity of $|\xi'|^\alpha$ on $[c,d]$ for $\alpha > 1$ with modulus $\mu > 0$, we have

$$\left| \frac{(\ln d)\xi(d) + (\ln c)\xi(c)}{\ln d + \ln c} \int_c^d \frac{(\ln x)\lambda(x)}{x} dx - \int_c^d \frac{(\ln x)\lambda(x)\xi(x)}{x} dx \right|$$

$$\leq \frac{(\ln d - \ln c)^2}{8} \|\lambda\|_\infty \left[\left(\frac{\alpha-1}{\alpha} \Delta_2(c^{\frac{\alpha}{\alpha-1}}, d^{\frac{\alpha}{\alpha-1}}) \right)^{1-\frac{1}{\alpha}} \right.$$

$$\times \left(\frac{|\xi'(c)|^\alpha}{2}(\Delta_1(c,d) - \Delta_5(c,d)) + \frac{|\xi'(d)|^\alpha}{2}(\Delta_1(c,d) + \Delta_5(c,d)) \right.$$

$$\left. - \frac{\mu}{4}\|\ln d - \ln c\|^2(\Delta_1(c,d) - \Delta_6(c,d)) \right)^{1/\alpha} + \left(\frac{\alpha-1}{\alpha}\Delta_2(d^{\frac{\alpha}{\alpha-1}}, c^{\frac{\alpha}{\alpha-1}}) \right)^{1-\frac{1}{\alpha}}$$

$$\times \left(\frac{|\xi'(c)|^\alpha}{2}(\Delta_1(d,c) + \Delta_5(d,c)) + \frac{|\xi'(d)|^\alpha}{2}(\Delta_1(d,c) - \Delta_5(d,c)) \right.$$

$$\left. \left. - \frac{\mu}{4}\|\ln d - \ln c\|^2(\Delta_1(d,c) - \Delta_6(d,c)) \right)^{1/\alpha} \right].$$

This completes the proof.

Corollary 2.4.3. *If* $\lambda(x) = \frac{1}{(\ln x)(\ln d - \ln c)}$, $\forall\, x \in [c,d]$ *with* $1 < c < d < \infty$ *in Theorem 2.4.2, then*

$$\left| \frac{(\ln d)\xi(d) + (\ln c)\xi(c)}{\ln d + \ln c} - \frac{1}{\ln d - \ln c}\int_c^d \frac{\xi(x)}{x} dx \right|$$

$$\leq \frac{(\ln d - \ln c)}{8\ln c} \left(\frac{\alpha-1}{\alpha} \right)^{1-\frac{1}{\alpha}} \left[(\Delta_2(c^{\frac{\alpha}{\alpha-1}}, d^{\frac{\alpha}{\alpha-1}}))^{1-\frac{1}{\alpha}} \right.$$

$$\times \left(\frac{|\xi'(c)|^\alpha}{2}(\Delta_1(c,d) - \Delta_5(c,d)) + \frac{|\xi'(d)|^\alpha}{2}(\Delta_1(c,d) + \Delta_5(c,d)) \right.$$

$$\left. - \frac{\mu}{4}\|\ln d - \ln c\|^2(\Delta_1(c,d) - \Delta_6(c,d)) \right)^{1/\alpha} + (\Delta_2(d^{\frac{\alpha}{\alpha-1}}, c^{\frac{\alpha}{\alpha-1}}))^{1-\frac{1}{\alpha}}$$

$$\times \left(\frac{|\xi'(c)|^\alpha}{2}(\Delta_1(d,c) + \Delta_5(d,c)) + \frac{|\xi'(d)|^\alpha}{2}(\Delta_1(d,c) - \Delta_5(d,c)) \right.$$

$$\left. \left. - \frac{\mu}{4}\|\ln d - \ln c\|^2(\Delta_1(d,c) - \Delta_6(d,c)) \right)^{1/\alpha} \right].$$

Corollary 2.4.4. *If $\mu = 0$ in Theorem 2.4.2, then*

$$\left| \frac{(\ln d)\xi(d) + (\ln c)\xi(c)}{\ln d + \ln c} \int_c^d \frac{(\ln x)\lambda(x)}{x} dx - \int_c^d \frac{(\ln x)\lambda(x)\xi(x)}{x} dx \right|$$

$$\leq \frac{(\ln d - \ln c)^2}{8} \|\lambda\|_\infty \left(\frac{\alpha - 1}{\alpha} \right)^{1 - \frac{1}{\alpha}} \left[\left(\Delta_2(c^{\frac{\alpha}{\alpha-1}}, d^{\frac{\alpha}{\alpha-1}}) \right)^{1 - \frac{1}{\alpha}} \right.$$

$$\times \left(\frac{|\xi'(c)|^\alpha}{2}(\Delta_1(c,d) - \Delta_5(c,d)) + \frac{|\xi'(d)|^\alpha}{2}(\Delta_1(c,d) + \Delta_5(c,d)) \right)^{1/\alpha}$$

$$+ \left(\Delta_2(d^{\frac{\alpha}{\alpha-1}}, c^{\frac{\alpha}{\alpha-1}}) \right)^{1 - \frac{1}{\alpha}}$$

$$\left. \times \left(\frac{|\xi'(c)|^\alpha}{2}(\Delta_1(d,c) + \Delta_5(d,c)) + \frac{|\xi'(d)|^\alpha}{2}(\Delta_1(d,c) - \Delta_5(d,c)) \right)^{1/\alpha} \right],$$

where $\|\lambda\|_\infty = sup_{x \in [c,d]} |\lambda(x)|$.

Remark 2.4.2. *If $|\xi'|^\alpha$ is geometrically quasi convex, then above theorem reduces to Theorem 2 of [128]: i.e.*

$$\left| \frac{(\ln d)\xi(d) + (\ln c)\xi(c)}{\ln d + \ln c} \int_c^d \frac{(\ln x)\lambda(x)}{x} dx - \int_c^d \frac{(\ln x)\lambda(x)\xi(x)}{x} dx \right|$$

$$\leq \frac{(\ln d - \ln c)^2}{8} \|\lambda\|_\infty \left(\frac{\alpha - 1}{\alpha} \right)^{1 - \frac{1}{\alpha}} \left[\left(\Delta_2(c^{\frac{\alpha}{\alpha-1}}, d^{\frac{\alpha}{\alpha-1}}) \right)^{1 - \frac{1}{\alpha}} (\Delta_1(c,d))^{\frac{1}{\alpha}} \right.$$

$$\times (sup\{|\xi'(\sqrt{cd})|, |\xi'(d)|\}) + \left(\Delta_2(d^{\frac{\alpha}{\alpha-1}}, c^{\frac{\alpha}{\alpha-1}}) \right)^{1 - \frac{1}{\alpha}} (\Delta_1(d,c))^{\frac{1}{\alpha}}$$

$$\left. \times (sup\{|\xi'(c)|, |\xi'(\sqrt{cd})|\}) \right].$$

Theorem 2.4.3. *Let $\xi : K \subseteq \mathbb{R}_+ = (0, \infty) \to \mathbb{R}$ be a differentiable function on K^0 and $c, d \in K^0$ with $c < d$, and let $\lambda : [c,d] \to [0, \infty)$ be a continuous positive mapping and geometrically symmetric to $\sqrt{cd}$. If $\xi' \in L^1[c,d]$, $\xi : K \subseteq \mathbb{R}_+ = (0, \infty) \to \mathbb{R}$ is geometrically symmetric with respect to $\sqrt{cd}$ and $|\xi'|^\alpha$ is strongly GA-convex on $[c,d]$ for $\alpha > 1$ with modulus $\mu > 0$ and $\alpha > l > 0$, then*

$$\left| \frac{(\ln d)\xi(d) + (\ln c)\xi(c)}{\ln d + \ln c} \int_c^d \frac{(\ln x)\lambda(x)}{x} dx - \int_c^d \frac{(\ln x)\lambda(x)\xi(x)}{x} dx \right|$$

$$\leq \frac{(\ln d - \ln c)^2}{8} \|\lambda\|_\infty \left(\frac{\alpha - 1}{\alpha - l} \right)^{1 - \frac{1}{\alpha}} \left(\frac{1}{l} \right)^{1/\alpha} \left[\left(\Delta_2(c^{\frac{\alpha-l}{\alpha-1}}, d^{\frac{\alpha-l}{\alpha-1}}) \right)^{1 - \frac{1}{\alpha}} \right.$$

$$\times \left(\frac{|\xi'(c)|^\alpha}{2}(\Delta_2(c^l, d^l) - \Delta_3(c^l, d^l)) + \frac{|\xi'(d)|^\alpha}{2}(\Delta_2(c^l, d^l) + \Delta_3(c^l, d^l)) \right.$$

$$\left. - \frac{\mu}{4}\|\ln d - \ln c\|^2(\Delta_2(c^l, d^l) - \Delta_4(c^l, d^l)) \right)^{1/\alpha} + \left(\Delta_2(d^{\frac{\alpha-l}{\alpha-1}}, c^{\frac{\alpha-l}{\alpha-1}}) \right)^{1-\frac{1}{\alpha}}$$

$$\times \left(\frac{|\xi'(c)|^\alpha}{2}(\Delta_2(d^l, c^l) + \Delta_3(d^l, c^l)) + \frac{|\xi'(d)|^\alpha}{2}(\Delta_2(d^l, c^l) - \Delta_3(d^l, c^l)) \right.$$

$$\left. - \frac{\mu}{4}\|\ln d - \ln c\|^2(\Delta_2(d^l, c^l) - \Delta_4(d^l, c^l)) \right)^{1/\alpha} \Bigg],$$

where $\|\lambda\|_\infty = sup_{x\in[c,d]}|\lambda(x)|$.

Proof From Lemma 2.2.8, we have

$$\left| \frac{(\ln d)\xi(d) + (\ln c)\xi(c)}{\ln d + \ln c} \int_c^d \frac{(\ln x)\lambda(x)}{x}dx - \int_c^d \frac{(\ln x)\lambda(x)\xi(x)}{x}dx \right|$$

$$\leq \frac{(\ln d - \ln c)}{2(\ln d + \ln c)}\|\lambda\|_\infty \left[\int_0^1 \left(\int_{\varrho_2(\delta)}^{\varrho_1(\delta)} \frac{\ln x}{x}dx \right) \varrho_1(\delta)|\ln(\varrho_1(\delta))||\xi'(\varrho_1(\delta))|d\delta \right.$$

$$\left. + \int_0^1 \left(\int_{\varrho_2(\delta)}^{\varrho_1(\delta)} \frac{\ln x}{x}dx \right) \varrho_2(\delta)|\ln(\varrho_2(\delta))||\xi'(\varrho_2(\delta))|d\delta \right]. \tag{2.17}$$

Applying Hölder's inequality in (2.17), we have

$$\left| \frac{(\ln d)\xi(d) + (\ln c)\xi(c)}{\ln d + \ln c} \int_c^d \frac{(\ln x)\lambda(x)}{x}dx - \int_c^d \frac{(\ln x)\lambda(x)\xi(x)}{x}dx \right|$$

$$\leq \frac{(\ln d - \ln c)^2}{8}\|\lambda\|_\infty \left[\left(\frac{\alpha-1}{\alpha-l} \int_0^1 \varrho_1^{\frac{\alpha-l}{\alpha-1}}(\delta)|\ln(\varrho_1^{\frac{\alpha-l}{\alpha-1}}(\delta))|d\delta \right)^{1-\frac{1}{\alpha}} \right.$$

$$\times \left(\frac{1}{l} \int_0^1 \varrho_1^l(\delta)|\ln(\varrho_1^l(\delta))||\xi'(\varrho_1(\delta))|^\alpha d\delta \right)^{1/\alpha}$$

$$+ \left(\frac{\alpha-1}{\alpha-l} \int_0^1 \varrho_2^{\frac{\alpha-l}{\alpha-1}}(\delta)|\ln(\varrho_2^{\frac{\alpha-l}{\alpha-1}}(\delta))|d\delta \right)^{1-\frac{1}{\alpha}}$$

$$\left. \times \left(\frac{1}{l} \int_0^1 \varrho_2^l(\delta)|\ln(\varrho_2^l(\delta))||\xi'(\varrho_2(\delta))|^\alpha d\delta \right)^{1/\alpha} \right].$$

Using Lemmas 2.2.6 and 2.2.7, and strong GA-convexity of $|\xi'|^\alpha$ on $[c, d]$ for $\alpha > 1$ with modulus $\mu > 0$, we have

$$\left| \frac{(\ln d)\xi(d) + (\ln c)\xi(c)}{\ln d + \ln c} \int_c^d \frac{(\ln x)\lambda(x)}{x}dx - \int_c^d \frac{(\ln x)\lambda(x)\xi(x)}{x}dx \right|$$

$$\leq \frac{(\ln d - \ln c)^2}{8} \|\lambda\|_\infty \left(\frac{\alpha - 1}{\alpha - l}\right)^{1 - \frac{1}{\alpha}} \left(\frac{1}{l}\right)^{1/\alpha} \left[(\Delta_2(c^{\frac{\alpha - l}{\alpha - 1}}, d^{\frac{\alpha - l}{\alpha - 1}}))^{1 - \frac{1}{\alpha}} \right.$$

$$\times \left(\frac{|\xi'(c)|^\alpha}{2} (\Delta_2(c^l, d^l) - \Delta_3(c^l, d^l)) + \frac{|\xi'(d)|^\alpha}{2} (\Delta_2(c^l, d^l) + \Delta_3(c^l, d^l)) \right.$$

$$\left. - \frac{\mu}{4} \|\ln d - \ln c\|^2 (\Delta_2(c^l, d^l) - \Delta_4(c^l, d^l)) \right)^{1/\alpha} + \left(\Delta_2(d^{\frac{\alpha - l}{\alpha - 1}}, c^{\frac{\alpha - l}{\alpha - 1}})\right)^{1 - \frac{1}{\alpha}}$$

$$\times \left(\frac{|\xi'(c)|^\alpha}{2} (\Delta_2(d^l, c^l) + \Delta_3(d^l, c^l)) + \frac{|\xi'(d)|^\alpha}{2} (\Delta_2(d^l, c^l) - \Delta_3(d^l, c^l)) \right.$$

$$\left. \left. - \frac{\mu}{4} \|\ln d - \ln c\|^2 (\Delta_2(d^l, c^l) - \Delta_4(d^l, c^l)) \right)^{1/\alpha} \right].$$

This completes the proof.

Corollary 2.4.5. *If* $\lambda(x) = \frac{1}{(\ln x)(\ln d - \ln c)}$, $\forall\, x \in [c, d]$ *with* $1 < c < d < \infty$ *in Theorem 2.4.3, then*

$$\left| \frac{(\ln d)\xi(d) + (\ln c)\xi(c)}{\ln d + \ln c} - \frac{1}{\ln d - \ln c} \int_c^d \frac{\xi(x)}{x} dx \right|$$

$$\leq \frac{(\ln d - \ln c)}{8(\ln c)} \|\lambda\|_\infty \left(\frac{\alpha - 1}{\alpha - l}\right)^{1 - \frac{1}{\alpha}} \left(\frac{1}{l}\right)^{1/\alpha} \left[(\Delta_2(c^{\frac{\alpha - l}{\alpha - 1}}, d^{\frac{\alpha - l}{\alpha - 1}}))^{1 - \frac{1}{\alpha}} \right.$$

$$\times \left(\frac{|\xi'(c)|^\alpha}{2} (\Delta_2(c^l, d^l) - \Delta_3(c^l, d^l)) + \frac{|\xi'(d)|^\alpha}{2} (\Delta_2(c^l, d^l) + \Delta_3(c^l, d^l)) \right.$$

$$\left. - \frac{\mu}{4} \|\ln d - \ln c\|^2 (\Delta_2(c^l, d^l) - \Delta_4(c^l, d^l)) \right)^{1/\alpha} + \left(\Delta_2(d^{\frac{\alpha - l}{\alpha - 1}}, c^{\frac{\alpha - l}{\alpha - 1}})\right)^{1 - \frac{1}{\alpha}}$$

$$\times \left(\frac{|\xi'(c)|^\alpha}{2} (\Delta_2(d^l, c^l) + \Delta_3(d^l, c^l)) + \frac{|\xi'(d)|^\alpha}{2} (\Delta_2(d^l, c^l) - \Delta_3(d^l, c^l)) \right.$$

$$\left. \left. - \frac{\mu}{4} \|\ln d - \ln c\|^2 (\Delta_2(d^l, c^l) - \Delta_4(d^l, c^l)) \right)^{1/\alpha} \right].$$

Corollary 2.4.6. *If* $\mu = 0$ *in Theorem 2.4.3, then*

$$\left| \frac{(\ln d)\xi(d) + (\ln c)\xi(c)}{\ln d + \ln c} \int_c^d \frac{(\ln x)\lambda(x)}{x} dx - \int_c^d \frac{(\ln x)\lambda(x)\xi(x)}{x} dx \right|$$

$$\leq \frac{(\ln d - \ln c)^2}{8} \|\lambda\|_\infty \left(\frac{\alpha - 1}{\alpha - l}\right)^{1 - \frac{1}{\alpha}} \left(\frac{1}{l}\right)^{1/\alpha} \left[(\Delta_2(c^{\frac{\alpha - l}{\alpha - 1}}, d^{\frac{\alpha - l}{\alpha - 1}}))^{1 - \frac{1}{\alpha}} \right.$$

$$\times \left(\frac{|\xi'(c)|^\alpha}{2} (\Delta_2(c^l, d^l) - \Delta_3(c^l, d^l)) + \frac{|\xi'(d)|^\alpha}{2} (\Delta_2(c^l, d^l) + \Delta_3(c^l, d^l)) \right)^{1/\alpha}$$

$$+ \left(\Delta_2(d^{\frac{\alpha - l}{\alpha - 1}}, c^{\frac{\alpha - l}{\alpha - 1}})\right)^{1 - \frac{1}{\alpha}}$$

$$\times \left(\frac{|\xi'(c)|^\alpha}{2} (\Delta_2(d^l, c^l) + \Delta_3(d^l, c^l)) + \frac{|\xi'(d)|^\alpha}{2} (\Delta_2(d^l, c^l) - \Delta_3(d^l, c^l)) \right)^{1/\alpha} \right],$$

where $\|\lambda\|_\infty = \sup_{x \in [c,d]} |\lambda(x)|$.

Remark 2.4.3. *If $|\xi'|^\alpha$ is geometrically quasi convex, then above theorem reduces to Theorem 3 of [128]: i.e.*

$$\left| \frac{(\ln d)\xi(d) + (\ln c)\xi(c)}{\ln d + \ln c} \int_c^d \frac{(\ln x)\lambda(x)}{x} dx - \int_c^d \frac{(\ln x)\lambda(x)\xi(x)}{x} dx \right|$$

$$\leq \frac{(\ln d - \ln c)^2}{8} \|\lambda\|_\infty \left(\frac{\alpha - 1}{\alpha - l} \right)^{1 - \frac{1}{\alpha}} \left(\frac{1}{l} \right)^{1/\alpha} \left[\left(\Delta_2(c^{\frac{\alpha-l}{\alpha-1}}, d^{\frac{\alpha-l}{\alpha-1}}) \right)^{1 - \frac{1}{\alpha}} \right.$$

$$\times \left(\Delta_2(c^l, d^l) \right)^{\frac{1}{\alpha}} (sup\{|\xi'(\sqrt{cd})|, |\xi'(d)|\}) + \left(\Delta_2(d^{\frac{\alpha-l}{\alpha-1}}, c^{\frac{\alpha-l}{\alpha-1}}) \right)^{1 - \frac{1}{\alpha}}$$

$$\left. \times \left(\Delta_2(d^l, c^l) \right)^{\frac{1}{\alpha}} (sup\{|\xi'(c)|, |\xi'(\sqrt{cd})|\}) \right].$$

2.5 Hermite–Hadamard Type Integral Inequalities for the Class of Strongly Convex Functions on Time Scales

In this section, first, we define a class of strongly convex function with respect to two auxiliary functions ψ_1 and ψ_2 on time scales $\mathbb{T}$ [92].

Definition 2.5.1. *Consider a time scale $\mathbb{T}$ and let $\psi_1, \psi_2 : (0,1) \to \mathbb{R}$ be two nonnegative funcions. A function $\xi : K = [c,d]_\mathbb{T} \to \mathbb{R}$ is said to be a (ψ_1, ψ_2)−strongly convex function with respect to two nonnegative functions ψ_1 and ψ_2 if there exists a constant $\mu > 0$ such that*

$$\xi((1-\delta)c + \delta d) \leq \psi_1(1-\delta)\psi_2(\delta)\xi(c) + \psi_2(1-\delta)\psi_1(\delta)\xi(d)$$
$$- \mu\delta(1-\delta)(d-c)^2, \quad \forall \, c,d \in K, \, \delta \in [0,1].$$

Now, we discuss some new special cases of Definition 2.5.1.

(I). If $\psi_1(\delta) = \psi_2(\delta) = \delta^s$ in Definition 2.5.1, then we get Breckner type of s−strongly convex functions.

Definition 2.5.2. *Consider a time scale $\mathbb{T}$ and $s \in [0,1]$ be a real number. A function $\xi : K = [c,d]_\mathbb{T} \to \mathbb{R}$ is a Breckner type s−strongly convex function, if*

$$\xi((1-\delta)c + \delta d) \leq (1-\delta)^s\delta^s[\xi(c) + \xi(d)] - \mu\delta(1-\delta)(d-c)^2,$$
$$\forall \, c,d \in K, \, \delta \in [0,1].$$

(II). If $\psi_1(\delta) = \psi_2(\delta) = 1$ in Definition 2.5.1, then we get P−strongly convex functions.

Definition 2.5.3. *Consider a time scale* $\mathbb{T}$, *then* $\xi : K = [c, d]_{\mathbb{T}} \rightarrow \mathbb{R}$ *is a* $P-$*strongly convex function, if*

$$\xi((1 - \delta)c + \delta d) \leq \xi(c) + \xi(d) - \mu\delta(1 - \delta)(d - c)^2, \quad \forall \, c, d \in K, \; \delta \in [0, 1].$$

Next, we shall present Hermite–Hadamard type inequalities for $(\psi_1, \psi_2)-$ strongly convex functions on Time-scales. [92]

Theorem 2.5.1. *Consider a time scale* $\mathbb{T}$ *and* $K = [c, d]_{\mathbb{T}}$ *such that* $c < d$ *and* $c, d \in \mathbb{T}$. *Suppose that there is a delta differentiable function* $\xi : K \rightarrow \mathbb{R}$ *on* K^0. *If* $|\xi^\Delta|$ *is* $(\psi_1, \psi_2)-$*strongly convex function with respect to two nonnegative functions* ψ_1 *and* ψ_2, *then*

$$\left| \frac{\xi(c) + \xi(d)}{2} - \frac{1}{d - c} \int_c^d \xi^\diamond(x)\Delta x \right|$$

$$\leq \frac{d - c}{2}[(A^*(\delta) + B^*(\delta))(|\xi^\Delta(c)| + |\xi^\Delta(d)|) - 2\mu(d - c)^2 C^*(\delta)],$$

where

$$A^*(\delta) = \int_0^1 \delta\psi_1(\delta)\psi_2(1 - \delta)\Delta\delta,$$

$$B^*(\delta) = \int_0^1 \delta\psi_2(\delta)\psi_1(1 - \delta)\Delta\delta$$

and

$$C^*(\delta) = \int_0^1 \delta^2(1 - \delta)\Delta\delta.$$

Proof Using Corollary 2.2.1, modulus property and $(\psi_1, \psi_2)-$strong convexity of $|\xi^\Delta|$, we obtain

$$\left| \frac{\xi(c) + \xi(d)}{2} - \frac{1}{d - c} \int_c^d \xi^\diamond(x)\Delta x \right|$$

$$\leq \frac{d - c}{2}\left[\int_0^1 \delta|\xi^\Delta(\delta d + (1 - \delta)c)|\Delta\delta + \int_0^1 \delta|\xi^\Delta(\delta c + (1 - \delta)d)|\Delta\delta \right]$$

$$\leq \frac{d - c}{2}\left[\int_0^1 \delta\{\psi_1(\delta)\psi_2(1 - \delta)|\xi^\Delta(d)| + \psi_2(\delta)\psi_1(1 - \delta)|\xi^\Delta(c)| \right.$$

$$- \mu\delta(1 - \delta)(d - c)^2\}\Delta\delta + \int_0^1 \delta\{\psi_1(\delta)\psi_2(1 - \delta)|\xi^\Delta(c)|$$

$$\left. + \psi_2(\delta)\psi_1(1 - \delta)|\xi^\Delta(d)| - \mu\delta(1 - \delta)(d - c)^2\}\Delta\delta \right]$$

$$= \frac{d - c}{2}\left[\int_0^1 \delta\{\psi_1(\delta)\psi_2(1 - \delta)(|\xi^\Delta(c)| + |\xi^\Delta(d)|) + \psi_2(\delta)\psi_1(1 - \delta)(|\xi^\Delta(c)| \right.$$

$$\left. + |\xi^\Delta(d)|)\}\Delta\delta - 2\mu(d - c)^2 \int_0^1 \delta^2(1 - \delta)\Delta\delta \right]$$

$$= \frac{d - c}{2}[(A^*(\delta) + B^*(\delta))(|\xi^\Delta(c)| + |\xi^\Delta(d)|) - 2\mu(d - c)^2 C^*(\delta)].$$

This completes the proof.

Corollary 2.5.1. *In Theorem 2.5.1, if* $|\xi^\Delta|$ *is a Breckner type* $s-$*strongly convex function, then*

$$\left| \frac{\xi(c) + \xi(d)}{2} - \frac{1}{d-c} \int_c^d \xi^\diamond(x) \Delta x \right|$$

$$\leq (d-c)[H_1(\delta)(|\xi^\Delta(c)| + |\xi^\Delta(d)|) - 2\mu(d-c)^2 \int_0^1 \delta^2(1-\delta)\Delta\delta],$$

where

$$H_1(\delta) = \int_0^1 \delta^{s+1}(1-\delta)^s \Delta\delta.$$

Corollary 2.5.2. *In Theorem 2.5.1, if* $|\xi^\Delta|$ *is a* $P-$*strongly convex function, then*

$$\left| \frac{\xi(c) + \xi(d)}{2} - \frac{1}{d-c} \int_c^d \xi^\diamond(x) \Delta x \right|$$

$$\leq (d-c)[H_2(\delta)(|\xi^\Delta(c)| + |\xi^\Delta(d)|) - 2\mu(d-c)^2 \int_0^1 \delta^2(1-\delta)\Delta\delta],$$

where

$$H_2(\delta) = \int_0^1 \delta\Delta\delta.$$

Corollary 2.5.3. *If* $\mathbb{T} = \mathbb{R}$, *then delta integral reduces to the usual Riemann integral from calculus. Hence, Theorem 2.5.1 becomes*

$$\left| \frac{\xi(c) + \xi(d)}{2} - \frac{1}{d-c} \int_c^d \xi(x) dx \right|$$

$$\leq \frac{d-c}{2}\left[(A^*(\delta) + B^*(\delta))(|\xi'(c)| + |\xi'(d)|) - 2\mu(d-c)^2 C^*(\delta)\right],$$

where

$$A^*(\delta) = \int_0^1 \delta\psi_1(\delta)\psi_2(1-\delta)d\delta,$$

$$B^*(\delta) = \int_0^1 \delta\psi_2(\delta)\psi_1(1-\delta)d\delta$$

and

$$C^*(\delta) = \int_0^1 \delta^2(1-\delta)d\delta.$$

Remark 2.5.1. *When* $\mu = 0$ *then above theorem reduces to Theorem 3 of [140], i.e.,*

$$\left| \frac{\xi(c) + \xi(d)}{2} - \frac{1}{d-c} \int_c^d \xi^\diamond(x) \Delta x \right|$$

$$\leq \frac{d-c}{2}[(A^*(\delta) + B^*(\delta))(|\xi^\Delta(c)| + |\xi^\Delta(d)|)].$$

Remark 2.5.2. *Letting* $\mathbb{T} = \mathbb{R}$ *along with* $\psi_1(\delta) = \delta^3$, $\psi_2(\delta) = 1$ *and* $\mu = 0$ *then above theorem reduces to Theorem 2.2 of [47], i.e.,*

$$\left| \frac{\xi(c) + \xi(d)}{2} - \frac{1}{d-c} \int_c^d \xi(x)dx \right| \leq \frac{d-c}{8} \left(|\xi'(c)| + |\xi'(d)| \right).$$

Theorem 2.5.2. *Consider a time scale* $\mathbb{T}$ *and* $K = [c,d]_{\mathbb{T}}$ *such that* $c < d$ *and* $c, d \in \mathbb{T}$. *Suppose that there is a delta differentiable function* $\xi : K \to \mathbb{R}$ *on* K^0. *If* $|\xi^\Delta|^b$ *is* $(\psi_1, \psi_2)-$*strongly convex function with respect to two nonnegative functions* ψ_1 *and* ψ_2, *where* $\frac{1}{a} + \frac{1}{b} = 1$ *with* $b > 1$. *Then, we have*

$$\left| \frac{\xi(c) + \xi(d)}{2} - \frac{1}{d-c} \int_c^d \xi^\diamond(x)\Delta x \right|$$

$$\leq \frac{d-c}{2} \left(\int_0^1 \delta^a \Delta\delta \right)^{\frac{1}{a}} \left[\left(\int_0^1 \{ \psi_1(\delta)\psi_2(1-\delta)|\xi^\Delta(d)|^b \right.\right.$$

$$+ \; \psi_2(\delta)\psi_1(1-\delta)|\xi^\Delta(c)|^b - \mu\delta(1-\delta)(d-c)^2\}\Delta\delta \Big)^{\frac{1}{b}}$$

$$+ \left(\int_0^1 \{ \psi_1(\delta)\psi_2(1-\delta)|\xi^\Delta(c)|^b + \psi_2(\delta)\psi_1(1-\delta)|\xi^\Delta(d)|^b \right.$$

$$\left.\left. -\mu\delta(1-\delta)(d-c)^2\}\Delta\delta \Big)^{\frac{1}{b}} \right].$$

Proof Using Corollary 2.2.1, modulus property, Hölder's integral inequality and $(\psi_1, \psi_2)-$strong convexity of $|\xi^\Delta|$, we get

$$\left| \frac{\xi(c) + \xi(d)}{2} - \frac{1}{d-c} \int_c^d \xi^\diamond(x)\Delta x \right|$$

$$\leq \frac{d-c}{2} \left[\left| \int_0^1 \delta\xi^\Delta(\delta d + (1-\delta)c)\Delta\delta \right| + \left| \int_0^1 \delta\xi^\Delta(\delta c + (1-\delta)d)\Delta\delta \right| \right]$$

$$\leq \frac{d-c}{2} \left(\int_0^1 \delta^a \Delta\delta \right)^{\frac{1}{a}} \left[\left(\int_0^1 |\xi^\Delta(\delta d + (1-\delta)c)|^b \Delta\delta \right)^{\frac{1}{b}} \right.$$

$$+ \left. \left(\int_0^1 |\xi^\Delta(\delta c + (1-\delta)d)|^b \Delta\delta \right)^{\frac{1}{b}} \right]$$

$$\leq \frac{d-c}{2} \left(\int_0^1 \delta^a \Delta\delta \right)^{\frac{1}{a}} \left[\left(\int_0^1 \{ \psi_1(\delta)\psi_2(1-\delta)|\xi^\Delta(d)|^b \right.\right.$$

$$+ \; \psi_2(\delta)\psi_1(1-\delta)|\xi^\Delta(c)|^b - \mu\delta(1-\delta)(d-c)^2\}\Delta\delta \Big)^{\frac{1}{b}}$$

$$+ \left(\int_0^1 \{ \psi_1(\delta)\psi_2(1-\delta)|\xi^\Delta(c)|^b + \psi_2(\delta)\psi_1(1-\delta)|\xi^\Delta(d)|^b \right.$$

$$\left.\left. -\mu\delta(1-\delta)(d-c)^2\}\Delta\delta \Big)^{\frac{1}{b}} \right].$$

This completes the proof.

Corollary 2.5.4. *In Theorem 2.5.2, if* $|\xi^\Delta|^b$ *is a Breckner type* $s-$*strongly convex function, then*

$$\left| \frac{\xi(c) + \xi(d)}{2} - \frac{1}{d-c} \int_c^d \xi^\diamond(x)\Delta x \right|$$

$$\leq (d-c) \left(\int_0^1 \delta^a \Delta\delta \right)^{\frac{1}{a}} \left[\left(\int_0^1 \{\delta^s(1-\delta)^s)(|\xi^\Delta(c)|^b + |\xi^\Delta(d)|^b) \right. \right.$$

$$\left. \left. - \mu\delta(1-\delta)(d-c)^2\}\Delta\delta \right)^{\frac{1}{b}} \right].$$

Corollary 2.5.5. *In Theorem 2.5.2, if* $|\xi^\Delta|^b$ *is a P-strongly convex function, then*

$$\left| \frac{\xi(c) + \xi(d)}{2} - \frac{1}{d-c} \int_c^d \xi^\diamond(x)\Delta x \right|$$

$$\leq (d-c) \left(\int_0^1 \delta^a \Delta\delta \right)^{\frac{1}{a}} \left[\left(\int_0^1 \{|\xi^\Delta(c)|^b + |\xi^\Delta(d)|^b \right. \right.$$

$$\left. \left. - \mu\delta(1-\delta)(d-c)^2\}\Delta\delta \right)^{\frac{1}{b}} \right].$$

Corollary 2.5.6. *If* $\mathbb{T} = \mathbb{R}$, *then delta integral reduces to the usual Riemann integral from calculus. Hence, Theorem 2.5.2 becomes*

$$\left| \frac{\xi(c) + \xi(d)}{2} - \frac{1}{d-c} \int_c^d \xi(x)dx \right|$$

$$\leq \frac{d-c}{2} \left(\int_0^1 \delta^a d\delta \right)^{\frac{1}{a}} \left[\left(\int_0^1 \{\psi_1(\delta)\psi_2(1-\delta)|\xi'(d)|^b + \psi_2(\delta)\psi_1(1-\delta)|\xi'(c)|^b \right. \right.$$

$$\left. - \mu\delta(1-\delta)(d-c)^2\}d\delta \right)^{\frac{1}{b}} + \left(\int_0^1 \{(\psi_1(\delta)\psi_2(1-\delta))|\xi'(c)|^b \right.$$

$$\left. \left. + \psi_2(\delta)\psi_1(1-\delta)|\xi'(d)|^b - \mu\delta(1-\delta)(d-c)^2\}d\delta \right)^{\frac{1}{b}} \right].$$

Remark 2.5.3. *When* $\mu = 0$, *then above theorem reduces to Theorem 4 of [140], i.e.,*

$$\left| \frac{\xi(c) + \xi(d)}{2} - \frac{1}{d-c} \int_c^d \xi^\diamond(x)\Delta x \right|$$

$$\leq \frac{d-c}{2} \left(\int_0^1 \delta^a \Delta\delta \right)^{\frac{1}{a}} \left[\left(\int_0^1 \{\psi_1(\delta)\psi_2(1-\delta)|\xi^\Delta(d)|^b \right. \right.$$

$$\left. + \psi_2(\delta)\psi_1(1-\delta)|\xi^\Delta(c)|^b\}\Delta\delta \right)^{\frac{1}{b}} + \left(\int_0^1 \{\psi_1(\delta)\psi_2(1-\delta)|\xi^\Delta(c)|^b \right.$$

$$\left. \left. + \psi_2(\delta)\psi_1(1-\delta)|\xi^\Delta(d)|^b\}\Delta\delta \right)^{\frac{1}{b}} \right].$$

Remark 2.5.4. *Letting* $\mathbb{T} = \mathbb{R}$ *along with* $\psi_1(\delta) = \delta^3$, $\psi_2(\delta) = 1$ *and* $\mu = 0$ *then above theorem reduces to Theorem 2.3 of [47], i.e.,*

$$\left| \frac{\xi(c) + \xi(d)}{2} - \frac{1}{d-c} \int_c^d \xi(x)dx \right| \leq \frac{d-c}{2(a+1)^{\frac{1}{a}}} \left(\frac{|\xi'(d)|^b + |\xi'(c)|^b}{2} \right)^{\frac{1}{b}}.$$

Theorem 2.5.3. *Consider a time scale* $\mathbb{T}$ *and* $K = [c, d]_{\mathbb{T}}$ *such that* $c < d$ *and* $c, d \in \mathbb{T}$. *Suppose that there is a delta differentiable function* $\xi : K \to \mathbb{R}$ *on* K^0. *If* $|\xi^\Delta|$ *is* $(\psi_1, \psi_2)-$*strongly convex function with respect to two nonnegative functions* ψ_1 *and* ψ_2, *then*

$$\left| \xi\left(\frac{c+d}{2} \right) - \frac{1}{d-c} \int_c^d \xi^\diamond(x)\Delta x \right| \leq (d-c) \left[A^{**}(\delta)|\xi^\Delta(d)| + B^{**}(\delta)|\xi^\Delta(c)| \right. $$
$$\left. - \mu(d-c)^2 C^{**}(\delta) \right],$$

where

$$A^{**}(\delta) = \int_0^{1/2} \delta\psi_1(\delta)\psi_2(1-\delta)\Delta\delta + \int_{1/2}^1 (1-\delta)\psi_1(\delta)\psi_2(1-\delta)\Delta\delta,$$

$$B^{**}(\delta) = \int_0^{1/2} \delta\psi_2(\delta)\psi_1(1-\delta)\Delta\delta + \int_{1/2}^1 (1-\delta)\psi_2(\delta)\psi_1(1-\delta)\Delta\delta$$

and

$$C^{**}(\delta) = \int_0^{1/2} \delta^2(1-\delta)\Delta\delta + \int_{1/2}^1 \delta(1-\delta)^2\Delta\delta.$$

Proof Using Corollary 2.2.2, modulus property and $(\psi_1, \psi_2)-$strong convexity of $|\xi^\Delta|$, we get

$$\left| \xi\left(\frac{c+d}{2} \right) - \frac{1}{d-c} \int_c^d \xi^\diamond(x)\Delta x \right|$$

$$\leq (d-c) \left[\int_0^{1/2} \delta|\xi^\Delta(\delta d + (1-\delta)c)|\Delta\delta + \int_{1/2}^1 |\delta - 1||\xi^\Delta(\delta d + (1-\delta)c)|\Delta\delta \right]$$

$$\leq (d-c) \left[\int_0^{1/2} \delta\{\psi_1(\delta)\psi_2(1-\delta)|\xi^\Delta(d)| + \psi_2(\delta)\psi_1(1-\delta)|\xi^\Delta(c)| \right.$$

$$- \mu\delta(1-\delta)(d-c)^2\}\Delta\delta + \int_{1/2}^1 (1-\delta)\{\psi_1(\delta)\psi_2(1-\delta)|\xi^\Delta(d)|$$

$$\left. + \psi_2(\delta)\psi_1(1-\delta)|\xi^\Delta(c)| - \mu\delta(1-\delta)(d-c)^2\}\Delta\delta \right]$$

$$= (d-c) \left[\left\{ \int_0^{1/2} \delta \psi_1(\delta) \psi_2(1-\delta) \Delta \delta + \int_{1/2}^1 (1-\delta) \psi_1(\delta) \psi_2(1-\delta) \Delta \delta \right\} \right.$$

$$\times |\xi^\Delta(d)| + \left\{ \int_0^{1/2} \delta \psi_2(\delta) \psi_1(1-\delta) \Delta \delta + \int_{1/2}^1 (1-\delta) \psi_2(\delta) \psi_1(1-\delta) \Delta \delta \right\}$$

$$\left. \times |\xi^\Delta(c)| - \mu(d-c)^2 \left\{ \int_0^{1/2} \delta^2(1-\delta) \Delta \delta + \int_{1/2}^1 \delta(1-\delta)^2 \Delta \delta \right\} \right]$$

$$= (d-c) \left[A^{**}(\delta) |\xi^\Delta(d)| + B^{**}(\delta) |\xi^\Delta(c)| - \mu(d-c)^2 C^{**}(\delta) \right].$$

This completes the proof.

Corollary 2.5.7. *If* $\mathbb{T} = \mathbb{R}$, *then our delta integral reduces to the usual Riemann integral from calculus. Hence, Theorem 2.5.3 becomes*

$$\left| \xi \left(\frac{c+d}{2} \right) - \frac{1}{d-c} \int_c^d \xi(x) dx \right|$$
$$\leq (d-c) \left[A^{**}(\delta) |\xi'(c)| + B^{**}(\delta) |\xi'(d)| \right.$$
$$\left. - \mu(d-c)^2 C^{**}(\delta) \right],$$

where

$$A^{**}(\delta) = \int_0^{1/2} \delta \psi_1(\delta) \psi_2(1-\delta) d\delta + \int_{1/2}^1 (1-\delta) \psi_1(\delta) \psi_2(1-\delta) d\delta,$$

$$B^{**}(\delta) = \int_0^{1/2} \delta \psi_2(\delta) \psi_1(1-\delta) d\delta + \int_{1/2}^1 (1-\delta) \psi_2(\delta) \psi_1(1-\delta) d\delta$$

and

$$C^{**}(\delta) = \int_0^{1/2} \delta^2(1-\delta) d\delta + \int_{1/2}^1 \delta(1-\delta)^2 d\delta.$$

Remark 2.5.5. *When* $\mu = 0$ *then above theorem reduces to Theorem 5 of [140], i.e.,*

$$\left| \xi \left(\frac{c+d}{2} \right) - \frac{1}{d-c} \int_c^d \xi^\diamond(x) \Delta x \right| \leq (d-c) \left[A^{**}(\delta) |\xi^\Delta(d)| + B^{**}(\delta) |\xi^\Delta(c)| \right].$$

Remark 2.5.6. *Letting* $\mathbb{T} = \mathbb{R}$ *along with* $\psi_1(\delta) = \delta$, $\psi_2(\delta) = 1$ *and* $\mu = 0$ *then above theorem reduces to Theorem 2.2 of [86], i.e.,*

$$\left| \xi \left(\frac{c+d}{2} \right) - \frac{1}{d-c} \int_c^d \xi(x) dx \right| \leq \frac{d-c}{8} \left[|\xi'(c)| + |\xi'(d)| \right].$$

Theorem 2.5.4. *Consider a time scale* $\mathbb{T}$ *and* $K = [c, d]_\mathbb{T}$ *such that* $c < d$ *and* $c, d \in \mathbb{T}$. *Suppose that there is a delta differentiable function* $\xi : K \to \mathbb{R}$ *on* K^0. *If* $|\xi^\Delta|^b$ *is* $(\psi_1, \psi_2)-$*strongly convex function with respect to two nonnegative functions* ψ_1 *and* ψ_2, *where* $\frac{1}{a} + \frac{1}{b} = 1$ *with* $b > 1$. *Then, we have*

$$\left| \xi\left(\frac{c+d}{2}\right) - \frac{1}{d-c} \int_c^d \xi^\diamond(x) \Delta x \right|$$

$$\leq (d-c) \left[\left(\int_0^{1/2} \delta^a \Delta\delta \right)^{\frac{1}{a}} \left(\int_0^{1/2} \{\psi_1(\delta)\psi_2(1-\delta)|\xi^\Delta(d)|^b \right. \right.$$

$$+ \psi_2(\delta)\psi_1(1-\delta)|\xi^\Delta(c)|^b - \mu\delta(1-\delta)(d-c)^2\} \Delta\delta)^{\frac{1}{b}}$$

$$+ \left(\int_{1/2}^1 |1-\delta|^a \Delta\delta \right)^{\frac{1}{a}} \left(\int_{1/2}^1 \{\psi_1(\delta)\psi_2(1-\delta)|\xi^\Delta(d)|^b \right.$$

$$\left. \left. + \psi_2(\delta)\psi_1(1-\delta)|\xi^\Delta(c)|^b - \mu\delta(1-\delta)(d-c)^2\} \Delta\delta)^{\frac{1}{b}} \right].$$

Proof Using Corollary 2.2.2, modulus property, Hölder's integral inequality and $(\psi_1, \psi_2)-$strong convexity of $|\xi^\Delta|$, we get

$$\left| \xi\left(\frac{c+d}{2}\right) - \frac{1}{d-c} \int_c^d \xi^\diamond(x) \Delta x \right|$$

$$\leq (d-c) \left[\left| \int_0^{1/2} \delta\xi^\Delta(\delta d + (1-\delta)c) \Delta\delta \right| + \left| \int_{1/2}^1 (\delta-1)\xi^\Delta(\delta d + (1-\delta)c) \Delta\delta \right| \right]$$

$$\leq (d-c) \left[\left(\int_0^{1/2} \delta^a \Delta\delta \right)^{\frac{1}{a}} \left(\int_0^{1/2} |\xi^\Delta(\delta d + (1-\delta)c|^b \Delta\delta \right)^{\frac{1}{b}} \right.$$

$$\left. + \left(\int_{1/2}^1 |1-\delta|^a \Delta\delta \right)^{\frac{1}{a}} \left(\int_{1/2}^1 |\xi^\Delta(\delta d + (1-\delta)c|^b \Delta\delta \right)^{\frac{1}{b}} \right]$$

$$\leq (d-c) \left[\left(\int_0^{1/2} \delta^a \Delta\delta \right)^{\frac{1}{a}} \left(\int_0^{1/2} \{\psi_1(\delta)\psi_2(1-\delta)|\xi^\Delta(d)|^b \right. \right.$$

$$+ \psi_2(\delta)\psi_1(1-\delta)|\xi^\Delta(c)|^b - \mu\delta(1-\delta)(d-c)^2\} \Delta\delta)^{\frac{1}{b}}$$

$$+ \left(\int_{1/2}^1 |1-\delta|^a \Delta\delta \right)^{\frac{1}{a}} \left(\int_{1/2}^1 \{\psi_1(\delta)\psi_2(1-\delta)|\xi^\Delta(d)|^b \right.$$

$$\left. \left. + \psi_2(\delta)\psi_1(1-\delta)|\xi^\Delta(c)|^b - \mu\delta(1-\delta)(d-c)^2\} \Delta\delta)^{\frac{1}{b}} \right].$$

This completes the proof.

Corollary 2.5.8. *If* $\mathbb{T} = \mathbb{R}$*, then our delta integral reduces to the usual Riemann integral from calculus. Hence, Theorem 2.5.4 becomes*

$$\left| \xi\left(\frac{c+d}{2}\right) - \frac{1}{d-c} \int_c^d \xi(x)dx \right|$$

$$\leq (d-c)\left[\left(\int_0^{1/2} \delta^a d\delta\right)^{\frac{1}{a}} \left(\int_0^{1/2} \{\psi_1(\delta)\psi_2(1-\delta)|\xi'(d)|^b\right.\right.$$

$$+\psi_2(\delta)\psi_1(1-\delta)|\xi'(c)|^b - \mu\delta(1-\delta)(d-c)^2\}d\delta\right)^{\frac{1}{b}}$$

$$+\left(\int_{1/2}^1 |1-\delta|^a d\delta\right)^{\frac{1}{a}} \left(\int_{1/2}^1 \{\psi_1(\delta)\psi_2(1-\delta)|\xi'(d)|^b\right.$$

$$\left.\left. + \psi_2(\delta)\psi_1(1-\delta)|\xi'(c)|^b - \mu\delta(1-\delta)(d-c)^2\}d\delta\right)^{\frac{1}{b}}\right].$$

Remark 2.5.7. *When* $\mu = 0$ *then above theorem reduces to Theorem 6 of [140], i.e.,*

$$\left| \xi\left(\frac{c+d}{2}\right) - \frac{1}{d-c} \int_c^d \xi^\diamond(x)\Delta x \right| \leq (d-c)\left[\left(\int_0^{1/2} \delta^a \Delta\delta\right)^{\frac{1}{a}}\right.$$

$$\times\left(\int_0^{1/2} \{\psi_1(\delta)\psi_2(1-\delta)|\xi^\Delta(d)|^b + \psi_2(\delta)\psi_1(1-\delta)|\xi^\Delta(c)|^b\}\Delta\delta\right)^{\frac{1}{b}}$$

$$+\left(\int_{1/2}^1 |1-\delta|^a \Delta\delta\right)^{\frac{1}{a}} \left(\int_{1/2}^1 \{\psi_1(\delta)\psi_2(1-\delta)|\xi^\Delta(d)|^b\right.$$

$$\left.\left. + \psi_2(\delta)\psi_1(1-\delta)|\xi^\Delta(c)|^b\}\Delta\delta)^{\frac{1}{b}}\right].$$

Remark 2.5.8. *Letting* $\mathbb{T} = \mathbb{R}$ *along with* $\psi_1(\delta) = \delta$*,* $\psi_2(\delta) = 1$ *and* $\mu = 0$ *then above theorem reduces to Theorem 2.3 of [86], i.e.,*

$$\left| \xi\left(\frac{c+d}{2}\right) - \frac{1}{d-c} \int_c^d \xi(x)dx \right|$$

$$\leq \frac{d-c}{16}\left(\frac{4}{a+1}\right)^{\frac{1}{a}} \left[\left(|\xi'(d)|^b + 3|\xi'(c)|^b\right)^{\frac{1}{b}}\right.$$

$$\left. + \left(|\xi'(c)|^b + 3|\xi'(d)|^b\right)^{\frac{1}{b}}\right].$$

Theorem 2.5.5. *Let* $\xi : \mathbb{T} \to \mathbb{R}$ *be a differntiable mapping and* $c, d \in \mathbb{T}$ *with* $c < d$*. Let* $|\xi^\Delta|$ *be* (ψ_1, ψ_2)−*strongly convex function with respect to two* nonnegative functions ψ_1 and ψ_2, then

$$\left| \xi(c)\{1 - \gamma_2(1,0)\} + \xi(d)\gamma_2(1,0) - \frac{1}{d-c} \int_c^d \xi^\diamond(x)\Delta x \right|$$

$$\leq \frac{d-c}{2}[A^{***}(\delta, \delta_1)|\xi^\Delta(c)| + B^{***}(\delta, \delta_1)|\xi^\Delta(d)| - \mu(d-c)^2 C^{***}(\delta, \delta_1)],$$

where

$$A^{***}(\delta, \delta_1) = \int_0^1 \int_0^1 \{\psi_1(\delta)\psi_2(1-\delta) + \psi_1(\delta_1)\psi_2(1-\delta_1)\}(\delta + \delta_1)\Delta\delta\Delta\delta_1,$$

$$B^{***}(\delta, \delta_1) = \int_0^1 \int_0^1 \{\psi_2(\delta)\psi_1(1-\delta) + \psi_2(\delta_1)\psi_1(1-\delta_1)\}(\delta + \delta_1)\Delta\delta\Delta\delta_1,$$

and

$$C^{***}(\delta, \delta_1) = \int_0^1 \int_0^1 \{\delta(1-\delta) + \delta_1(1-\delta_1)\}(\delta + \delta_1)\Delta\delta\Delta\delta_1.$$

Proof Using Lemma 2.2.9, property of modulus and (ψ_1, ψ_2)−strong convexity of $|\xi^\Delta|$, we obtain

$$\left| \xi(c)\{1 - \gamma_2(1,0)\} + \xi(d)\gamma_2(1,0) - \frac{1}{d-c} \int_c^d \xi^\diamond(x)\Delta x \right|$$

$$\leq \frac{d-c}{2} \int_0^1 \int_0^1 |\xi^\Delta(\delta c + (1-\delta)d) - \xi^\Delta(\delta_1 c + (1-\delta_1)d)||\delta - \delta_1|\Delta\delta\Delta\delta_1$$

$$\leq \frac{d-c}{2} \int_0^1 \int_0^1 \{|\xi^\Delta(\delta c + (1-\delta)d)| + |\xi^\Delta(\delta_1 c + (1-\delta_1)d)|\}(\delta + \delta_1)\Delta\delta\Delta\delta_1$$

$$\leq \frac{d-c}{2} \int_0^1 \int_0^1 \{\psi_1(\delta)\psi_2(1-\delta)|\xi^\Delta(c)| + \psi_2(\delta)\psi_1(1-\delta)|\xi^\Delta(d)|$$

$$- \mu\delta(1-\delta)(d-c)^2 + \psi_1(\delta_1)\psi_2(1-\delta_1)|\xi^\Delta(c)| + \psi_2(\delta_1)\psi_1(1-\delta_1)|\xi^\Delta(d)|$$

$$- \mu\delta_1(1-\delta_1)(d-c)^2\}(\delta + \delta_1)\Delta\delta\Delta\delta_1$$

$$= \frac{d-c}{2} \int_0^1 \int_0^1 [\{\psi_1(\delta)\psi_2(1-\delta) + \psi_1(\delta_1)\psi_2(1-\delta_1)\}|\xi^\Delta(c)|$$

$$+ \{\psi_2(\delta)\psi_1(1-\delta) + \psi_2(\delta_1)\psi_1(1-\delta_1)\}|\xi^\Delta(d)|$$

$$- \mu(d-c)^2\{\delta(1-\delta) + \delta_1(1-\delta_1)\}](\delta + \delta_1)\Delta\delta\Delta\delta_1$$

$$= \frac{d-c}{2}[A^{***}(\delta, \delta_1)|\xi^\Delta(c)| + B^{***}(\delta, \delta_1)|\xi^\Delta(d)| - \mu(d-c)^2 C^{***}(\delta, \delta_1)].$$

This completes the proof.

Corollary 2.5.9. *If* $\mathbb{T} = \mathbb{R}$, *then our delta integral reduces to the usual Riemann integral from calculus. Hence, Theorem 2.5.5 becomes*

$$\left| \frac{\xi(c) + \xi(d)}{2} - \frac{1}{d-c} \int_c^d \xi(x)dx \right|$$

$$\leq \frac{d-c}{2}[A^{***}(\delta, \delta_1)|\xi'(c)| + B^{***}(\delta, \delta_1)|\xi'(d)| - \mu(d-c)^2 C^{***}(\delta, \delta_1)],$$

where

$$A^{***}(\delta, \delta_1) = \int_0^1 \int_0^1 \{\psi_1(\delta)\psi_2(1-\delta) + \psi_1(\delta_1)\psi_2(1-\delta_1)\}(\delta + \delta_1)d\delta d\delta_1,$$

$$B^{***}(\delta, \delta_1) = \int_0^1 \int_0^1 \{\psi_2(\delta)\psi_1(1-\delta) + \psi_2(\delta_1)\psi_1(1-\delta_1)\}(\delta + \delta_1)d\delta d\delta_1,$$

$$C^{***}(\delta, \delta_1) = \int_0^1 \int_0^1 \{\delta(1-\delta) + (\delta_1(1-\delta_1)\}(\delta + \delta_1)d\delta d\delta_1$$

and

$$\gamma_2(1,0) = \int_0^1 (1-\delta)d\delta = 1/2.$$

Remark 2.5.9. *When* $\mu = 0$ *then above theorem reduces to Theorem 7 of [140], i.e.,*

$$\left| \xi(c)\{1 - \gamma_2(1,0)\} + \xi(d)\gamma_2(1,0) - \frac{1}{d-c} \int_c^d \xi^\diamond(x)\Delta x \right|$$

$$\leq \frac{d-c}{2}[A^{***}(\delta, \delta_1)|\xi^\Delta(c)| + B^{***}(\delta, \delta_1)|\xi^\Delta(d)|].$$

Remark 2.5.10. *Letting* $\mathbb{T} = \mathbb{R}$ *along with* $\psi_1(\delta) = \frac{1}{2}$, $\psi_2(\delta) = \frac{1}{4}$ *and* $\mu = 0$ *then above theorem reduces to Theorem 2.2 of [47], i.e.,*

$$\left| \frac{\xi(c) + \xi(d)}{2} - \frac{1}{d-c} \int_c^d \xi(x)dx \right| \leq \frac{d-c}{8}\left(|\xi'(c)| + |\xi'(d)| \right).$$

Theorem 2.5.6. *Let* $\xi : [c,d]_\mathbb{T} \to \mathbb{R}$ *be a delta differentiable mapping on* $\mathbb{T}^0$ *such that* $c < d$. *If* $|\xi^\Delta|$ *is* (ψ_1, ψ_2)−*strongly convex function with respect to two nonnegative functions* ψ_1 *and* ψ_2, *then*

$$\left| \xi\left(\frac{c+d}{2}\right) - \frac{1}{d-c} \int_c^d \xi^\diamond(x)\Delta x \right|$$

$$\leq \frac{d-c}{2}[A^{****}(\delta, \delta_1)|\xi^\Delta(c)| + B^{****}(\delta, \delta_1)|\xi^\Delta(d)| - \mu(d-c)^2 C^{****}(\delta, \delta_1)],$$

where

$$A^{****}(\delta, \delta_1) = \int_0^1 \int_0^1 |\psi(\delta_1) - \psi(\delta)| \left[\psi_1(\delta)\psi_2(1-\delta) + \psi_1(\delta_1)\psi_2(1-\delta_1)\right] \Delta\delta\Delta\delta_1,$$

$$B^{****}(\delta, \delta_1) = \int_0^1 \int_0^1 |\psi(\delta_1) - \psi(\delta)| \left[\psi_1(1-\delta)\psi_2(\delta) + \psi_1(1-\delta_1)\psi_2(\delta_1)\right] \Delta\delta\Delta\delta_1$$

and

$$C^{****}(\delta, \delta_1) = \int_0^1 \int_0^1 |\psi(\delta_1) - \psi(\delta)| \left[\delta(1-\delta) + \delta_1(1-\delta_1)\right] \Delta\delta\Delta\delta_1.$$

Proof Using Lemma 2.2.10, property of modulus and (ψ_1, ψ_2)−strong convexity of $|\xi^\Delta|$, we obtain

$$\left| \xi\left(\frac{c+d}{2}\right) - \frac{1}{d-c}\int_c^d \xi^\diamond(x)\Delta x \right|$$

$$\leq \frac{d-c}{2} \int_0^1 \int_0^1 |\xi^\Delta(\delta c + (1-\delta)d) - \xi^\Delta(\delta_1 c + (1-\delta_1)d)|$$
$$\times |\psi(\delta_1) - \psi(\delta)| \Delta\delta\Delta\delta_1$$

$$\leq \frac{d-c}{2} \int_0^1 \int_0^1 \{|\xi^\Delta(\delta c + (1-\delta)d)| + |\xi^\Delta(\delta_1 c + (1-\delta_1)d)|\}$$
$$\times |\psi(\delta_1) - \psi(\delta)| \Delta\delta\Delta\delta_1$$

$$\leq \frac{d-c}{2} \int_0^1 \int_0^1 \{\psi_1(\delta)\psi_2(1-\delta)|\xi^\Delta(c)| + \psi_2(\delta)\psi_1(1-\delta)|\xi^\Delta(d)|$$
$$- \mu\delta(1-\delta)(d-c)^2 + \psi_1(\delta_1)\psi_2(1-\delta_1)|\xi^\Delta(c)| + \psi_2(\delta_1)\psi_1(1-\delta_1)|\xi^\Delta(d)|$$
$$- \mu\delta_1(1-\delta_1)|d-c|^2\}|\psi(\delta_1) - \psi(\delta)| \Delta\delta\Delta\delta_1$$

$$= \frac{d-c}{2} \int_0^1 \int_0^1 [\{\psi_1(\delta)\psi_2(1-\delta) + \psi_1(\delta_1)\psi_2(1-\delta_1)\}|\xi^\Delta(c)|$$
$$+ \{\psi_2(\delta)\psi_1(1-\delta) + \{\psi_2(\delta_1)\psi_1(1-\delta_1)\}|\xi^\Delta(d)|$$
$$- \mu(d-c)^2\{\delta(1-\delta) + \delta_1(1-\delta_1)\}]|\psi(\delta_1) - \psi(\delta)| \Delta\delta\Delta\delta_1$$

$$= \frac{d-c}{2}[A^{****}(\delta, \delta_1)|\xi^\Delta(c)| + B^{****}(\delta, \delta_1)|\xi^\Delta(d)| - \mu(d-c)^2 C^{****}(\delta, \delta_1)].$$

This completes the proof.

Remark 2.5.11. *If* $\mathbb{T} = \mathbb{R}$, *then our delta integral reduces to the usual Riemann integral from calculus. Hence, Theorem 2.5.6 becomes*

$$\left| \xi\left(\frac{c+d}{2}\right) - \frac{1}{d-c}\int_c^d \xi(x)dx \right|$$

$$\leq \frac{d-c}{2}[A^{****}(\delta, \delta_1)|\xi'(c)| + B^{****}(\delta, \delta_1)|\xi'(d)| - \mu(d-c)^2 C^{****}(\delta, \delta_1)],$$

where

$$A^{****}(\delta,\delta_1) = \int_0^1 \int_0^1 |\psi(\delta_1)-\psi(\delta)| \left[\psi_1(\delta)\psi_2(1-\delta) + \psi_1(\delta_1)\psi_2(1-\delta_1)\right] d\delta d\delta_1,$$

$$B^{****}(\delta,\delta_1) = \int_0^1 \int_0^1 |\psi(\delta_1)-\psi(\delta)| \left[\psi_1(1-\delta)\psi_2(\delta) + \psi_1(1-\delta_1)\psi_2(\delta_1)\right] d\delta d\delta_1$$

and

$$C^{****}(\delta,\delta_1) = \int_0^1 \int_0^1 |\psi(\delta_1)-\psi(\delta)| \left[\delta(1-\delta) + \delta_1(1-\delta_1)\right] d\delta d\delta_1.$$

Remark 2.5.12. *When* $\mu = 0$ *then above theorem reduces to Theorem 8 of [140], i.e.,*

$$\left| \xi\left(\frac{c+d}{2}\right) - \frac{1}{d-c}\int_c^d \xi^\diamond(x)\Delta x \right|$$
$$\leq \frac{d-c}{2}[A^{****}(\delta,\delta_1)|\xi^\Delta(c)| + B^{****}(\delta,\delta_1)|\xi^\Delta(d)|].$$

Remark 2.5.13. *Letting* $\mathbb{T} = \mathbb{R}$ *along with* $\psi_1(\delta) = \frac{1}{2}$, $\psi_2(\delta) = \frac{1}{4}$ *and* $\mu = 0$ *then above theorem reduces to Theorem 2.2 of [86], i.e.,*

$$\left| \xi\left(\frac{c+d}{2}\right) - \frac{1}{d-c}\int_c^d \xi(x)dx \right| \leq \frac{d-c}{8}\left[|\xi'(c)| + |\xi'(d)|\right].$$

Example 2.5.1. *Let* $\mathbb{T} = \mathbb{R}$. *Obviously,* $\xi(x) = x$ *is a strongly convex function with* $\psi_1(\delta) = 2-\delta$, $\psi_2(\delta) = 1$, $\mu = 1$ *and continuous on* $(0,\infty)$, *so we may apply Theorem 2.5.1 with* $c = 1/2$ *and* $d = 1$. *Clearly*

$$\left| \frac{\xi(c)+\xi(d)}{2} - \frac{1}{d-c}\int_c^d \xi(x)dx \right|$$
$$= \frac{3}{4} - 2\int_{1/2}^1 x dx$$
$$= 0. \tag{2.18}$$

On the other hand

$$\frac{d-c}{2}[(A^*(\delta) + B^*(\delta))(|\xi'(c)| + |\xi'(d)|) - 2\mu(d-c)^2 C^*(\delta)]$$
$$= \frac{1}{4}\left[\frac{9}{6} \times 2 - 2 \times 1 \times \frac{1}{4} \times \frac{1}{12}\right]$$
$$\approx 0.7395, \tag{2.19}$$

where

$$A^*(\delta) = \int_0^1 \delta(2 - \delta)d\delta = \frac{2}{3},$$

$$B^*(\delta) = \int_0^1 \delta(1 + \delta)d\delta = \frac{5}{6}$$

and

$$C^*(\delta) = \int_0^1 \delta^2(1 - \delta)d\delta = \frac{1}{12}.$$

From (2.18) *and* (2.19), *we see that*

$$0 < 0.7395.$$

Example 2.5.2. *Let* $\mathbb{T} = \mathbb{R}$. *Obviously,* $\xi(x) = x + 1$ *is strongly convex with* $\psi_1(\delta) = 2$, $\psi_2(\delta) = 4$, $\mu = 1/4$ *and continuous on* $(0, \infty)$, *so we may apply Theorem 2.5.3 with* $c = 0$ *and* $d = 1/4$. *Clearly*

$$\left| \xi\left(\frac{c + d}{2}\right) - \frac{1}{d - c} \int_c^d \xi(x)dx \right|$$

$$= \left| \frac{9}{8} - 4 \int_0^{1/2} (x + 1)dx \right|$$

$$= 0. \tag{2.20}$$

On the other hand

$$(d - c)\left[A^{**}(\delta)|\xi'(c)| + B^{**}(\delta)|\xi'(d)| - c(d - c)^2 C^{**} \right]$$

$$= \frac{1}{4}\left[(2 \times 1 + 2 \times 1) - \frac{1}{4} \times \frac{1}{16} \times \frac{5}{96} \right]$$

$$\approx 0.9997, \tag{2.21}$$

where

$$A^{**}(\delta) = B^{**}(\delta) = 8\int_0^{1/2} \delta d\delta + 8\int_{1/2}^1 (1 - \delta)d\delta = 2,$$

$$C^{**}(\delta) = \int_0^{1/2} \delta^2(1 - \delta)d\delta + \int_{1/2}^1 \delta(1 - \delta)^2 d\delta = \frac{5}{96}.$$

From (2.20) *and* (2.21), *we see that*

$$0 < 0.9997.$$

Example 2.5.3. *Let* $\mathbb{T} = \mathbb{R}$*. Obviously,* $\xi(x) = \sqrt{x}$ *is strongly convex with* $\psi_1(\delta) = 4$*,* $\psi_2(\delta) = \delta$*,* $\mu = 1/4$*, and continuous on* $(0, \infty)$*, so we may apply Theorem 2.5.5 with* $c = 2$ *and* $d = 4$*. Clearly*

$$\left| \xi\left(\frac{c+d}{2}\right) - \frac{1}{d-c} \int_c^d \xi(x)dx \right|$$

$$= \left| \sqrt{\frac{2+4}{2}} - \frac{1}{2} \int_2^4 \sqrt{x}\,dx \right|$$

$$\approx 0.0081. \tag{2.22}$$

On the other hand

$$\frac{d-c}{2}[A^{***}(\delta,\delta_1)|\xi^{\Delta}(c)| + B^{***}(\delta,\delta_1)|\xi^{\Delta}(d)| - \mu(d-c)^2 C^{***}(\delta,\delta_1)]$$

$$= \frac{4-2}{2}\left[\frac{10}{3} \times \frac{1}{2\sqrt{2}} + \frac{14}{3} \times \frac{1}{4} - \frac{1}{4} \times 4 \times \frac{1}{3}\right]$$

$$\approx 2.0118, \tag{2.23}$$

where

$$A^{***}(\delta,\delta_1) = \int_0^1 \int_0^1 \{4(1-\delta) + 4(1-\delta_1)\}(\delta + \delta_1)d\delta d\delta_1 = \frac{10}{3},$$

$$B^{***}(\delta,\delta_1) = \int_0^1 \int_0^1 (4\delta + 4\delta_1)(\delta + \delta_1)d\delta d\delta_1 = \frac{14}{3}$$

and

$$C^{***}(\delta,\delta_1) = \int_0^1 \int_0^1 \{\delta(1-\delta) + \delta_1(1-\delta_1)\}(\delta + \delta_1)d\delta d\delta_1 = \frac{1}{3}.$$

From (2.22) *and* (2.23)*, we see that*

$$0.0081 < 2.0118.$$

Chapter 3

Integral Inequalities for Strongly Generalized Convex Functions of Higher Order

3.1 Introduction

Lin and Fukushima [97] introduced strongly convex functions of higher order to simplify the study of mathematical programs with equilibrium constraints. Obviously, strong convexity of higher order is a generalization of strong convexity, the function $\xi(x) = x^4$ is strongly convex of order 4, but not strongly convex of order 2 on $\mathbb{R}$, see [97]. Lin and Fukushima [97] have established that the optimal solution of MPEC under strong convexity of higher order is same as the optimal solution of penalized problem. Further, Lin and Fukushima [97] have shown that the higher order strong convexity of a function is equivalent to higher order strong monotonicity of the gradient map of the function.

Antczak [6] introduced the class of exponentially convex functions which can be considered as a significant extension of the convex functions. Exponentially convex functions have applications in various fields such as Mathematical programming, Information geometry, Big data analysis, Machine learning, Statistics, Sequential prediction and Stochastic optimization, see [2, 122, 123, 131]. Awan *et al.* [11] investigated some other kinds of exponentially convex functions and established several new Hermite–Hadamard type integral inequalities via exponentially convex functions. Noor and Noor [123] defined and introduced some new concepts of the strongly exponentially convex functions with respect to an auxiliary non-negative bifunction and investigated the optimality conditions for the strongly exponentially convex functions.

Recently, Rashied *et al.* [140] established some Trapezoid type inequalities for generalized fractional integral and related inequalities via exponentially convex functions. Further, Rashid *et al.* [142] derived a new integral identity involving Riemann–Liouville fractional integral and obtained new fractional bounds involving the functions having exponential convexity property.

The organization of this chapter is as follows: In Section 3.2, we recall some basic results that are necessary for our main results. In Section 3.3, we

DOI: 10.1201/9781003408284-3

introduce the concept of strongly η-convex functions of higher order, as a generalization of the strongly η-convex functions. We investigate the Hermite–Hadamard and Hermite–Hadamard-Fejér type inequalities for strongly η-convex functions of higher order. Some special cases of these results are also investigated in this section. In Section 3.4, we derive Hermite–Hadamard integral inequality for higher order strongly exponentially convex functions. Also, we discuss Reimann-Liouville fractional estimates via strongly exponentially convex functions of higher order. Moreover, some particular cases of the main results are discussed.

3.2   Preliminaries

Lin and Fukushima [97] gave the following concept of strongly convex functions of higher order.

Definition 3.2.1. *A function $\xi : K \subseteq \mathbb{R}^n \to \mathbb{R}$ is said to be strongly convex with order $\sigma > 0$ on a convex set $K \subseteq \mathbb{R}^n$ if there exist a constant $\mu > 0$, such that*

$$\xi(\delta x + (1 - \delta)y) \le \delta \xi(x) + (1 - \delta)\xi(y) - \mu\delta(1 - \delta)\|x - y\|^\sigma,$$

for any $x, y \in K$ and any $\delta \in [0, 1]$.

For $\sigma = 2$, the above definition reduces to the strong convexity.
The following concept of exponentially convex function is given by Antczak [6].

Definition 3.2.2. *A positive function $\xi : K \longrightarrow \mathbb{R}$ is said to be an exponentially convex function, if*

$$e^{\xi(\delta x + (1-\delta)y)} \le \delta e^{\xi(x)} + (1 - \delta)e^{\xi(y)}, \quad \forall\, x, y \in K,\ \delta \in [0, 1].$$

Noor and Noor [123] gave the following concept of strongly exponentially convex function of higher order.

Definition 3.2.3. *A positive function ξ on the convex set K is said to be higher order strongly exponentially convex functon of order $\sigma > 1$ if there exists a constant $\mu > 0$, such that*

$$e^{\xi(\delta x + (1-\delta)y)} \le \delta e^{\xi(x)} + (1-\delta)e^{\xi(y)} - \mu\delta(1-\delta)\|y - x\|^\sigma, \quad \forall\, x, y \in K,\ \delta \in [0, 1].$$

Rashid *et al.* [140] obtained the following corollary.

Corollary 3.2.1. *Let* $\xi : K = [c,d] \to \mathbb{R}$ *be an absolutely continuous mapping on* (c,d) *such that* $(e^\xi)' \in L^1[c,d]$. *Then the following equality holds:*

$$\frac{(x-c)^\alpha e^{\xi(c)} + (d-x)^\alpha e^{\xi(d)}}{d-c} - \frac{\Gamma(\alpha+1)[J^\alpha_{x-} e^{\xi(c)} + J^\alpha_{x+} e^{\xi(d)}]}{d-c}$$

$$= \frac{(x-c)^{\alpha+1}}{d-c} \int_0^1 (\delta^\alpha - 1)e^{\xi(\delta x+(1-\delta)c)}\xi'(\delta x + (1-\delta)c)d\delta$$

$$+ \frac{(d-x)^{\alpha+1}}{d-c} \int_0^1 (1-\delta^\alpha)e^{\xi(\delta x+(1-\delta)d)}\xi'(\delta x + (1-\delta)d)\, d\delta.$$

Rashid *et al.* [142] proposed the following lemma.

Lemma 3.2.1. *Let* $\alpha > 0$ *be a number and let* $\xi : K = [c,d] \longrightarrow \mathbb{R}$ *be a differentiable function on* (c,d), *then*

$$\Gamma_\xi(c,d,\alpha) = \frac{d-c}{16}\left[\int_0^1 \delta^\alpha e^{\xi\left(\delta \frac{3c+d}{4}+(1-\delta)c\right)}\xi'\left(\delta\frac{3c+d}{4} + (1-\delta)c\right)d\delta \right.$$

$$+ \int_0^1 (\delta^\alpha - 1)e^{\xi\left(\delta\frac{c+d}{2}+(1-\delta)\frac{3c+d}{4}\right)}\xi'\left(\delta\frac{c+d}{2} + (1-\delta)\frac{3c+d}{4}\right)d\delta$$

$$+ \int_0^1 \delta^\alpha e^{\xi\left(\delta\frac{c+3d}{4}+(1-\delta)\frac{c+d}{2}\right)}\xi'\left(\delta\frac{c+3d}{4} + (1-\delta)\frac{c+d}{2}\right)d\delta$$

$$\left. + \int_0^1 (\delta^\alpha - 1)e^{\xi\left(\delta d+(1-\delta)\frac{c+3d}{4}\right)}\xi'\left(\delta d + (1-\delta)\frac{c+3d}{4}\right)d\delta \right],$$

where

$$\Gamma_\xi(c,d,\alpha) = \frac{1}{2}\left[e^{\xi\left(\frac{3c+d}{4}\right)} + e^{\xi\left(\frac{c+3d}{4}\right)} \right] - \frac{4^{(\alpha-1)}\Gamma(\alpha+1)}{(d-c)^\alpha}$$

$$\left[J^\alpha_{\left(\frac{3c+d}{4}\right)-} e^{\xi(c)} + J^\alpha_{\left(\frac{c+d}{2}\right)-} e^{\xi\left(\frac{3c+d}{4}\right)} + J^\alpha_{\left(\frac{c+3d}{4}\right)-} e^{\xi\left(\frac{c+d}{2}\right)} + J^\alpha_{d-} e^{\xi\left(\frac{c+3d}{4}\right)} \right].$$

3.3 Strongly Generalized Convex Functions of Higher Order

Now we are ready to discuss our main results. We define the η-convex and strongly η-convex function of higher order in $\mathbb{R}^n$ [106].

Definition 3.3.1. *A function* $\xi : K \subseteq \mathbb{R}^n \to \mathbb{R}$ *is said to be* η-*convex function with respect to* $\eta : \xi(K) \times \xi(K) \to \mathbb{R}$, *if*

$$\xi(\delta x + (1-\delta)y) \le \xi(y) + \delta\eta(\xi(x),\xi(y)), \quad \forall x,y \in K, \quad \delta \in [0,1].$$

Definition 3.3.2. *A function $\xi : K \subseteq \mathbb{R}^n \to \mathbb{R}$ is said to be strongly η-convex function of order $\sigma > 0$ with respect to $\eta : \xi(K) \times \xi(K) \to \mathbb{R}$ and modulus $\mu > 0$, if*

$$\xi(\delta x + (1-\delta)y) \le \xi(y) + \delta\eta(\xi(x),\xi(y)) - \mu\delta(1-\delta)\|x-y\|^\sigma, \ \ \forall x, y \in K, \delta \in [0,1]. \tag{3.1}$$

Example 3.3.1. $X = \mathbb{R}^+$, $\xi(x) = 4x$, $\eta(x,y) = \exp(x-y)^4 + x$. *Then, ξ is strongly η-convex function of order 4 with modulus 1.*

For $\sigma = 2$, the above definition reduces to the strongly η-convex function but if σ is not equal to 2, they are different.

Example 3.3.2. $X = \mathbb{R}^+ \cup \{0\}$, $\xi(x) = x$, $\eta(x,y) = (x-y)^4 + x + y$. *Then, ξ is strongly η-convex function of order 4 with modulus 1 and is not strongly η-convex function of order 2.*

Remark 3.3.1. *If $K \subset \mathbb{R}$ and $x = y$ in (3.1), then (3.1) reduces to Remark 1.4 of [12].*

Theorem 3.3.1. *Let $\xi : K \subseteq \mathbb{R}^n \to \mathbb{R}$ be a differentiable strongly η-convex function of order $\sigma > 0$. If ξ has minimum at y, then*

$$\eta(\xi(x),\xi(y)) - \mu\|x-y\|^\sigma \ge 0. \tag{3.2}$$

Proof Since ξ has minimum at y, then $\nabla\xi(y) = 0$ and the condition

$$\langle \nabla\xi(y), x-y \rangle \ge 0 \tag{3.3}$$

is satisfied automatically.

We know that ξ is strongly η-convex of order $\sigma > 0$, then

$$\xi(\delta x + (1-\delta)y) \le \xi(y) + \delta\eta(\xi(x),\xi(y)) - \mu\delta(1-\delta)\|x-y\|^\sigma.$$

Dividing above inequality by δ and taking limit $\delta \to 0$ on both sides, we have

$$\langle \nabla\xi(y), x-y \rangle \le \eta(\xi(x),\xi(y)) - \mu\|x-y\|^\sigma. \tag{3.4}$$

From (3.3) and (3.4), we have $\eta(\xi(x),\xi(y)) - \mu\|x-y\|^\sigma \ge 0$. This completes the proof.

Remark 3.3.2. *When $K \subset \mathbb{R}$ and $\sigma = 2$, then above theorem reduces to Theorem 1.6 of [12], i.e.,*

$$\eta(\xi(x),\xi(y)) - \mu(x-y)^2 \ge 0.$$

Theorem 3.3.2. *Let $\xi : [c, d] \to \mathbb{R}$ be strongly η-convex function of order $\sigma > 0$ with modulus $\mu > 0$. If $\eta(.,.)$ is bounded from above on $\xi([c, d]) \times \xi([c, d])$, then*

$$\xi\left(\frac{c+d}{2}\right) - \frac{M_\eta}{2} + \frac{\mu}{4(\sigma+1)}\|d - c\|^\sigma \leq \frac{1}{d-c}\int_c^d \xi(x)dx$$

$$\leq \frac{\xi(c) + \xi(d)}{2} + \frac{\eta(\xi(c), \xi(d)) + \eta(\xi(d), \xi(c))}{4} - \frac{\mu}{6}\|d - c\|^\sigma$$

$$\leq \frac{\xi(c) + \xi(d)}{2} + \frac{M_\eta}{2} - \frac{\mu}{6}\|d - c\|^\sigma,$$

where M_η is upper bound of η.

Proof Since ξ is strongly η-convex of order σ, then

$$\xi\left(\frac{c+d}{2}\right)$$

$$= \xi\left(\frac{1}{2}\left(\frac{c+d-\delta(d-c)}{2}\right) + \frac{1}{2}\left(\frac{c+d+\delta(d-c)}{2}\right)\right)$$

$$\leq \xi\left(\frac{c+d+\delta(d-c)}{2}\right)$$

$$+ \frac{1}{2}\eta\left(\xi\left(\frac{c+d-\delta(d-c)}{2}\right), \xi\left(\frac{c+d+\delta(d-c)}{2}\right)\right) - \frac{\mu}{4}\delta^\sigma\|d - c\|^\sigma$$

$$\leq \xi\left(\frac{c+d+\delta(d-c)}{2}\right) + \frac{M_\eta}{2} - \frac{\mu}{4}\delta^\sigma\|d - c\|^\sigma.$$

This implies

$$\xi\left(\frac{c+d}{2}\right) - \frac{M_\eta}{2} + \frac{\mu}{4}\delta^\sigma\|d - c\|^\sigma \leq \xi\left(\frac{c+d+\delta(d-c)}{2}\right). \tag{3.5}$$

Similarly,

$$\xi\left(\frac{c+d}{2}\right) - \frac{M_\eta}{2} + \frac{\mu}{4}\delta^\sigma\|d - c\|^\sigma \leq \xi\left(\frac{c+d-\delta(d-c)}{2}\right). \tag{3.6}$$

By using the change of variable technique, we have

$$\frac{1}{d-c}\int_c^d \xi(x)dx = \frac{1}{d-c}\left[\int_c^{\frac{c+d}{2}} \xi(x)dx + \int_{\frac{c+d}{2}}^d \xi(x)dx\right]$$

$$= \frac{1}{2}\int_0^1 \xi\left(\frac{c+d-\delta(d-c)}{2}\right)d\delta$$

$$+ \frac{1}{2}\int_0^1 \xi\left(\frac{c+d+\delta(d-c)}{2}\right)d\delta$$

$$\geq \int_0^1 \left[\xi\left(\frac{c+d}{2}\right) - \frac{M_\eta}{2} + \frac{\mu}{4}\delta^\sigma\|d - c\|^\sigma\right]d\delta$$

$$= \xi\left(\frac{c+d}{2}\right) - \frac{M_\eta}{2} + \frac{\mu}{4(\sigma+1)}\|d - c\|^\sigma.$$

We now prove the right hand side of the theorem. Since ξ is strongly η-convex function of order $\sigma > 0$, we have

$$\xi(\delta c + (1-\delta)d) \le \xi(d) + \delta\eta(\xi(c),\xi(d)) - \mu\delta(1-\delta)\|d-c\|^\sigma.$$

Integrating above inequality with respect to δ on $[0,1]$, we have

$$\int_0^1 \xi(\delta c + (1-\delta)d)d\delta \le \int_0^1 [\xi(d) + \delta\eta(\xi(c),\xi(d)) - \mu\delta(1-\delta)\|d-c\|^\sigma]d\delta,$$

$$\frac{1}{d-c}\int_c^d \xi(x)dx \le \xi(d) + \frac{1}{2}\eta(\xi(c),\xi(d)) - \frac{\mu}{6}\|d-c\|^\sigma = P.$$

Similarly,

$$\frac{1}{d-c}\int_c^d \xi(x)dx \le \xi(c) + \frac{1}{2}\eta(\xi(d),\xi(c)) - \frac{\mu}{6}\|d-c\|^\sigma = Q.$$

Therefore,

$$\frac{1}{d-c}\int_c^d \xi(x)dx \le Min\{P,Q\}$$

$$\le \frac{\xi(c)+\xi(d)}{2} + \frac{\eta(\xi(c)+\xi(d)) + \eta(\xi(d),\xi(c))}{4} - \frac{\mu}{6}\|d-c\|^\sigma$$

$$\le \frac{\xi(c)+\xi(d)}{2} + \frac{M_\eta}{2} - \frac{\mu}{6}\|d-c\|^\sigma.$$

This completes the proof.

Remark 3.3.3. *When $\sigma = 2$, then above theorem reduces to Theorem 2.1 of [12], i.e.,*

$$\xi\left(\frac{c+d}{2}\right) - \frac{M_\eta}{2} + \frac{\mu}{12}\|d-c\|^2 \le \frac{1}{d-c}\int_c^d \xi(x)dx$$

$$\le \frac{\xi(c)+\xi(d)}{2} + \frac{\eta(\xi(c),\xi(d)) + \eta(\xi(d),\xi(c))}{4} - \frac{\mu}{6}\|d-c\|^2$$

$$\le \frac{\xi(c)+\xi(d)}{2} + \frac{M_\eta}{2} - \frac{\mu}{6}\|d-c\|^2.$$

Remark 3.3.4. *If we consider $\mu = 0$, then above theorem reduces to Theorem 5 of [54], i.e.,*

$$\xi\left(\frac{c+d}{2}\right) - \frac{M_\eta}{2} \le \frac{1}{d-c}\int_c^d \xi(x)dx$$

$$\le \frac{\xi(c)+\xi(d)}{2} + \frac{\eta(\xi(c),\xi(d)) + \eta(\xi(d),\xi(c))}{4}$$

$$\le \frac{\xi(c)+\xi(d)}{2} + \frac{M_\eta}{2}.$$

Now we establish the result on Fejér type inequality for strongly η-convex function of order $\sigma > 0$ [106].

Theorem 3.3.3. *Let $\xi : [c, d] \to \mathbb{R}$ be a strongly η-convex function of order σ, such that $\eta(.,.)$ is bounded above on $\xi([c,d]) \times \xi([c,d])$. Also suppose that $\psi : [c,d] \to \mathbb{R}^+$ is integrable and symmetric with respect to $\frac{c+d}{2}$, then*

$$\xi\left(\frac{c+d}{2}\right) \int_c^d \psi(x)dx + \frac{\mu}{4}\int_c^d \|(c+d-2x)\|^\sigma \psi(x)dx - L_\eta(c,d)$$

$$\leq \int_c^d \xi(x)\psi(x)dx \leq \frac{\xi(c)+\xi(d)}{2}\int_c^d \psi(x)dx$$

$$- \mu\|d-c\|^{(\sigma-2)}\int_c^d (d-x)(x-c)\psi(x)dx + R_\eta(c,d),$$

where
$L_\eta(c,d) = \frac{1}{2}\int_c^d \eta(\xi(c+d-x), \xi(x))\psi(x)dx$
and $R_\eta(c,d) = \frac{\eta(\xi(c),\xi(d))+\eta(\xi(d),\xi(c))}{2(d-c)}\int_c^d(d-x)\psi(x)dx$,
respectively.

Proof First, we prove the first pair inequality of the theorem.
Since ξ is strongly η-convex function of order σ, then

$$\xi\left(\frac{c+d}{2}\right) = \xi\left(\frac{1}{2}((1-\delta)d+\delta c) + \frac{1}{2}((1-\delta)c+\delta d)\right)$$

$$\leq \xi((1-\delta)c+\delta d) + \frac{1}{2}\eta\left(\xi((1-\delta)d+\delta c), \xi((1-\delta)c+\delta d)\right)$$

$$- \frac{\mu}{4}\|(d-c)(1-2\delta)\|^\sigma.$$

Since $\psi : [c,d] \to \mathbb{R}^+$ is integrable and symmetric with respect to $\frac{c+d}{2}$, then

$$\xi\left(\frac{c+d}{2}\right)\int_c^d \psi(x)dx = (d-c)\psi\left(\frac{c+d}{2}\right)\int_0^1 \psi\left((1-\delta)c+\delta d\right)d\delta$$

$$\leq (d-c)\int_0^1 \xi\left((1-\delta)c+\delta d\right)\psi\left((1-\delta)c+\delta d\right)d\delta$$

$$+ \frac{(d-c)}{2}\int_0^1 \eta\left(\xi((1-\delta)d+\delta c), \xi((1-\delta)c+\delta d)\right)\psi((1-\delta)c+\delta d)d\delta$$

$$- \frac{\mu}{4}(d-c)\int_0^1 \|(d-c)(1-2\delta)\|^\sigma \psi((1-\delta)c+\delta d)d\delta$$

$$= \int_c^d \xi(x)\psi(x)dx + \frac{1}{2}\int_c^d \eta\left(\xi(c+d-x), \xi(x)\right)\psi(x)dx$$

$$- \frac{\mu}{4}\int_c^d \|c+d-2x\|^\sigma \psi(x)dx.$$

Next, we prove the second pair inequality of the theorem,

$$\int_c^d \xi(x)\psi(x)dx = (d-c)\int_0^1 \xi(\delta c + (1-\delta)d)\psi(\delta c + (1-\delta)d)d\delta.$$

Using the definition of strongly η-convex function of order σ, we have

$$\int_c^d \xi(x)\psi(x)dx \le (d-c)\left[\xi(d)\int_0^1 \psi(\delta c + (1-\delta)d)d\delta \right.$$

$$+ \eta(\xi(c),\xi(d))\int_0^1 \delta\psi(\delta c + (1-\delta)d)d\delta \qquad (3.7)$$

$$\left. - \mu\|d-c\|^\sigma \int_0^1 \delta(1-\delta)\psi(\delta c + (1-\delta)d)d\delta\right].$$

Similarly,

$$\int_c^d \xi(x)\psi(x)dx \le (d-c)\left[\xi(c)\int_0^1 \psi(\delta c + (1-\delta)d)d\delta \right.$$

$$+ \eta(\psi(d),\psi(c))\int_0^1 \delta\psi(\delta c + (1-\delta)d)d\delta \qquad (3.8)$$

$$\left. - \mu\|d-c\|^\sigma \int_0^1 \delta(1-\delta)\psi(\delta c + (1-\delta)d)d\delta\right].$$

From (3.7) and (3.8), we have

$$2\int_c^d \xi(x)\psi(x)dx \le (d-c)[\xi(c)+\xi(d)]\int_0^1 \psi(\delta c + (1-\delta)d)d\delta$$

$$+ (d-c)[\eta(\xi(c),\xi(d))+\eta(\xi(d),\xi(c))]\int_0^1 \delta\psi(\delta c + (1-\delta)d)d\delta$$

$$- 2\mu\|d-c\|^{(\sigma+1)}\int_0^1 \delta(1-\delta)\psi(\delta c + (1-\delta)d)d\delta.$$

Applying change of variable technique in above inequality, we have

$$\int_c^d \xi(x)\psi(x)dx \le \frac{[\xi(c)+\xi(d)]}{2}\int_c^d \psi(x)dx$$

$$+ \frac{[\eta(\xi(c),\xi(d))+\eta(\xi(d),\xi(c))]}{2(d-c)}\int_c^d (d-x)\psi(x)dx$$

$$- \mu\|d-c\|^{(\sigma-2)}\int_c^d (d-x)(x-c)\psi(x)dx.$$

This completes the proof.

Remark 3.3.5. *When* $\mu = 0$, *then the first pair of inequality of the above theorem reduces to Theorem 7 of [54], i.e.,*

$$\xi\left(\frac{c+d}{2}\right)\int_c^d \psi(x)dx - \frac{1}{2}\int_c^d \eta(\xi(c+d-x),\xi(x))\psi(x)dx \le \int_c^d \xi(x)\psi(x)dx.$$

Remark 3.3.6. *When $\mu = 0$, then the second pair of inequality of the above theorem reduces to Theorem 6 of [54], i.e.,*

$$\int_c^d \xi(x)\psi(x)dx \leq \frac{\xi(c) + \xi(d)}{2} \int_c^d \psi(x)dx$$
$$+ \frac{\eta(\xi(c), \xi(d)) + \eta(\xi(d), \xi(c))}{2(d-c)} \int_c^d (d-x)\psi(x)dx.$$

Now, we discuss a new variant of Hermite–Hadamard inequality for differentiable strongly η-convex of order σ [106].

Theorem 3.3.4. *Let $\xi : K^0 \subset \mathbb{R} \to \mathbb{R}$ be n-times differentiable strongly η-convex function of order σ on K^0 where $c, d \in K^0$ with $c < d$ and $\xi' \in L^1[c,d]$. If $|\xi^{(n)}|^p$ is strongly η-convex function of order σ with $\mu \geq 1$, then for $n \geq 2$ and $p \geq 1$, we have*

$$\left| \frac{\xi(c) + \xi(d)}{2} - \frac{1}{d-c} \int_c^d \xi(x)dx - \sum_{k=2}^{n-1} \frac{(k-1)(d-c)^k}{2(k+1)!} \xi^{(k)}(c) \right|$$
$$\leq \frac{(d-c)^n}{2n!} \alpha^{(1-\frac{1}{p})}(n)[\alpha(n)|\xi^{(n)}(d)|^p + \beta(n)\eta(|\xi^{(n)}(c)|^p, |\xi^{(n)}(d)|^p)$$
$$- \mu\gamma(n)\|c - d\|^\sigma]^{\frac{1}{p}},$$

where
$$\alpha(n) := \frac{n-1}{n+1}, \quad \beta(n) := \frac{n^2-2}{(n+1)(n+2)}, \quad \gamma(n) := \frac{n-1}{(n+1)(n+3)}.$$

Proof Recall Lemma 2.2.5

$$\frac{\xi(c) + \xi(d)}{2} - \frac{1}{d-c} \int_c^d \xi(x)dx - \sum_{k=2}^{n-1} \frac{(k-1)(d-c)^k}{2(k+1)!} \xi^{(k)}(c)$$
$$= \frac{(d-c)^n}{2n!} \int_0^1 \delta^{n-1}(n - 2\delta)\xi^{(n)}(\delta c + (1-\delta)d)d\delta.$$

Case 1 When $p = 1$,
using the definition of strong η-convexity of order σ, we have

$$\left| \frac{\xi(c) + \xi(d)}{2} - \frac{1}{d-c} \int_c^d \xi(x)dx - \sum_{k=2}^{n-1} \frac{(k-1)(d-c)^k}{2(k+1)!} \xi^{(k)}(c) \right|$$
$$\leq \frac{(d-c)^n}{2n!} \left[\int_0^1 \delta^{n-1}(n - 2\delta)|\xi^{(n)}(\delta c + (1-\delta)d)|d\delta \right]$$

$$\leq \frac{(d-c)^n}{2n!} \left[|\xi^{(n)}(d)| \int_0^1 \delta^{n-1}(n-2\delta)d\delta \right.$$

$$+ \eta(|\xi^{(n)}(c)|, |\xi^{(n)}(d)|) \int_0^1 \delta^n (n-2\delta)d\delta$$

$$\left. - \mu \|c-d\|^\sigma \int_0^1 \delta^n (1-\delta)(n-2\delta)d\delta \right]$$

$$= \frac{(d-c)^n}{2n!} \left[\frac{n-1}{n+1} |\xi^{(n)}(d)| + \frac{n^2-2}{(n+1)(n+2)} \eta(|\xi^{(n)}(c)|, |\xi^{(n)}(d)|) \right.$$

$$\left. - \frac{\mu(n-1)}{(n+1)(n+3)} \|c-d\|^\sigma \right].$$

Case 2 When $p > 1$,

$$\left| \frac{\xi(c)+\xi(d)}{2} - \frac{1}{d-c} \int_c^d \xi(x)dx - \sum_{k=2}^{n-1} \frac{(k-1)(d-c)^k}{2(k+1)!} \xi^{(k)}(c) \right|$$

$$\leq \frac{(d-c)^n}{2n!} \left[\int_0^1 (\delta^{n-1}(n-2\delta))^{1-\frac{1}{p}} (\delta^{n-1}(n-2\delta))^{\frac{1}{p}} |\xi^{(n)}(\delta c + (1-\delta)d)| d\delta \right]$$

Using Hölder's inequality, we have

$$\left| \frac{\xi(c)+\xi(d)}{2} - \frac{1}{d-c} \int_c^d \xi(x)dx - \sum_{k=2}^{n-1} \frac{(k-1)(d-c)^k}{2(k+1)!} \xi^{(k)}(c) \right|$$

$$\leq \frac{(d-c)^n}{2n!} \left[\int_0^1 \delta^{n-1}(n-2\delta)d\delta \right]^{1-\frac{1}{p}}$$

$$\times \left[\int_0^1 \delta^{n-1}(n-2\delta)|\xi^{(n)}(\delta c + (1-\delta)d)|^p d\delta \right]^{\frac{1}{p}}$$

$$\leq \frac{(d-c)^n}{2n!} \left(\frac{n-1}{n+1} \right)^{1-\frac{1}{p}} \left[|\xi^{(n)}(d)|^p \int_0^1 \delta^{n-1}(n-2\delta)d\delta \right.$$

$$+ \eta(|\xi^{(n)}(c)|^p, |\xi^{(n)}(d)|^p) \int_0^1 \delta^n (n-2\delta)d\delta$$

$$\left. - \mu \|c-d\|^\sigma \int_0^1 \delta^n (1-\delta)(n-2\delta)d\delta \right]^{\frac{1}{p}}.$$

This implies

$$\left| \frac{\xi(c) + \xi(d)}{2} - \frac{1}{d-c} \int_c^d \xi(x)dx - \sum_{k=2}^{n-1} \frac{(k-1)(d-c)^k}{2(k+1)!} \xi^{(k)}(c) \right|$$

$$\leq \frac{(d-c)^n}{2n!} \left(\frac{n-1}{n+1} \right)^{1-\frac{1}{p}} \left[\frac{n-1}{n+1} |\xi^{(n)}(d)|^p \right.$$

$$+ \frac{n^2 - 2}{(n+1)(n+2)} \eta(|\xi^{(n)}(c)|^p, |\xi^{(n)}(d)|^p)$$

$$\left. - \frac{\mu(n-1)}{(n+1)(n+3)} ||c - d||^\sigma \right]^{\frac{1}{p}}.$$

This completes the proof.

3.4 Integral Inequalities for Higher Order Strongly Exponentially Convex Functions

We derive new Hermite–Hadamard inequality for higher order strongly exponentially convex functions.

Theorem 3.4.1. *Let $\xi : K = [c, d] \longrightarrow \mathbb{R}$ be a strongly exponentially convex function of order $\sigma > 1$ with modulus $\mu > 0$, then the function satisfies the following:*

$$e^{\xi(\frac{c+d}{2})} + \frac{\mu}{4} ||d - c||^\sigma \leq \frac{1}{d-c} \int_c^d e^{\xi(x)} dx \leq \frac{e^{\xi(c)} + e^{\xi(d)}}{2} - \frac{\mu}{6} ||d - c||^\sigma. \quad (3.9)$$

Proof Since ξ is a strongly exponentially convex function of order $\sigma > 1$ on K, we have

$$e^{\xi(\delta x + (1-\delta)y)} \leq \delta e^{\xi(x)} + (1-\delta)e^{\xi(y)} - \mu\delta(1-\delta)||y - x||^\sigma, \quad \forall\ x, y \in K, \ \delta \in [0, 1]. \quad (3.10)$$

For $\delta = \frac{1}{2}$, we get

$$e^{\xi(\frac{x+y}{2})} \leq \frac{e^{\xi(x)} + e^{\xi(y)}}{2} - \frac{\mu}{4} ||y - x||^\sigma.$$

Let $x = (1 - \delta)c + \delta d$ and $y = \delta c + (1 - \delta)d$, we have

$$e^{\xi(\frac{c+d}{2})} \leq \frac{e^{\xi[(1-\delta)c+\delta d]} + e^{\xi[(\delta c+(1-\delta)d)]}}{2} - \frac{\mu}{4} ||d - c||^\sigma.$$

Integrating above with respect to δ over $[0, 1]$ and using the change of variable technique, we have

$$e^{\xi\left(\frac{c+d}{2}\right)} \leq \frac{1}{d-c}\int_c^d e^{\xi(x)}dx - \frac{\mu}{4}\|d-c\|^\sigma. \tag{3.11}$$

Integrating (3.10) with respect to δ over $[0, 1]$, we have

$$\frac{1}{d-c}\int_c^d e^{\xi(x)}dx \leq \frac{e^{\xi(c)}+e^{\xi(d)}}{2} - \frac{\mu}{6}\|d-c\|^\sigma. \tag{3.12}$$

From (3.11) and (3.12), we obtain

$$e^{\xi\left(\frac{c+d}{2}\right)} + \frac{\mu}{4}\|d-c\|^\sigma \leq \frac{1}{d-c}\int_c^d e^{\xi(x)}dx \leq \frac{e^{\xi(c)}+e^{\xi(d)}}{2} - \frac{\mu}{6}\|d-c\|^\sigma.$$

This completes the proof.

Example 3.4.1. Let $K = [\frac{1}{2}, 1]$ and $\mu = \frac{1}{10}$. Let $\xi : K \longrightarrow \mathbb{R}$ be defined by $\xi(x) = x$ for all $x \in K$. Obviously, ξ is higher order strongly exponentially convex function for $\mu = \frac{1}{10}$. Then, the function ξ satisfies the above theorem.

Remark 3.4.1. When $\mu = 0$, then Theorem 3.4.1 reduces to the following:

$$e^{\xi\left(\frac{c+d}{2}\right)} \leq \frac{1}{d-c}\int_c^d e^{\xi(x)}dx \leq \frac{e^{\xi(c)}+e^{\xi(d)}}{2},$$

which is Hermite–Hadamard inequality for exponentially convex functions given by [46].

Now, we obtain some new Riemann–Liouville fractional estimates via strongly exponentially convex functions of higher order.

Theorem 3.4.2. Let $\xi : K = [c, d] \to \mathbb{R}$ be an absolutely continuous mapping on (c, d) such that $(e^\xi)' \in L^1[c, d]$. If the function $|\xi|$ is strongly exponentially convex function of order $\sigma_1 > 1$ with modulus $\mu_1 > 0$ and $|\xi'|$ is strongly convex function of order $\sigma_2 > 0$ with modulus $\mu_2 > 0$, then

$$\left| \frac{(x-c)^\alpha e^{\xi(c)} + (d-x)^\alpha e^{\xi(d)}}{d-c} - \frac{\Gamma(\alpha+1)[J_{x^-}^\alpha e^{\xi(c)} + J_{x^+}^\alpha e^{\xi(d)}]}{d-c} \right|$$

$$\leq \frac{\alpha}{3(\alpha+3)}\phi(x)\left(\frac{(x-c)^{\alpha+1}+(d-x)^{\alpha+1}}{d-c}\right)$$

$$+ \frac{\alpha^3+6\alpha^2+11\alpha}{3(\alpha+1)(\alpha+2)(\alpha+3)}\left(\frac{(x-c)^{\alpha+1}\phi(c)+(d-x)^{\alpha+1}\phi(d)}{d-c}\right)$$

$$+ \frac{\alpha^2+5\alpha}{6(\alpha+2)(\alpha+3)}\left(\frac{(x-c)^{\alpha+1}\Delta_1(x,c)+(d-x)^{\alpha+1}\Delta_1(x,d)}{d-c}\right)$$

$$- \frac{\alpha^2+7\alpha}{12(\alpha+3)(\alpha+4)}(\mu_1|\xi'(x)|\|d-c\|^{\sigma_1} + \mu_2|e^{\xi(x)}|\|d-c\|^{\sigma_2})$$

$$\times \left(\frac{(x-c)^{\alpha+1} + (d-x)^{\alpha+1}}{d-c} \right) - \frac{\alpha^3 + 9\alpha^2 + 26\alpha}{12(\alpha+2)(\alpha+3)(\alpha+4)}$$

$$\times \left\{ \mu_1 \|d-c\|^{\sigma_1} \left(\frac{|\xi'(c)|(x-c)^{\alpha+1} + |\xi'(d)|(d-x)^{\alpha+1}}{d-c} \right) \right.$$

$$+ \mu_2 \|d-c\|^{\sigma_2} \left(\frac{|e^{\xi(c)}|(x-c)^{\alpha+1} + |e^{\xi(d)}|(d-x)^{\alpha+1}}{d-c} \right) \right\}$$

$$+ \frac{\alpha^3 + 12\alpha^2 + 47\alpha}{30(\alpha+3)(\alpha+4)(\alpha+5)} \mu_1 \mu_2 \|d-c\|^{\sigma_1+\sigma_2} \left(\frac{(x-c)^{\alpha+1} + (d-x)^{\alpha+1}}{d-c} \right),$$

where

$$\Delta_1(x,c) = |e^{\xi(x)} \xi'(c)| + |e^{\xi(c)} \xi'(x)|, \quad \Delta_1(x,d) = |e^{\xi(x)} \xi'(d)| + |e^{\xi(d)} \xi'(x)|$$

and $\phi(x) = |e^{\xi(x)} \xi'(x)|, \quad \phi(c) = |e^{\xi(c)} \xi'(c)|, \quad \phi(d) = |e^{\xi(d)} \xi'(d)|.$

Proof Using Corollary 3.2.1, property of modulus and the given hypothesis of the theorem, we obtain

$$\left| \frac{(x-c)^{\alpha} e^{\xi(c)} + (d-x)^{\alpha} e^{\xi(d)}}{d-c} - \frac{\Gamma(\alpha+1)[J_{x-}^{\alpha} e^{\xi(c)} + J_{x+}^{\alpha} e^{\xi(d)}]}{d-c} \right|$$

$$\leq \frac{(x-c)^{\alpha+1}}{d-c} \int_0^1 |(\delta^{\alpha} - 1)| |e^{\xi(\delta x + (1-\delta)c)} \xi'(\delta x + (1-\delta)c)| d\delta$$

$$+ \frac{(d-x)^{\alpha+1}}{d-c} \int_0^1 |(1-\delta^{\alpha})| |e^{\xi(\delta x + (1-\delta)d)} \xi'(\delta x + (1-\delta)d)| d\delta$$

$$\leq \frac{(x-c)^{\alpha+1}}{d-c} \int_0^1 (1-\delta^{\alpha})(\delta|e^{\xi(x)}| + (1-\delta)|e^{\xi(c)}| - \mu_1 \delta(1-\delta)\|d-c\|^{\sigma_1})$$

$$\times (\delta|\xi'(x)| + (1-\delta)|\xi'(c)| - \mu_2 \delta(1-\delta)\|d-c\|^{\sigma_2}) d\delta + \frac{(d-x)^{\alpha+1}}{d-c}$$

$$\times \int_0^1 (1-\delta^{\alpha})(\delta|e^{\xi(x)}| + (1-\delta)|e^{\xi(d)}| - \mu_1 \delta(1-\delta)\|d-c\|^{\sigma_1})(\delta|\xi'(x)|$$

$$+ (1-\delta)|\xi'(d)| - \mu_2 \delta(1-\delta)\|d-c\|^{\sigma_2}) d\delta$$

$$= \frac{(x-c)^{\alpha+1}}{d-c} \left[|e^{\xi(x)} \xi'(x)| \int_0^1 (1-\delta^{\alpha}) \delta^2 d\delta + |e^{\xi(c)} \xi'(c)| \int_0^1 (1-\delta^{\alpha})(1-\delta)^2 d\delta \right.$$

$$+ (|e^{\xi(x)} \xi'(c)| + |e^{\xi(c)} \xi'(x)|) \int_0^1 (1-\delta^{\alpha}) \delta(1-\delta) d\delta - (\mu_1 |\xi'(x)|) \|d-c\|^{\sigma_1}$$

$$+ \mu_2 |e^{\xi(x)}| \|d-c\|^{\sigma_2}) \int_0^1 (1-\delta^{\alpha}) \delta^2 (1-\delta) d\delta - (\mu_1 |\xi'(c)|) \|d-c\|^{\sigma_1}$$

$$+ \mu_2 |e^{\xi(c)}| \|d-c\|^{\sigma_2}) \int_0^1 (1-\delta^{\alpha}) \delta(1-\delta)^2 d\delta + \mu_1 \mu_2 \|d-c\|^{\sigma_1+\sigma_2}$$

$$\left. \times \int_0^1 (1-\delta^{\alpha}) \delta^2 (1-\delta)^2 d\delta \right]$$

$$+ \frac{(d-x)^{\alpha+1}}{d-c} \Bigg[|e^{\xi(x)}\xi'(x)| \int_0^1 (1-\delta^\alpha)\delta^2 d\delta$$

$$+ |e^{\xi(d)}\xi'(d)| \int_0^1 (1-\delta^\alpha)(1-\delta)^2 d\delta + (|e^{\xi(x)}\xi'(d)| + |e^{\xi(d)}\xi'(x)|)$$

$$\times \int_0^1 (1-\delta^\alpha)\delta(1-\delta)d\delta - (\mu_1|\xi'(x)|\|d-c\|^{\sigma_1} + \mu_2|e^{\xi(x)}|\|d-c\|^{\sigma_2})$$

$$\times \int_0^1 (1-\delta^\alpha)\delta^2(1-\delta)d\delta - (\mu_1|\xi'(d)|\|d-c\|^{\sigma_1} + \mu_2|e^{\xi(d)}|\|d-c\|^{\sigma_2})$$

$$\times \int_0^1 (1-\delta^\alpha)\delta(1-\delta)^2 d\delta + \mu_1\mu_2\|d-c\|^{\sigma_1+\sigma_2} \int_0^1 (1-\delta^\alpha)\delta^2(1-\delta)^2 d\delta \Bigg]$$

$$= \frac{\alpha}{3(\alpha+3)}\phi(x)\left(\frac{(x-c)^{\alpha+1}+(d-x)^{\alpha+1}}{d-c}\right)$$

$$+ \frac{\alpha^3+6\alpha^2+11\alpha}{3(\alpha+1)(\alpha+2)(\alpha+3)}\left(\frac{(x-c)^{\alpha+1}\phi(c)+(d-x)^{\alpha+1}\phi(d)}{d-c}\right)$$

$$+ \frac{\alpha^2+5\alpha}{6(\alpha+2)(\alpha+3)}\left(\frac{(x-c)^{\alpha+1}\Delta_1(x,c)+(d-x)^{\alpha+1}\Delta_1(x,d)}{d-c}\right)$$

$$- \frac{\alpha^2+7\alpha}{12(\alpha+3)(\alpha+4)}(\mu_1|\xi'(x)|\|d-c\|^{\sigma_1} + \mu_2|e^{\xi(x)}|\|d-c\|^{\sigma_2})$$

$$\times \left(\frac{(x-c)^{\alpha+1}+(d-x)^{\alpha+1}}{d-c}\right) - \frac{\alpha^3+9\alpha^2+26\alpha}{12(\alpha+2)(\alpha+3)(\alpha+4)}$$

$$\times \left\{ \mu_1\|d-c\|^{\sigma_1}\left(\frac{|\xi'(c)|(x-c)^{\alpha+1}+|\xi'(d)|(d-x)^{\alpha+1}}{d-c}\right) \right.$$

$$\left. + \mu_2\|d-c\|^{\sigma_2}\left(\frac{|e^{\xi(c)}|(x-c)^{\alpha+1}+|e^{\xi(d)}|(d-x)^{\alpha+1}}{d-c}\right)\right\}$$

$$+ \frac{\alpha^3+12\alpha^2+47\alpha}{30(\alpha+3)(\alpha+4)(\alpha+5)}\mu_1\mu_2\|d-c\|^{\sigma_1+\sigma_2}\left(\frac{(x-c)^{\alpha+1}+(d-x)^{\alpha+1}}{d-c}\right).$$

This completes the proof.

Corollary 3.4.1. *If we choose* $\alpha = 1$, *then under the assumption of Theorem 3.4.2, we have a new result*

$$\left| \frac{(x-c)e^{\xi(c)}+(d-x)e^{\xi(d)}}{d-c} - \frac{1}{d-c}\int_c^d e^{\xi(x)}dx \right|$$

$$\leq \frac{1}{12}\phi(x)\left(\frac{(x-c)^2+(d-x)^2}{d-c}\right) + \frac{1}{4}\left(\frac{(x-c)^2\phi(c)+(d-x)^2\phi(d)}{d-c}\right)$$

$$+ \frac{1}{12}\left(\frac{(x-c)^2\Delta_1(x,c)+(d-x)^2\Delta_2(x,d)}{d-c}\right) - \frac{1}{30}(\mu_1|\xi'(x)|\|d-c\|^{\sigma_1}$$

$$+ \mu_2 |e^{\xi(x)}| \|d - c\|^{\sigma_2}) \left(\frac{(x-c)^2 + (d-x)^2}{d-c} \right) - \frac{1}{20} \{ \mu_1 \|d - c\|^{\sigma_1}$$

$$\times \left(\frac{|\xi'(c)|(x-c)^{\alpha+1} + |\xi'(d)|(d-x)^{\alpha+1}}{d-c} \right)$$

$$+ \mu_2 \|d - c\|^{\sigma_2} \left(\frac{|e^{\xi(c)}|(x-c)^{\alpha+1} + |e^{\xi(d)}|(d-x)^{\alpha+1}}{d-c} \right) \}$$

$$+ \frac{1}{60} \mu_1 \mu_2 \|d - c\|^{\sigma_1 + \sigma_2} \left(\frac{(x-c)^2 + (d-x)^2}{d-c} \right).$$

Remark 3.4.2. *When $\mu_1, \mu_2 = 0$, then above theorem reduces to Corollary 2.5 of [146], i.e.,*

$$\left| \frac{(x-c)^\alpha e^{\xi(c)} + (d-x)^\alpha e^{\xi(d)}}{d-c} - \frac{\Gamma(\alpha+1)[J_{x^-}^\alpha e^{\xi(c)} + J_{x^+}^\alpha e^{\xi(d)}]}{d-c} \right|$$

$$\leq \frac{\alpha}{3(\alpha+3)} \phi(x) \left(\frac{(x-c)^{\alpha+1} + (d-x)^{\alpha+1}}{d-c} \right) + \frac{\alpha^3 + 6\alpha^2 + 11\alpha}{3(\alpha+1)(\alpha+2)(\alpha+3)}$$

$$\times \left(\frac{(x-c)^{\alpha+1}\phi(c) + (d-x)^{\alpha+1}\phi(d)}{d-c} \right) + \frac{\alpha^2 + 5\alpha}{6(\alpha+2)(\alpha+3)}$$

$$\times \left(\frac{(x-c)^{\alpha+1}\Delta_1(x,c) + (d-x)^{\alpha+1}\Delta_1(x,d)}{d-c} \right).$$

Remark 3.4.3. *When $\mu_1, \mu_2 = 0$ and $\alpha = 1$ then above theorem reduces to Corollary 2.4 of [146], i.e.,*

$$\left| \frac{(x-c)e^{\xi(c)} + (d-x)e^{\xi(d)}}{d-c} - \frac{1}{d-c} \int_c^d e^{\xi(x)} dx \right|$$

$$\leq \frac{1}{12} \phi(x) \left(\frac{(x-c)^2 + (d-x)^2}{d-c} \right) + \frac{1}{4} \left(\frac{(x-c)^2\phi(c) + (d-x)^2\phi(d)}{d-c} \right) + \frac{1}{12}$$

$$\times \left(\frac{(x-c)^2\Delta_1(x,c) + (d-x)^2\Delta_1(x,d)}{d-c} \right).$$

Theorem 3.4.3. *Let $\xi : K \to \mathbb{R}$ be an absolutely continuous mapping on (c,d) such that $(e^\xi)' \in L^1[c,d]$, where $c,d \in K$ with $c < d$. If the function $|\xi|^q$ is strongly exponentially convex function of order $\sigma_1 > 1$ with modulus $\mu_1 > 0$ and $|\xi'|^q$ is strongly convex function of order $\sigma_2 > 0$ with modulus $\mu_2 > 0$,* *where $\frac{1}{p} + \frac{1}{q} = 1$ with $q > 1$. Then, we have*

$$\left| \frac{(x-c)^\alpha e^{\xi(c)} + (d-x)^\alpha e^{\xi(d)}}{d-c} - \frac{\Gamma(\alpha+1)[J^\alpha_{x^-} e^{\xi(c)} + J^\alpha_{x^+} e^{\xi(d)}]}{d-c} \right|$$

$$\leq \left(\frac{1}{\alpha}\beta\left(p+1, \frac{1}{\alpha}\right) \right)^{\frac{1}{p}} \left[\frac{(x-c)^{\alpha+1}}{d-c} \left(\frac{\Delta_2(x,c)}{3} + \frac{\Delta_3(x,c)}{6} \right. \right.$$

$$- \frac{\mu_1 \|d-c\|^{\sigma_1}(|\xi'(x)|^q + |\xi'(c)|^q)}{12} - \frac{\mu_2 \|d-c\|^{\sigma_2}(|e^{\xi(x)}|^q + |e^{\xi(c)}|^q)}{12}$$

$$+ \left. \frac{\mu_1\mu_2\|d-c\|^{\sigma_1+\sigma_2}}{30} \right)^{\frac{1}{q}} + \frac{(d-x)^{\alpha+1}}{d-c} \left(\frac{\Delta_2(x,d)}{3} + \frac{\Delta_3(x,d)}{6} \right.$$

$$- \frac{\mu_1 \|d-c\|^{\sigma_1}(|\xi'(x)|^q + |\xi'(d)|^q)}{12} - \frac{\mu_2 \|d-c\|^{\sigma_2}(|e^{\xi(x)}|^q + |e^{\xi(d)}|^q)}{12}$$

$$+ \left. \left. \frac{\mu_1\mu_2\|d-c\|^{\sigma_1+\sigma_2}}{30} \right)^{\frac{1}{q}} \right],$$

where
$\Delta_2(x,c) = |e^{\xi(x)}\xi'(x)|^q + |e^{\xi(c)}\xi'(c)|^q$, $\Delta_2(x,d) = |e^{\xi(x)}\xi'(x)|^q + |e^{\xi(d)}\xi'(d)|^q$
$\Delta_3(x,c) = |e^{\xi(x)}\xi'(c)|^q + |e^{\xi(c)}\xi'(x)|^q$ *and* $\Delta_3(x,d) = |e^{\xi(x)}\xi'(d)|^q + |e^{\xi(d)}\xi'(x)|^q$.

Proof Using Corollary 3.2.1, Hölder's inequality and the given hypothesis of the theorem, we obtain

$$\left| \frac{(x-c)^\alpha e^{\xi(c)} + (d-x)^\alpha e^{\xi(d)}}{d-c} - \frac{\Gamma(\alpha+1)[J^\alpha_{x^-} e^{\xi(c)} + J^\alpha_{x^+} e^{\xi(d)}]}{d-c} \right|$$

$$\leq \frac{(x-c)^{\alpha+1}}{d-c} \left(\int_0^1 |(\delta^\alpha - 1)|^p d\delta \right)^{\frac{1}{p}} \left(\int_0^1 |e^{\xi(\delta x + (1-\delta)c)}\xi'(\delta x + (1-\delta)c)|^q d\delta \right)^{\frac{1}{q}}$$

$$+ \frac{(d-x)^{\alpha+1}}{d-c} \left(\int_0^1 |(1-\delta^\alpha)|^p d\delta \right)^{\frac{1}{p}} \left(\int_0^1 |e^{\xi(\delta x + (1-\delta)d)}\xi'(\delta x + (1-\delta)d)|^q d\delta \right)^{\frac{1}{q}}$$

$$\leq \frac{(x-c)^{\alpha+1}}{d-c} \left(\int_0^1 |(1-\delta^\alpha)|^p d\delta \right)^{\frac{1}{p}} \left(\int_0^1 (1-\delta^\alpha)(\delta|e^{\xi(x)}|^q + (1-\delta)|e^{\xi(c)}|^q \right.$$

$$\left. - \mu_1\delta(1-\delta)\|d-c\|^{\sigma_1})(\delta|\xi'(x)|^q + (1-\delta)|\xi'(c)|^q - \mu_2\delta(1-\delta)\|d-c\|^{\sigma_2})d\delta \right)^{\frac{1}{q}}$$

$$+ \frac{(d-x)^{\alpha+1}}{d-c} \left(\int_0^1 |(1-\delta^\alpha)|^p \right)^{\frac{1}{p}} \left(\int_0^1 (\delta|e^{\xi(x)}|^q + (1-\delta)|e^{\xi(d)}|^q \right.$$

$$\left. - \mu_1\delta(1-\delta)\|d-c\|^{\sigma_1})(\delta|\xi'(x)|^q + (1-\delta)|\xi'(d)|^q - \mu_2\delta(1-\delta)\|d-c\|^{\sigma_2})d\delta \right)^{\frac{1}{q}}$$

$$= \frac{(x-c)^{\alpha+1}}{d-c} \left(\int_0^1 (1-\delta^\alpha)^p d\delta \right)^{\frac{1}{p}} \left[|e^{\xi(x)}\xi'(x)|^q \int_0^1 \delta^2 d\delta + |e^{\xi(c)}\xi'(c)|^q \right.$$

$$\times \int_0^1 (1-\delta)^2 d\delta + (|e^{\xi(x)}\xi'(c)|^q + |e^{\xi(c)}\xi'(x)|^q) \int_0^1 \delta(1-\delta)d\delta$$

$$- \left(\mu_1 |\xi'(x)|^q \|d - c\|^{\sigma_1} + \mu_2 |e^{\xi(x)}|^q \|d - c\|^{\sigma_2} \right) \int_0^1 \delta^2 (1 - \delta) d\delta$$

$$- \left(\mu_1 |\xi'(c)|^q \|d - c\|^{\sigma_1} + \mu_2 |e^{\xi(c)}|^q \|d - c\|^{\sigma_2} \right) \int_0^1 \delta (1 - \delta)^2 d\delta$$

$$+ \mu_1 \mu_2 \|d - c\|^{\sigma_1 + \sigma_2} \int_0^1 \delta^2 (1 - \delta)^2 d\delta \Big]^{\frac{1}{q}} + \frac{(d - x)^{\alpha+1}}{d - c} \left(\int_0^1 (1 - \delta^\alpha)^p d\delta \right)^{\frac{1}{p}}$$

$$\times \left[|e^{\xi(x)} \xi'(x)|^q \int_0^1 \delta^2 d\delta + |e^{\xi(d)} \xi'(d)|^q \int_0^1 (1 - \delta)^2 d\delta + (|e^{\xi(x)} \xi'(d)|^q \right.$$

$$+ |e^{\xi(d)} \xi'(x)|^q) \times \int_0^1 \delta (1 - \delta) d\delta - \left(\mu_1 |\xi'(x)|^q \|d - c\|^{\sigma_1} + \mu_2 |e^{\xi(x)}|^q \|d - c\|^{\sigma_2} \right)$$

$$\times \int_0^1 \delta^2 (1 - \delta) d\delta - \left(\mu_1 |\xi'(d)|^q \|d - c\|^{\sigma_1} + \mu_2 |e^{\xi(d)}|^q \|d - c\|^{\sigma_2} \right)$$

$$\times \int_0^1 \delta (1 - \delta)^2 d\delta + \mu_1 \mu_2 \|d - c\|^{\sigma_1 + \sigma_2} \int_0^1 \delta^2 (1 - \delta)^2 d\delta \Big]^{\frac{1}{q}}$$

$$= \left(\int_0^1 (1 - \delta^\alpha)^p d\delta \right)^{\frac{1}{p}} \left[\frac{(x - c)^{\alpha+1}}{d - c} \left(\frac{\Delta_2(x, c)}{3} + \frac{\Delta_3(x, c)}{6} \right. \right.$$

$$- \frac{(\mu_1 |\xi'(x)|^q \|d - c\|^{\sigma_1} + \mu_2 |e^{\xi(x)}|^q \|d - c\|^{\sigma_2})}{12}$$

$$- \frac{(\mu_1 |\xi'(c)|^q \|d - c\|^{\sigma_1} + \mu_2 |e^{\xi(c)}|^q \|d - c\|^{\sigma_2})}{12} + \frac{\mu_1 \mu_2 \|d - c\|^{\sigma_1 + \sigma_2}}{30} \Big)^{\frac{1}{q}}$$

$$+ \frac{(d - x)^{\alpha+1}}{d - c} \left(\frac{\Delta_2(x, d)}{3} + \frac{\Delta_3(x, d)}{6} \right.$$

$$- \frac{(\mu_1 |\xi'(x)|^q \|d - c\|^{\sigma_1} + \mu_2 |e^{\xi(x)}|^q \|d - c\|^{\sigma_2})}{12}$$

$$\left. \left. - \frac{(\mu_1 |\xi'(d)|^q \|d - c\|^{\sigma_1} + \mu_2 |e^{\xi(d)}|^q \|d - c\|^{\sigma_2})}{12} + \frac{\mu_1 \mu_2 \|d - c\|^{\sigma_1 + \sigma_2}}{30} \Big)^{\frac{1}{q}} \right]$$

$$= \left(\frac{1}{\alpha} \beta \left(p + 1, \frac{1}{\alpha} \right) \right)^{\frac{1}{p}} \left[\frac{(x - c)^{\alpha+1}}{d - c} \left(\frac{\Delta_2(x, c)}{3} + \frac{\Delta_3(x, c)}{6} \right. \right.$$

$$- \frac{\mu_1 \|d - c\|^{\sigma_1} (|\xi'(x)|^q + |\xi'(c)|^q)}{12} - \frac{\mu_2 \|d - c\|^{\sigma_2} (|e^{\xi(x)}|^q + |e^{\xi(c)}|^q)}{12}$$

$$+ \frac{\mu_1 \mu_2 \|d - c\|^{\sigma_1 + \sigma_2}}{30} \Big)^{\frac{1}{q}} + \frac{(d - x)^{\alpha+1}}{d - c} \left(\frac{\Delta_2(x, d)}{3} + \frac{\Delta_3(x, d)}{6} \right.$$

$$- \frac{\mu_1 \|d - c\|^{\sigma_1} (|\xi'(x)|^q + |\xi'(d)|^q)}{12} - \frac{\mu_2 \|d - c\|^{\sigma_2} (|e^{\xi(x)}|^q + |e^{\xi(d)}|^q)}{12}$$

$$\left. \left. + \frac{\mu_1 \mu_2 \|d - c\|^{\sigma_1 + \sigma_2}}{30} \Big)^{\frac{1}{q}} \right].$$

Corollary 3.4.2. *If we choose* $\alpha = 1$, *then under the assumption of Theorem 3.4.3, we have a new result*

$$
\left| \frac{(x-c)e^{\xi(c)} + (d-x)e^{\xi(d)}}{d-c} - \frac{1}{d-c}\int_c^d e^{\xi(x)}dx \right|
$$

$$
\leq \left(\frac{1}{p+1}\right)^{\frac{1}{p}} \left[\frac{(x-c)^2}{d-c} \left(\frac{\Delta_2(x,c)}{3} + \frac{\Delta_3(x,c)}{6} \right. \right.
$$

$$
- \frac{\mu_1 \|d-c\|^{\sigma_1}(|\xi'(x)|^q + |\xi'(c)|^q)}{12} - \frac{\mu_2\|d-c\|^{\sigma_2}(|e^{\xi(x)}|^q + |e^{\xi(c)}|^q)}{12}
$$

$$
+ \left. \frac{\mu_1\mu_2\|d-c\|^{\sigma_1+\sigma_2}}{30} \right)^{\frac{1}{q}} + \frac{(d-x)^2}{d-c}\left(\frac{\Delta_2(x,d)}{3} + \frac{\Delta_3(x,d)}{6} \right.
$$

$$
- \frac{\mu_1\|d-c\|^{\sigma_1}(|\xi'(x)|^q + |\xi'(d)|^q)}{12} - \frac{\mu_2\|d-c\|^{\sigma_2}(|e^{\xi(x)}|^q + |e^{\xi(d)}|^q)}{12}
$$

$$
+ \left. \left. \frac{\mu_1\mu_2\|d-c\|^{\sigma_1+\sigma_2}}{30} \right)^{\frac{1}{q}} \right].
$$

Remark 3.4.4. *When* $\mu_1, \mu_2 = 0$, *then above theorem reduces to Corollary 2.8 of [146], i.e.,*

$$
\left| \frac{(x-c)^{\alpha}e^{\xi(c)} + (d-x)^{\alpha}e^{\xi(d)}}{d-c} - \frac{\Gamma(\alpha+1)[J_{x^-}^{\alpha}e^{\xi(c)} + J_{x^+}^{\alpha}e^{\xi(d)}]}{d-c} \right|
$$

$$
\leq \left(\frac{1}{\alpha}\beta\left(p+1, \frac{1}{\alpha}\right)\right)^{\frac{1}{p}} \left[\frac{(x-c)^{\alpha+1}}{d-c}\left(\frac{\Delta_2(x,c)}{3} + \frac{\Delta_3(x,c)}{6} \right)^{\frac{1}{q}} \right.
$$

$$
+ \left. \frac{(d-x)^{\alpha+1}}{d-c}\left(\frac{\Delta_2(x,d)}{3} + \frac{\Delta_3(x,d)}{6} \right)^{\frac{1}{q}} \right].
$$

Remark 3.4.5. *When* $\mu_1, \mu_2 = 0$ *and* $\alpha = 1$, *then above theorem reduces to Corollary 2.7 of [146], i.e.,*

$$
\left| \frac{(x-c)e^{\xi(c)} + (d-x)e^{\xi(d)}}{d-c} - \frac{1}{d-c}\int_c^d e^{\xi(x)}dx \right|
$$

$$
\leq \left(\frac{1}{p+1}\right)^{\frac{1}{p}} \left[\frac{(x-c)^2}{d-c}\left(\frac{\Delta_2(x,c)}{3} + \frac{\Delta_3(x,c)}{6} \right)^{\frac{1}{q}} \right.
$$

$$
+ \left. \frac{(d-x)^2}{d-c}\left(\frac{\Delta_2(x,d)}{3} + \frac{\Delta_3(x,d)}{6} \right)^{\frac{1}{q}} \right].
$$

Theorem 3.4.4. *Let* $\alpha > 0$ *be a number and let* $\xi : K = [c,d] \longrightarrow \mathbb{R}$ *be a differentiable function on* (c,d). *If the function* $|\xi|$ *is strongly exponentially* *convex function of order $\sigma_1 > 1$ with modulus $\mu_1 > 0$ and $|\xi'|$ is strongly convex function of order $\sigma_2 > 0$ with modulus $\mu_2 > 0$. Then, we have*

$$
\begin{aligned}
&|\Gamma_\xi(c,d,\alpha)| \\
&\leq \frac{d-c}{16}\left[\frac{\alpha^3 + 9\alpha^2 + 20\alpha + 6}{3(\alpha+1)(\alpha+2)(\alpha+3)}\left(\left|e^{\xi\left(\frac{3c+d}{4}\right)}\xi'\left(\frac{3c+d}{4}\right)\right|\right.\right. \\
&\left.\left.+\left|e^{\xi\left(\frac{c+3d}{4}\right)}\xi'\left(\frac{c+3d}{4}\right)\right|\right)\right] \\
&+\frac{d-c}{16}\left[\frac{2}{(\alpha+1)(\alpha+2)(\alpha+3)}\left|e^{\xi(c)}\xi'(c)\right|\right. \\
&+\frac{\alpha^3 + 3\alpha^2 + 2\alpha + 6}{3(\alpha+1)(\alpha+2)(\alpha+3)}\left|e^{\xi\left(\frac{c+d}{2}\right)}\xi'\left(\frac{c+d}{2}\right)\right| + \frac{\alpha}{3(\alpha+3)}\left|e^{\xi(d)}\xi'(d)\right| \\
&+\frac{(A_1(c,d) + A_6(c,d))}{(\alpha+2)(\alpha+3)} - \frac{\alpha^3 + 9\alpha^2 + 38\alpha + 24}{12(\alpha+2)(\alpha+3)(\alpha+4)}(A_2(c,d) \\
&+A_7(c,d)) - \frac{2A_3(c,d)}{(\alpha+2)(\alpha+3)(\alpha+4)} - \frac{\alpha^3 + 9\alpha^2 + 14\alpha + 24}{12(\alpha+2)(\alpha+3)(\alpha+4)}A_5(c,d) \\
&+\frac{\alpha(\alpha+5)}{6(\alpha+2)(\alpha+3)}(A_4(c,d) + A_8(c,d)) - \frac{\alpha(\alpha+7)}{12(\alpha+3)(\alpha+4)}A_9(c,d) \\
&+\frac{4(\alpha^3 + 12\alpha^2 + 47\alpha + 30)}{30(\alpha+3)(\alpha+4)(\alpha+5)}\mu_1\mu_2\left\|\frac{(d-c)}{4}\right\|^{\sigma_1+\sigma_2}\right],
\end{aligned}
$$

where

$$
A_1(c,d) = \left|e^{\xi(c)}\xi'\left(\frac{3c+d}{4}\right)\right| + \left|e^{\xi\left(\frac{3c+d}{4}\right)}\xi'(c)\right|,
$$

$$
A_2(c,d) = \mu_1\left\|\frac{(d-c)}{4}\right\|^{\sigma_1}\left|\xi'\left(\frac{3c+d}{4}\right)\right| + \mu_2\left\|\frac{(d-c)}{4}\right\|^{\sigma_2}|e^{\xi\left(\frac{3c+d}{4}\right)}|,
$$

$$
A_3(c,d) = \mu_1\left\|\frac{(d-c)}{4}\right\|^{\sigma_1}|\xi'(c)| + \mu_2\left\|\frac{(d-c)}{4}\right\|^{\sigma_2}|e^{\xi(c)}|,
$$

$$
A_4(c,d) = \left|e^{\xi\left(\frac{3c+d}{4}\right)}\xi'\left(\frac{c+d}{2}\right)\right| + \left|e^{\xi\left(\frac{c+d}{2}\right)}\xi'\left(\frac{3c+d}{4}\right)\right|,
$$

$$
A_5(c,d) = \mu_1\left\|\frac{(d-c)}{4}\right\|^{\sigma_1}\left|\xi'\left(\frac{c+d}{2}\right)\right| + \mu_2\left\|\frac{(d-c)}{4}\right\|^{\sigma_2}|e^{\xi\left(\frac{c+d}{2}\right)}|,
$$

$$
A_6(c,d) = \left|e^{\xi\left(\frac{c+d}{2}\right)}\xi'\left(\frac{c+3d}{4}\right)\right| + \left|e^{\xi\left(\frac{c+3d}{4}\right)}\xi'\left(\frac{c+d}{2}\right)\right|,
$$

$$
A_7(c,d) = \mu_1\left\|\frac{(d-c)}{4}\right\|^{\sigma_1}\left|\xi'\left(\frac{c+3d}{4}\right)\right| + \mu_2\left\|\frac{(d-c)}{4}\right\|^{\sigma_2}|e^{\xi\left(\frac{c+3d}{4}\right)}|,
$$

$$
A_8(c,d) = |e^{\xi\left(\frac{c+3d}{4}\right)}\xi'(d)| + \left|e^{\xi(d)}\xi'\left(\frac{c+3d}{4}\right)\right|,
$$

$$
A_9(c,d) = \mu_1\left\|\frac{(d-c)}{4}\right\|^{\sigma_1}|\xi'(d)| + \mu_2\left\|\frac{(d-c)}{4}\right\|^{\sigma_2}|e^{\xi(d)}|.
$$

Proof Using Lemma 3.2.1, property of modulus and the given hypothesis of the theorem,

$$|\Gamma_\xi(c,d,\alpha)| \le \frac{d-c}{16} \sum_{i=1}^{4} I_i,$$

where

$$I_1 = \int_0^1 \delta^\alpha \left| e^{\xi\left(\delta\frac{3c+d}{4}+(1-\delta)c\right)}\xi'\left(\delta\frac{3c+d}{4}+(1-\delta)c\right)\right| d\delta$$

$$\le \int_0^1 \delta^\alpha \left(\delta\left|e^{\xi\left(\frac{3c+d}{4}\right)}\right| + (1-\delta)|e^{\xi(c)}| - \mu_1\delta(1-\delta)\left\|\frac{d-c}{4}\right\|^{\sigma_1}\right)$$

$$\times \left(\delta\left|\xi'\left(\frac{3c+d}{4}\right)\right| + (1-\delta)|\xi'(c)| - \mu_2\delta(1-\delta)\left\|\frac{d-c}{4}\right\|^{\sigma_2}\right) d\delta$$

$$= \left|e^{\xi\left(\frac{3c+d}{4}\right)}\xi'\left(\frac{3c+d}{4}\right)\right| \int_0^1 \delta^{\alpha+2}d\delta + |e^{\xi(c)}\xi'(c)| \int_0^1 \delta^\alpha(1-\delta)^2 d\delta$$

$$+ \left(\left|e^{\xi(c)}\xi'\left(\frac{3c+d}{4}\right)\right| + |e^{\xi\left(\frac{3c+d}{4}\right)}\xi'(c)|\right) \int_0^1 \delta^{\alpha+1}(1-\delta)d\delta$$

$$- \left(\mu_1\left\|\frac{(d-c)}{4}\right\|^{\sigma_1}\left|\xi'\left(\frac{3c+d}{4}\right)\right| + \mu_2\left\|\frac{(d-c)}{4}\right\|^{\sigma_2}|e^{\xi\left(\frac{3c+d}{4}\right)}|\right)$$

$$\times \int_0^1 \delta^{\alpha+2}(1-\delta)d\delta - \left(\mu_1\left\|\frac{(d-c)}{4}\right\|^{\sigma_1}|\xi'(c)| + \mu_2\left\|\frac{(d-c)}{4}\right\|^{\sigma_2}|e^{\xi(c)}|\right)$$

$$\times \int_0^1 \delta^{\alpha+1}(1-\delta)^2 d\delta + \mu_1\mu_2\left\|\frac{(d-c)}{4}\right\|^{\sigma_1+\sigma_2}\int_0^1 \delta^{\alpha+2}(1-\delta)^2 d\delta$$

$$= \frac{1}{\alpha+3}\left|e^{\xi\left(\frac{3c+d}{4}\right)}\xi'\left(\frac{3c+d}{4}\right)\right| + \frac{2}{(\alpha+1)(\alpha+2)(\alpha+3)}\left|e^{\xi(c)}\xi'(c)\right|$$

$$+ \frac{A_1(c,d)}{(\alpha+2)(\alpha+3)} - \frac{A_2(c,d)}{(\alpha+3)(\alpha+4)} - \frac{2A_3(c,d)}{(\alpha+2)(\alpha+3)(\alpha+4)}$$

$$+ \frac{2}{(\alpha+3)(\alpha+4)(\alpha+5)}\mu_1\mu_2\left\|\frac{(d-c)}{4}\right\|^{\sigma_1+\sigma_2},$$

$$I_2 = \int_0^1 (1-\delta^\alpha)\left|e^{\xi\left(\delta\frac{c+d}{2}+(1-\delta)\frac{3c+d}{4}\right)}\xi'\left(\delta\frac{c+d}{2}+(1-\delta)\frac{3c+d}{4}\right)\right| d\delta$$

$$\le \frac{\alpha}{3(\alpha+3)}\left|e^{\xi\left(\frac{c+d}{2}\right)}\xi'\left(\frac{c+d}{2}\right)\right| + \frac{\alpha(\alpha^2+6\alpha+11)}{3(\alpha+1)(\alpha+2)(\alpha+3)}$$

$$\times \left|e^{\xi\left(\frac{3c+d}{4}\right)}\xi'\left(\frac{3c+d}{4}\right)\right| + \frac{\alpha(\alpha+5)}{6(\alpha+2)(\alpha+3)}A_4(c,d)$$

$$- \frac{\alpha(\alpha+7)}{12(\alpha+3)(\alpha+4)}A_5(c,d) - \frac{\alpha(\alpha^2+9\alpha+26)}{12(\alpha+2)(\alpha+3)(\alpha+4)}A_2(c,d)$$

$$+ \frac{2\alpha(\alpha^2+12\alpha+47)}{30(\alpha+3)(\alpha+4)(\alpha+5)}\mu_1\mu_2\left\|\frac{(d-c)}{4}\right\|^{\sigma_1+\sigma_2},$$

$$I_3 = \int_0^1 \delta^\alpha \left| e^{\xi\left(\delta\frac{c+3d}{4}+(1-\delta)\frac{c+d}{2}\right)}\xi'\left(\delta\frac{c+3d}{4}+(1-\delta)\frac{c+d}{2}\right)\right| d\delta$$

$$\leq \frac{1}{\alpha+3}\left|e^{\xi\left(\frac{c+3d}{4}\right)}\xi'\left(\frac{c+3d}{4}\right)\right| + \frac{2}{(\alpha+1)(\alpha+2)(\alpha+3)}\left|e^{\xi\left(\frac{c+d}{2}\right)}\xi'\left(\frac{c+d}{2}\right)\right|$$

$$+ \frac{A_6(c,d)}{(\alpha+2)(\alpha+3)} - \frac{A_7(c,d)}{(\alpha+3)(\alpha+4)} - \frac{2A_5(c,d)}{(\alpha+2)(\alpha+3)(\alpha+4)}$$

$$+ \frac{2}{(\alpha+3)(\alpha+4)(\alpha+5)}\mu_1\mu_2\left\|\frac{(d-c)}{4}\right\|^{\sigma_1+\sigma_2}$$

and

$$I_4 = \int_0^1 (1-\delta^\alpha)\left|e^{\xi\left(\delta d+(1-\delta)\frac{c+3d}{4}\right)}\xi'\left(\delta d+(1-\delta)\frac{c+3d}{4}\right)\right| d\delta$$

$$\leq \frac{\alpha}{3(\alpha+3)}\left|e^{\xi(d)}\xi'(d)\right| + \frac{\alpha(\alpha^2+6\alpha+11)}{3(\alpha+1)(\alpha+2)(\alpha+3)}\left|e^{\xi\left(\frac{c+3d}{4}\right)}\xi'\left(\frac{c+3d}{4}\right)\right|$$

$$+ \frac{\alpha(\alpha+5)}{6(\alpha+2)(\alpha+3)}A_8(c,d) - \frac{\alpha(\alpha+7)}{12(\alpha+3)(\alpha+4)}A_9(c,d)$$

$$- \frac{\alpha(\alpha^2+9\alpha+26)}{12(\alpha+2)(\alpha+3)(\alpha+4)}A_7(c,d)$$

$$+ \frac{2\alpha(\alpha^2+12\alpha+47)}{30(\alpha+3)(\alpha+4)(\alpha+5)}\mu_1\mu_2\left\|\frac{(d-c)}{4}\right\|^{\sigma_1+\sigma_2}.$$

This completes the proof.

Corollary 3.4.3. *If we choose* $\alpha = 1$, *then under the assumption of Theorem 3.4.4, we have a new result*

$$\left|\frac{1}{2}\left[e^{\xi\left(\frac{3c+d}{4}\right)}+e^{\xi\left(\frac{c+3d}{4}\right)}\right] - \frac{1}{d-c}\int_c^d e^{\xi(x)}\,dx\right]\right|$$

$$\leq \frac{d-c}{16}\left[\frac{1}{2}\left(\left|e^{\xi\left(\frac{3c+d}{4}\right)}\xi'\left(\frac{3c+d}{4}\right)\right| + \left|e^{\xi\left(\frac{c+3d}{4}\right)}\xi'\left(\frac{c+3d}{4}\right)\right|\right)\right.$$

$$+ \frac{1}{12}\left|e^{\xi(c)}\xi'(c)\right| + \frac{1}{6}\left|e^{\xi\left(\frac{c+d}{2}\right)}\xi'\left(\frac{c+d}{2}\right)\right| + \frac{1}{12}\left|e^{\xi(d)}\xi'(d)\right|$$

$$+ \frac{1}{12}(A_1(c,d)+A_6(c,d)) - \frac{1}{10}(A_2(c,d)+A_7(c,d)) - \frac{A_3(c,d)}{30}$$

$$- \frac{1}{15}A_5(c,d) + \frac{1}{12}(A_4(c,d)+A_8(c,d)) - \frac{1}{30}A_9(c,d)$$

$$\left. + \frac{1}{10}\mu_1\mu_2\left\|\frac{(d-c)}{4}\right\|^{\sigma_1+\sigma_2}\right].$$

Theorem 3.4.5. *Let $\alpha > 0$ be a number and let $\xi : K = [c,d] \longrightarrow \mathbb{R}$ be a differentiable function on (c,d). If the function $|\xi|^q$ is strongly exponentially convex function of order $\sigma_1 > 1$ with modulus $\mu_1 > 0$ and $|\xi'|^q$ is strongly convex function of order $\sigma_2 > 0$ with modulus $\mu_2 > 0$, where $\frac{1}{p} + \frac{1}{q} = 1$, $q > 1$, then*

$$|\Gamma_\xi(c,d,\alpha)|$$

$$\leq \frac{d-c}{(60^{\frac{1}{q}})16} \left(\frac{1}{\alpha}\right)^{\frac{1}{p}} \left[\left(\frac{\alpha}{1+p\alpha}\right)^{\frac{1}{p}} \left\{ \left(20 \left(\left| e^{\xi(\frac{3c+d}{4})} \xi'\left(\frac{3c+d}{4}\right) \right|^q \right.\right.\right.$$

$$+ \left| e^{\xi(c)} \xi'(c) \right|^q \right) + 10B_1(c,d) - 5B_2(c,d) + 2\mu_1\mu_2 \left\| \frac{(d-c)}{4} \right\|^{\sigma_1+\sigma_2} \right)^{\frac{1}{q}}$$

$$+ \left(20 \left(\left| e^{\xi(\frac{c+3d}{4})} \xi'\left(\frac{c+3d}{4}\right) \right|^q + \left| e^{\xi(\frac{c+d}{2})} \xi'\left(\frac{c+d}{2}\right) \right|^q \right) + 10B_3(c,d) \right.$$

$$\left. -5B_4(c,d) + 2\mu_1\mu_2 \left\| \frac{(d-c)}{4} \right\|^{\sigma_1+\sigma_2} \right)^{\frac{1}{q}} \right\} + \left(\beta\left(p+1, \frac{1}{\alpha}\right) \right)^{\frac{1}{p}}$$

$$\times \left\{ \left(20 \left(\left| e^{\xi(\frac{c+d}{2})} \xi'\left(\frac{c+d}{2}\right) \right|^q + \left| e^{\xi(\frac{3c+d}{4})} \xi'\left(\frac{3c+d}{4}\right) \right|^q \right) + 10B_5(c,d) \right.\right.$$

$$\left. -5B_6(c,d) + 2\mu_1\mu_2 \left\| \frac{(d-c)}{4} \right\|^{\sigma_1+\sigma_2} \right)^{\frac{1}{q}} + \left(20 \left(\left| e^{\xi(d)} \xi'(d) \right|^q \right.\right.$$

$$\left. + \left| e^{\xi(\frac{c+3d}{4})} \xi'\left(\frac{c+3d}{4}\right) \right|^q \right) + 10B_7(c,d) - 5B_8(c,d)$$

$$\left.\left. + 2\mu_1\mu_2 \left\| \frac{(d-c)}{4} \right\|^{\sigma_1+\sigma_2} \right)^{\frac{1}{q}} \right\} \right],$$

where

$$B_1(c,d) = \left| e^{\xi(c)} \xi'\left(\frac{3c+d}{4}\right) \right|^q + \left| e^{\xi(\frac{3c+d}{4})} \xi'(c) \right|^q ,$$

$$B_2(c,d) = \mu_1 \left\| \frac{(d-c)}{4} \right\|^{\sigma_1} \left(\left| \xi'\left(\frac{3c+d}{4}\right) \right|^q + |\xi'(c)|^q \right)$$

$$+ \mu_2 \left\| \frac{(d-c)}{4} \right\|^{\sigma_2} \left(\left| e^{\xi(\frac{3c+d}{4})} \right|^q + \left| e^{\xi(c)} \right|^q \right),$$

$$B_3(c,d) = \left| e^{\xi(\frac{c+d}{2})} \xi'\left(\frac{c+3d}{4}\right) \right|^q + \left| e^{\xi(\frac{c+3d}{4})} \xi'\left(\frac{c+d}{2}\right) \right|^q ,$$

$$B_4(c,d) = \mu_1 \left\|\frac{(d-c)}{4}\right\|^{\sigma_1} \left(\left|\xi'\left(\frac{c+3d}{4}\right)\right|^q + \left|\xi'\left(\frac{c+d}{2}\right)\right|^q\right)$$

$$+ \mu_2 \left\|\frac{(d-c)}{4}\right\|^{\sigma_2} \left(\left|e^{\xi\left(\frac{c+3d}{4}\right)}\right|^q + \left|e^{\xi\left(\frac{c+d}{2}\right)}\right|^q\right),$$

$$B_5(c,d) = \left|e^{\xi\left(\frac{3c+d}{4}\right)}\xi'\left(\frac{c+d}{2}\right)\right|^q + \left|e^{\xi\left(\frac{c+d}{2}\right)}\xi'\left(\frac{3c+d}{4}\right)\right|^q,$$

$$B_6(c,d) = \mu_1 \left\|\frac{(d-c)}{4}\right\|^{\sigma_1} \left(\left|\xi'\left(\frac{c+d}{2}\right)\right|^q + \left|\xi'\left(\frac{3c+d}{4}\right)\right|^q\right)$$

$$+ \mu_2 \left\|\frac{(d-c)}{4}\right\|^{\sigma_2} \left(\left|e^{\xi\left(\frac{c+d}{2}\right)}\right|^q + \left|e^{\xi\left(\frac{3c+d}{4}\right)}\right|^q\right),$$

$$B_7(c,d) = \left|e^{\xi\left(\frac{c+3d}{4}\right)}\xi'(d)\right|^q + \left|e^{\xi(d)}\xi'\left(\frac{c+3d}{4}\right)\right|^q,$$

$$B_8(c,d) = \mu_1 \left\|\frac{(d-c)}{4}\right\|^{\sigma_1} \left(|\xi'(d)|^q + \left|\xi'\left(\frac{c+3d}{4}\right)\right|^q\right)$$

$$+ \mu_2 \left\|\frac{(d-c)}{4}\right\|^{\sigma_2} \left(|e^{\xi(d)}|^q + |e^{\xi\left(\frac{c+3d}{4}\right)}|^q\right).$$

Proof Using Lemma 3.2.1, Hölder's inequality and the given hypothesis of the theorem, we obtain

$$|\Gamma_\xi(c,d,\alpha)|$$

$$\leq \frac{d-c}{16}\left[\left(\int_0^1 (\delta^\alpha)^p d\delta\right)^{\frac{1}{p}} \left(\sum_{r=1}^2 J_r^{\frac{1}{q}}\right) + \left(\int_0^1 (1-\delta^\alpha)^p d\delta\right)^{\frac{1}{p}} \left(\sum_{r=3}^4 J_r^{\frac{1}{q}}\right)\right],$$

where

$$J_1 = \int_0^1 \left|e^{\xi\left(\delta\frac{3c+d}{4}+(1-\delta)c\right)}\xi'\left(\delta\frac{3c+d}{4}+(1-\delta)c\right)\right|^q d\delta$$

$$\leq \left|e^{\xi\left(\frac{3c+d}{4}\right)}\xi'\left(\frac{3c+d}{4}\right)\right|^q \int_0^1 \delta^2 d\delta + \left|e^{\xi(c)}\xi'(c)\right|^q \int_0^1 (1-\delta)^2 d\delta$$

$$+ \left(\left|e^{\xi(c)}\xi'\left(\frac{3c+d}{4}\right)\right|^q + |e^{\xi\left(\frac{3c+d}{4}\right)}\xi'(c)|^q\right)\int_0^1 \delta(1-\delta)d\delta$$

$$- \left(\mu_1 \left\|\frac{(d-c)}{4}\right\|^{\sigma_1}\left|\xi'\left(\frac{3c+d}{4}\right)\right|^q + \mu_2 \left\|\frac{(d-c)}{4}\right\|^{\sigma_2}|e^{\xi\left(\frac{3c+d}{4}\right)}|^q\right)$$

$$\times \int_0^1 \delta^2(1-\delta)d\delta - \left(\mu_1 \left\|\frac{(d-c)}{4}\right\|^{\sigma_1}|\xi'(c)|^q + \mu_2 \left\|\frac{(d-c)}{4}\right\|^{\sigma_2}|e^{\xi(c)}|^q\right)$$

$$\times \int_0^1 \delta(1-\delta)^2 d\delta + \mu_1\mu_2 \left\|\frac{(d-c)}{4}\right\|^{\sigma_1+\sigma_2} \int_0^1 \delta^2(1-\delta)^2 d\delta$$

$$= \frac{1}{3} \left| e^{\xi\left(\frac{3c+d}{4}\right)} \xi'\left(\frac{3c+d}{4}\right) \right|^q + \frac{1}{3} \left| e^{\xi(c)} \xi'(c) \right|^q + \frac{1}{6} \left(\left| e^{\xi(c)} \xi'\left(\frac{3c+d}{4}\right) \right|^q \right.$$

$$+ \left| e^{\xi\left(\frac{3c+d}{4}\right)} \xi'(c) \right|^q \right) - \frac{1}{12} \left(\mu_1 \left\| \frac{(d-c)}{4} \right\|^{\sigma_1} \left| \xi'\left(\frac{3c+d}{4}\right) \right|^q \right.$$

$$+ \mu_2 \left\| \frac{(d-c)}{4} \right\|^{\sigma_2} \left| e^{\xi\left(\frac{3c+d}{4}\right)} \right|^q \right) - \frac{1}{12} \left(\mu_1 \left\| \frac{(d-c)}{4} \right\|^{\sigma_1} \left| \xi'(c) \right|^q \right.$$

$$+ \mu_2 \left\| \frac{(d-c)}{4} \right\|^{\sigma_2} \left| e^{\xi(c)} \right|^q \right) + \frac{1}{30} \mu_1 \mu_2 \left\| \frac{(d-c)}{4} \right\|^{\sigma_1+\sigma_2}$$

$$= \frac{1}{60} \left[20 \left(\left| e^{\xi\left(\frac{3c+d}{4}\right)} \xi'\left(\frac{3c+d}{4}\right) \right|^q + \left| e^{\xi(c)} \xi'(c) \right|^q \right) + 10 B_1(c,d) \right.$$

$$\left. - 5 B_2(c,d) + 2\mu_1\mu_2 \left\| \frac{(d-c)}{4} \right\|^{\sigma_1+\sigma_2} \right],$$

$$J_2 = \int_0^1 \left| e^{\xi\left(\delta\frac{c+3d}{4} + (1-\delta)\frac{c+d}{2}\right)} \xi'\left(\delta\frac{c+3d}{4} + (1-\delta)\frac{c+d}{2}\right) \right|^q d\delta$$

$$\leq \frac{1}{60} \left[20 \left(\left| e^{\xi\left(\frac{c+3d}{4}\right)} \xi'\left(\frac{c+3d}{4}\right) \right|^q + \left| e^{\xi\left(\frac{c+d}{2}\right)} \xi'\left(\frac{c+d}{2}\right) \right|^q \right) + 10 B_3(c,d) \right.$$

$$\left. - 5 B_4(c,d) + 2\mu_1\mu_2 \left\| \frac{(d-c)}{4} \right\|^{\sigma_1+\sigma_2} \right],$$

$$J_3 = \int_0^1 \left| e^{\xi\left(\delta\frac{c+d}{2} + (1-\delta)\frac{3c+d}{4}\right)} \xi'\left(\delta\frac{c+d}{2} + (1-\delta)\frac{3c+d}{4}\right) \right|^q d\delta$$

$$\leq \frac{1}{60} \left[20 \left(\left| e^{\xi\left(\frac{c+d}{2}\right)} \xi'\left(\frac{c+d}{2}\right) \right|^q + \left| e^{\xi\left(\frac{3c+d}{4}\right)} \xi'\left(\frac{3c+d}{4}\right) \right|^q \right) + 10 B_5(c,d) \right.$$

$$\left. - 5 B_6(c,d) + 2\mu_1\mu_2 \left\| \frac{(d-c)}{4} \right\|^{\sigma_1+\sigma_2} \right]$$

and

$$J_4 = \int_0^1 \left| e^{\xi\left(\delta d + (1-\delta)\frac{c+3d}{4}\right)} \xi'\left(\delta d + (1-\delta)\frac{c+3d}{4}\right) \right|^q d\delta$$

$$\leq \frac{1}{60} \left[20 \left(\left| e^{\xi(d)} \xi'(d) \right|^q + \left| e^{\xi\left(\frac{c+3d}{4}\right)} \xi'\left(\frac{c+3d}{4}\right) \right|^q \right) + 10 B_7(c,d) \right.$$

$$\left. - 5 B_8(c,d) + 2\mu_1\mu_2 \left\| \frac{(d-c)}{4} \right\|^{\sigma_1+\sigma_2} \right].$$

This completes the proof.

Corollary 3.4.4. *If we choose* $\alpha = 1$, *then under the assumption of Theorem 3.4.5, we have a new result*

$$\left| \frac{1}{2} \left[e^{\xi\left(\frac{3c+d}{4}\right)} + e^{\xi\left(\frac{c+3d}{4}\right)} \right] - \frac{1}{d-c} \int_c^d e^{\xi(x)} dx \right|$$

$$\leq \frac{d-c}{(60^{\frac{1}{q}})16} \left(\frac{1}{1+p} \right)^{\frac{1}{p}} \left[\left\{ 20 \left(\left| e^{\xi\left(\frac{3c+d}{4}\right)} \xi'\left(\frac{3c+d}{4}\right) \right|^q + \left| e^{\xi(c)} \xi'(c) \right|^q \right) \right. \right.$$

$$\left. + 10B_1(c,d) - 5B_2(c,d) + 2\mu_1\mu_2 \left\| \frac{(d-c)}{4} \right\|^{\sigma_1+\sigma_2} \right\}^{\frac{1}{q}}$$

$$+ \left\{ 20 \left(\left| e^{\xi\left(\frac{c+3d}{4}\right)} \xi'\left(\frac{c+3d}{4}\right) \right|^q + \left| e^{\xi\left(\frac{c+d}{2}\right)} \xi'\left(\frac{c+d}{2}\right) \right|^q \right) + 10B_3(c,d) \right.$$

$$\left. - 5B_4(c,d) + 2\mu_1\mu_2 \left\| \frac{(d-c)}{4} \right\|^{\sigma_1+\sigma_2} \right\}^{\frac{1}{q}} + \left\{ 20 \left(\left| e^{\xi\left(\frac{c+d}{2}\right)} \xi'\left(\frac{c+d}{2}\right) \right|^q \right. \right.$$

$$\left. + \left| e^{\xi\left(\frac{3c+d}{4}\right)} \xi'\left(\frac{3c+d}{4}\right) \right|^q \right) + 10B_5(c,d)$$

$$\left. - 5B_6(c,d) + 2\mu_1\mu_2 \left\| \frac{(d-c)}{4} \right\|^{\sigma_1+\sigma_2} \right\}^{\frac{1}{q}}$$

$$+ \left\{ 20 \left(\left| e^{\xi(d)} \xi'(d) \right|^q + \left| e^{\xi\left(\frac{c+3d}{4}\right)} \xi'\left(\frac{c+3d}{4}\right) \right|^q \right) + 10B_7(c,d) \right.$$

$$\left. \left. - 5B_8(c,d) + 2\mu_1\mu_2 \left\| \frac{(d-c)}{4} \right\|^{\sigma_1+\sigma_2} \right\}^{\frac{1}{q}} \right].$$

Remark 3.4.6. *When* $\mu_1, \mu_2 = 0$, *then above theorem reduces to Theorem 2.2 of [142], i.e.,*

$$|\Gamma_\xi(c,d,\alpha)|$$

$$\leq \frac{d-c}{(60^{\frac{1}{q}})16} \left(\frac{1}{\alpha} \right)^{\frac{1}{p}} \left[\left(\frac{\alpha}{1+p\alpha} \right)^{\frac{1}{p}} \left\{ \left(20 \left(\left| e^{\xi\left(\frac{3c+d}{4}\right)} \xi'\left(\frac{3c+d}{4}\right) \right|^q \right. \right. \right. \right.$$

$$+ \left| e^{\xi(c)} \xi'(c) \right|^q \right) + 10B_1(c,d) \right)^{\frac{1}{q}} + \left(20 \left(\left| e^{\xi\left(\frac{c+3d}{4}\right)} \xi'\left(\frac{c+3d}{4}\right) \right|^q \right. \right.$$

$$\left. \left. + \left| e^{\xi\left(\frac{c+d}{2}\right)} \xi'\left(\frac{c+d}{2}\right) \right|^q \right) + 10B_3(c,d) \right)^{\frac{1}{q}} \right\} + \left(\beta\left(p+1, \frac{1}{\alpha} \right) \right)^{\frac{1}{p}}$$

$$\times \left\{ \left(20 \left(\left| e^{\xi\left(\frac{c+d}{2}\right)} \xi'\left(\frac{c+d}{2}\right) \right|^q + \left| e^{\xi\left(\frac{3c+d}{4}\right)} \xi'\left(\frac{3c+d}{4}\right) \right|^q \right) + 10B_5(c,d) \right)^{\frac{1}{q}} \right.$$

$$\left. \left. + \left(20 \left(\left| e^{\xi(d)} \xi'(d) \right|^q + \left| e^{\xi\left(\frac{c+3d}{4}\right)} \xi'\left(\frac{c+3d}{4}\right) \right|^q \right) + 10B_7(c,d) \right)^{\frac{1}{q}} \right\} \right].$$

Remark 3.4.7. *When $\mu_1, \mu_2 = 0$ and $\alpha = 1$, then above theorem reduces to Corollary 2.2 of [142], i.e.,*

$$\left| \frac{1}{2}\left[e^{\xi\left(\frac{3c+d}{4}\right)} + e^{\xi\left(\frac{c+3d}{4}\right)}\right] - \frac{1}{d-c}\int_c^d e^{\xi(x)}\,dx \right|$$

$$\leq \frac{d-c}{(60^{\frac{1}{q}})16}\left(\frac{1}{1+p}\right)^{\frac{1}{p}}\left[\left\{\left(20\left(\left|e^{\xi\left(\frac{3c+d}{4}\right)}\xi'\left(\frac{3c+d}{4}\right)\right|^q + \left|e^{\xi(c)}\xi'(c)\right|^q\right)\right.\right.\right.$$

$$+10B_1(c,d))^{\frac{1}{q}} + \left(20\left(\left|e^{\xi\left(\frac{c+3d}{4}\right)}\xi'\left(\frac{c+3d}{4}\right)\right|^q + \left|e^{\xi\left(\frac{c+d}{2}\right)}\xi'\left(\frac{c+d}{2}\right)\right|^q\right)\right.$$

$$\left.+10B_3(c,d))^{\frac{1}{q}}\right\} + \left\{\left(20\left(\left|e^{\xi\left(\frac{c+d}{2}\right)}\xi'\left(\frac{c+d}{2}\right)\right|^q + \left|e^{\xi\left(\frac{3c+d}{4}\right)}\xi'\left(\frac{3c+d}{4}\right)\right|^q\right)\right.\right.$$

$$+10B_5(c,d))^{\frac{1}{q}} + \left(20\left(\left|e^{\xi(d)}\xi'(d)\right|^q + \left|e^{\xi\left(\frac{c+3d}{4}\right)}\xi'\left(\frac{c+3d}{4}\right)\right|^q\right)\right.$$

$$\left.\left.+10B_7(c,d))^{\frac{1}{q}}\right\}\right].$$

Chapter 4

Integral Inequalities for Generalized Preinvex Functions

4.1 Introduction

Pini [135] established the relationship between invexity and generalized convexity and showed that $\xi(x) = x^3$ is quasi-convex but not invex, since $x = 0$ is a stationary point but not a minimum point and also given an example $\xi(x, y) = -x^2 + xy - e^y$ which is invex but not quasi-convex because it fails to satisfy the second-order necessary and sufficient condition for quasi-convexity.

Noor proved some Hermite–Hadamard type inequalities for preinvex [119], log-preinvex [118] and the product of two preinvex functions [120]. Further, İşcan [64] obtained some Hermite–Hadamard type inequalities using fractional integrals for preinvex functions.

Recently, Wang and Fečkan [171] investigated fractional integral identities for a differentiable mapping involving Riemann–Liouville fractional integrals and Hadamard fractional integrals and gave some inequalities via standard convex, r-convex, s-convex, m-convex, (s,m)-convex, (β, m)-convex functions, etc. Further, the Hermite–Hadamard type inequality for fractional integrals obtained by Liu et al. [98].

Bector et al. [15] introduced preunivex and univex functions with respect to a vector function η, scalar function ϕ and b as generalizations of preinvex, invex, b-preinvex, b-invex and b-vex functions. It is well known that if $\xi : K \to \mathbb{R}$ is differentiable and preunivex with respect to η, ϕ and b then ξ is univex with respect to η, ϕ and b, where $b(x, u) = \lim_{\lambda \to 0^+} b(x, u, \delta)$ [15].

Antczak [7] showed the relationships between different classes of (p, r)-invex functions with respect to the same function η. It is well known that if a real-valued function ξ defined on an invex set $K \subset \mathbb{R}^n$ with respect to η is preunivex with respect to η, ϕ and b then the level set L_α of ξ is invex with respect to η for each α [7].

Dragomir et al. [42] generalized the Hermite–Hadamard type integral inequalities and discussed their application in special means and numerical integration. Recently, Zhang et al. [182] established some new bounds on Hermite–Hadamard and Simpson's inequalities for mappings whose absolute values of second derivatives are generalized (m, h)-preinvex. They also derived a general

DOI: 10.1201/9781003408284-4

k-fractional integral identity along with multi parameters for twice differentiable mappings.

The organization of this chapter is as follows: In Section 4.2, we recall some basic results which are necessary for our main results. In Section 4.3, we prove a new form of Hermite–Hadamard inequality using left and right-sided ψ-Riemann–Liouville fractional integrals for preinvexity. We present two essential results of ψ-Riemann–Liouville fractional integral identities using the first-order derivative of a preinvex function. With the help of these results, we obtain some fractional Hermite–Hadamard inequalities and give some examples which satisfy the theorems. Further, in Section 4.4, we discuss some applications for special means with the help of these results. In Section 4.5, we propose the concept of generalized (m, h)-preunivex and establish some new bounds on Hermite–Hadamard and Simpson's inequalities for mappings whose absolute values of second derivatives are generalized (m, h)-preunivex. Some interesting results are also obtained by using the special parameter values for various suitable choices of function h.

4.2   Preliminaries

In this section, we recall some basic definitions and results required for this chapter.

Let $K \subseteq \mathbb{R}^n$ be non empty, $\eta : K \times K \to \mathbb{R}^n$, $b : K \times K \times [0,1] \to \mathbb{R}_+$, $\xi : K \to \mathbb{R}$ and $\phi : \mathbb{R} \to \mathbb{R}$. For $x \in K$, $y \in K$, $0 \le \delta \le 1$, we assume that b stands for $b(x, y, \delta) \ge 0$, and $\delta b \le 1$.

Noor [118] established the following Hermite–Hadamard inequalities for preinvex functions.

Theorem 4.2.1. *Let $\xi : K = [c, c + \eta(d, c)] \to (0, \infty)$ be a preinvex function on the interval of real numbers $int(K)$ and $c, d \in int(K)$ with $c < c + \eta(d, c)$. Then,*

$$\xi\left(\frac{2c + \eta(d, c)}{2}\right) \le \frac{1}{\eta(d, c)} \int_c^{c + \eta(d, c)} \xi(x) dx \le \frac{\xi(c) + \xi(d)}{2}. \qquad (4.1)$$

The following definition is taken by Kilbas *et al.* [85].

Definition 4.2.1. *Let (c, d) $(-\infty \le c < d \le \infty)$ be an interval of the real line $\mathbb{R}$ and $\alpha > 0$. Also let $\psi(x)$ be an increasing and positive monotone function on $(c, d]$, having a continuous derivative $\psi'(x)$ on (c, d). The left and right-sided ψ-Riemann–Liouville fractional integrals of a function ξ with respect to an other function ψ on $[c, d]$ are defined by*

$$J_{c^+}^{\alpha;\psi} \xi(x) = \frac{1}{\Gamma(\alpha)} \int_c^x \psi'(v)(\psi(x) - \psi(v))^{\alpha-1} \xi(v) dv,$$

$$J_{d-}^{\alpha:\psi}\xi(x) = \frac{1}{\Gamma(\alpha)}\int_x^d \psi'(v)(\psi(v) - \psi(x))^{\alpha-1}\xi(v)dv,$$

respectively.

The following concept of preunivex functions established by Bector *et al.* [15].

Definition 4.2.2. *At* $y \in K$, *where* K *is a invex set, the function* ξ *is said to be preunivex with respect to* η, ϕ *and* b, *if for every* $x \in K$ *and* $0 \leq \delta \leq 1$,

$$\xi[y + \delta\eta(x,y)] \leq \xi(y) + \delta b\phi[\xi(x) - \xi(y)].$$

Du *et al.* [48] introduced the following concept of m-invex set.

Definition 4.2.3. *A set* $K \subseteq \mathbb{R}^n$ *is called m-invex with respect to the mapping* $\eta : K \times K \times (0,1] \to \mathbb{R}^n$ *for some fixed* $m \in (0,1]$ *if* $mx + \delta\eta(y,x,m) \in K$ *holds for all* $x, y \in K$ *and* $\delta \in [0,1]$.

Zhang *et al.* [182] derived the following general k-fractional integral identity with multi-parameters for twice differentiable functions.

Lemma 4.2.1. *Let* $K \subseteq \mathbb{R}$ *be an open m-invex subset with respect to* η : $K \times K \times (0,1] \to \mathbb{R}\backslash\{0\}$ *for some fixed* $m \in (0,1]$, *and let* $c, d \in A$, $c < d$ *with* $\eta(d,c,m) > 0$. *Assume that* $\xi : K \to \mathbb{R}$ *is a twice differentiable function on* K *such that* ξ'' *is integrable on* $[mc, mc + \eta(d,c,m)]$. *Then the following identity for Riemann–Liouville k-fractional integrals along with* $x \in [c,d]$, $\lambda \in [0,1]$, $\alpha > 0$, *and* $k > 0$ *exists:*

$$I_{\xi,\eta}(\alpha, k; x, \lambda, m, c, d)$$

$$= \frac{\eta^{\frac{\alpha}{k}+2}(x,c,m)}{(\frac{\alpha}{k}+1)\eta(d,c,m)}\int_0^1 \delta\left[\left(\frac{\alpha}{k}+1\right)\lambda - \delta^{\frac{\alpha}{k}}\right]\xi''(mc + \delta\eta(x,c,m))d\delta$$

$$+ \frac{(-1)^{\frac{\alpha}{k}+2}\eta^{\frac{\alpha}{k}+2}(x,d,m)}{(\frac{\alpha}{k}+1)\eta(d,c,m)}\int_0^1 \delta\left[\left(\frac{\alpha}{k}+1\right)\lambda - \delta^{\frac{\alpha}{k}}\right]\xi''(md + \delta\eta(x,d,m))d\delta,$$

where

$$I_{\xi,\eta}(\alpha, k; x, \lambda, m, c, d) = \frac{1-\lambda}{\eta(d,c,m)}[\eta^{\frac{\alpha}{k}}(x,c,m)\xi(mc + \eta(x,c,m))$$

$$+ (-1)^{\frac{\alpha}{k}}\eta^{\frac{\alpha}{k}}(x,d,m)\xi(md + \eta(x,d,m))] + \frac{\lambda}{\eta(d,c,m)}[\eta^{\frac{\alpha}{k}}(x,c,m)\xi(mc)$$

$$+ (-1)^{\frac{\alpha}{k}}\eta^{\frac{\alpha}{k}}(x,d,m)\xi(md)]$$

$$+ \frac{\frac{1}{\frac{\alpha}{k}+1} - \lambda}{\eta(d,c,m)}[(-1)^{\frac{\alpha}{k}+1}\eta^{\frac{\alpha}{k}+1}(x,d,m)\xi'(md + \eta(x,d,m))$$

$$- \eta^{\frac{\alpha}{k}+1}(x,c,m)\xi'(mc + \eta(x,c,m))] - \frac{\Gamma_k(\alpha+k)}{\eta(d,c,m)}[{}_kJ^{\alpha}_{(mc+\eta(x,c,m))-}\xi(mc)$$

$$+ {}_kJ^{\alpha}_{(md+\eta(x,d,m))+}\xi(md)]$$

and Γ_k *is the k-gamma function.*

4.3   Hermite–Hadamard Type Inequalities via Preinvex Functions

In this section, first, we prove Hermite–Hadamard inequalities for ψ-Riemann–Liouville fractional integrals via preinvexity [158].

Theorem 4.3.1. *Let $K \subseteq \mathbb{R}$ be an open invex subset with respect to η : $K \times K \to \mathbb{R}$ and $c, d \in K$ with $c < c + \eta(d, c)$. If $\xi : [c, c + \eta(d, c)] \to (0, \infty)$ is a preinvex function, $\xi \in L^1[c, c + \eta(d, c)]$ and η satisfies Condition C. Also suppose $\psi(x)$ is an increasing and positive monotone function on $(c, c + \eta(d, c))$, having a continuous derivative $\psi'(x)$ on $(c, c + \eta(d, c))$ and $\alpha \in (0, 1)$. Then,*

$$\xi\left(c + \frac{1}{2}\eta(d, c)\right) \leq \frac{\Gamma(\alpha + 1)}{2\eta^\alpha(d, c)} \left[J^{\alpha:\psi}_{\psi^{-1}(c)^+}(\xi \circ \psi)\psi^{-1}(c + \eta(d, c)) \right.$$

$$\left. + J^{\alpha:\psi}_{\psi^{-1}(c+\eta(d,c))^-}(\xi \circ \psi)\psi^{-1}(c) \right]$$

$$\leq \frac{\xi(c) + \xi(c + \eta(d, c))}{2} \leq \frac{\xi(c) + \xi(d)}{2}.$$

Proof By definition of invex set $x, y \in K$, then $x + \delta\eta(y, x) \in K$, $\forall \delta \in [0, 1]$. From Theorem 4.2.1, we get

$$\xi\left(x + \frac{1}{2}\eta(y, x)\right) \leq \frac{\xi(x) + \xi(y)}{2}. \tag{4.2}$$

Using $x = c + (1 - \delta)\eta(d, c)$ and $y = c + \delta\eta(d, c)$ in (4.2), we have

$$\xi(c + (1 - \delta)\eta(d, c) + \frac{1}{2}\eta(c + \delta\eta(d, c), c + (1 - \delta)\eta(d, c)))$$

$$\leq \frac{\xi(c + (1 - \delta)\eta(d, c)) + \xi(c + \delta\eta(d, c))}{2}. \tag{4.3}$$

Applying Condition C in (4.3), we have

$$\xi\left(c + \frac{1}{2}\eta(d, c)\right) \leq \frac{\xi(c + (1 - \delta)\eta(d, c)) + \xi(c + \delta\eta(d, c))}{2}. \tag{4.4}$$

Multiplying (4.4) by $\delta^{\alpha-1}$ on both sides and integrating the resultant with respect to δ over $[0, 1]$, we have

$$\frac{2}{\alpha}\xi\left(c + \frac{1}{2}\eta(d, c)\right) \leq \int_0^1 \delta^{\alpha-1}\xi(c + (1 - \delta)\eta(d, c))d\delta$$

$$+ \int_0^1 \delta^{\alpha-1}\xi(c + \delta\eta(d, c))d\delta. \tag{4.5}$$

Next,

$$\frac{\Gamma(\alpha+1)}{2\eta^\alpha(d,c)}\left[J^{\alpha:\psi}_{\psi^{-1}(c)^+}(\xi\circ\psi)\psi^{-1}(c+\eta(d,c)) + J^{\alpha:\psi}_{\psi^{-1}(c+\eta(d,c))^-}(\xi\circ\psi)\psi^{-1}(c)\right]$$

$$= \frac{\alpha}{2\eta^\alpha(d,c)}\left[\int_{\psi^{-1}(c)}^{\psi^{-1}(c+\eta(d,c))}(c+\eta(d,c)-\psi(v))^{\alpha-1}(\xi\circ\psi)(v)\psi'(v)dv\right.$$

$$\left.+\int_{\psi^{-1}(c)}^{\psi^{-1}(c+\eta(d,c))}(\psi(v)-c)^{\alpha-1}(\xi\circ\psi)(v)\psi'(v)dv\right]$$

$$= \frac{\alpha}{2}\left[\int_0^1\delta^{\alpha-1}\xi(c+(1-\delta)\eta(d,c))d\delta + \int_0^1\delta^{\alpha-1}\xi(c+\delta\eta(d,c))d\delta\right]. \quad (4.6)$$

From (4.5) and (4.6), we have

$$\xi\left(c+\frac{1}{2}\eta(d,c)\right) \leq \frac{\Gamma(\alpha+1)}{2\eta^\alpha(d,c)}\left[J^{\alpha:\psi}_{\psi^{-1}(c)^+}(\xi\circ\psi)\psi^{-1}(c+\eta(d,c))\right.$$

$$\left.+J^{\alpha:\psi}_{\psi^{-1}(c+\eta(d,c))^-}(\xi\circ\psi)\psi^{-1}(c)\right].$$

Now, we prove the second pair inequality of the theorem

$$\xi(c+\delta\eta(d,c)) = \xi(c+\eta(d,c)+(1-\delta)\eta(c,c+\eta(d,c)))$$
$$\leq \delta\xi(c+\eta(d,c))+(1-\delta)\xi(c). \quad (4.7)$$

Similarly,

$$\xi(c+(1-\delta)\eta(d,c)) = \xi(c+\eta(d,c)+\delta\eta(c,c+\eta(d,c)))$$
$$\leq (1-\delta)\xi(c+\eta(d,c))+\delta\xi(c). \quad (4.8)$$

From (4.7) and (4.8), we have

$$\xi(c+\delta\eta(d,c))+\xi(c+(1-\delta)\eta(d,c)) \leq \xi(c)+\xi(c+\eta(d,c)). \quad (4.9)$$

Multiplying both sides by $\delta^{\alpha-1}$ in (4.9), then integrating with respect to δ over 0 to 1, we have

$$\int_0^1\delta^{\alpha-1}\xi(c+\delta\eta(d,c))d\delta + \int_0^1\delta^{\alpha-1}\xi(c+(1-\delta)\eta(d,c))d\delta$$

$$\leq \frac{\xi(c)+\xi(c+\eta(d,c))}{\alpha}. \quad (4.10)$$

From (4.6) and (4.10), we have

$$\frac{\Gamma(\alpha+1)}{2\eta^\alpha(d,c)}\left[J^{\alpha:\psi}_{\psi^{-1}(c)^+}(\xi\circ\psi)\psi^{-1}(c+\eta(d,c)) + J^{\alpha:\psi}_{\psi^{-1}(c+\eta(d,c))^-}(\xi\circ\psi)\psi^{-1}(c)\right]$$

$$\leq \frac{\xi(c)+\xi(c+\eta(d,c))}{2} \leq \frac{\xi(c)+\xi(d)}{2}.$$

This completes the proof.

Remark 4.3.1. *When $\eta(d,c) = d-c$, then above theorem reduces to Theorem 2.1 of [98], i.e.*

$$\xi\left(\frac{c+d}{2}\right) \leq \frac{\Gamma(\alpha+1)}{2(d-c)^\alpha}\left[J^{\alpha:\psi}_{\psi^{-1}(c)+}(\xi \circ \psi)(\psi^{-1}(d))\right.$$
$$\left. + J^{\alpha:\psi}_{\psi^{-1}(d)-}(\xi \circ \psi)(\psi^{-1}(c))\right] \leq \frac{\xi(c)+\xi(d)}{2}.$$

Now, we present results of ψ-Riemann–Liouville fractional integral identities including the first-order derivative of a preinvex function [158].

Lemma 4.3.1. *Let $K \subseteq \mathbb{R}$ be an open invex subset with respect to $\eta : K \times K \to \mathbb{R}$ and $c,d \in K$ with $c < c+\eta(d,c)$. Suppose that $\xi : K \to \mathbb{R}$ is a differentiable function. If ξ' is preinvex function on K and $\xi' \in L^1[c, c+\eta(d,c)]$, $\psi(x)$ is an increasing and positive monotone function on $(c, c+\eta(d,c))$, having a continuous derivative $\psi'(x)$ on $(c, c+\eta(d,c))$ and $\alpha \in (0,1)$. Then,*

$$\frac{\xi(c)+\xi(c+\eta(d,c))}{2} - \frac{\Gamma(\alpha+1)}{2\eta^\alpha(d,c)}\left[J^{\alpha:\psi}_{\psi^{-1}(c)+}(\xi \circ \psi)\psi^{-1}(c+\eta(d,c))\right.$$
$$\left. + J^{\alpha:\psi}_{\psi^{-1}(c+\eta(d,c))-}(\xi \circ \psi)\psi^{-1}(c)\right]$$
$$= \frac{1}{2\eta^\alpha(d,c)}\int_{\psi^{-1}(c)}^{\psi^{-1}(c+\eta(d,c))}[(\psi(v)-c)^\alpha - (c+\eta(d,c)-\psi(v))^\alpha]$$
$$\times (\xi' \circ \psi)(v)\psi'(v)dv$$
$$= \frac{\eta(d,c)}{2}\int_0^1 ((1-\delta)^\alpha - \delta^\alpha)\xi'(c+(1-\delta)\eta(d,c))d\delta.$$

Proof Let $M_1 = \frac{\Gamma(\alpha+1)}{2\eta^\alpha(d,c)}J^{\alpha:\psi}_{\psi^{-1}(c)+}(\xi \circ \psi)\psi^{-1}(c+\eta(d,c))$
and
$M_2 = \frac{\Gamma(\alpha+1)}{2\eta^\alpha(d,c)}J^{\alpha:\psi}_{\psi^{-1}(c+\eta(d,c))-}(\xi \circ \psi)\psi^{-1}(c)$.

$$M_1 = \frac{\alpha}{2\eta^\alpha(d,c)}\int_{\psi^{-1}(c)}^{\psi^{-1}(c+\eta(d,c))}(c+\eta(d,c)-\psi(v))^{\alpha-1}(\xi \circ \psi)(v)\psi'(v)dv$$
$$= -\frac{1}{2\eta^\alpha(d,c)}\int_{\psi^{-1}(c)}^{\psi^{-1}(c+\eta(d,c))}(\xi \circ \psi)(v)d(c+\eta(d,c)-\psi(v))^\alpha$$
$$= -\frac{1}{2\eta^\alpha(d,c)}\left[-\xi(c)\eta^\alpha(d,c) - \int_{\psi^{-1}(c)}^{\psi^{-1}(c+\eta(d,c))}(c+\eta(d,c)-\psi(v))^\alpha\right.$$
$$\times (\xi' \circ \psi)(v)\psi'(v)dv]$$
$$= \frac{\xi(c)}{2} + \frac{1}{2\eta^\alpha(d,c)}\int_{\psi^{-1}(c)}^{\psi^{-1}(c+\eta(d,c))}(c+\eta(d,c)-\psi(v))^\alpha(\xi' \circ \psi)(v)\psi'(v)dv.$$

$$M_2 = \frac{\alpha}{2\eta^\alpha(d,c)} \int_{\psi^{-1}(c)}^{\psi^{-1}(c+\eta(d,c))} (\psi(v) - c)^{\alpha-1} (\xi \circ \psi)(v) \psi'(v) dv$$

$$= \frac{1}{2\eta^\alpha(d,c)} \int_{\psi^{-1}(c)}^{\psi^{-1}(c+\eta(d,c))} (\xi \circ \psi)(v) d(\psi(v) - c)^\alpha$$

$$= \frac{1}{2\eta^\alpha(d,c)} \Big[\xi(c + \eta(d,c))\eta^\alpha(d,c)$$

$$- \int_{\psi^{-1}(c)}^{\psi^{-1}(c+\eta(d,c))} (\psi(v) - c)^\alpha (\xi' \circ \psi)(v) \psi'(v) dv\Big]$$

$$= \frac{\xi(c + \eta(d,c))}{2} - \frac{1}{2\eta^\alpha(d,c)} \int_{\psi^{-1}(c)}^{\psi^{-1}(c+\eta(d,c))} (\psi(v) - c)^\alpha (\xi' \circ \psi)(v) \psi'(v) dv.$$

It follows that

$$\frac{\xi(c) + \xi(c + \eta(d,c))}{2} - M_1 - M_2$$

$$= \frac{1}{2\eta^\alpha(d,c)} \int_{\psi^{-1}(c)}^{\psi^{-1}(c+\eta(d,c))} [(\psi(v) - c)^\alpha - (c + \eta(d,c) - \psi(v))^\alpha]$$

$$\times (\xi' \circ \psi)(v) \psi'(v) dv. \tag{4.11}$$

Next, we prove the second pair equality of the lemma.

Let $M_3 = \dfrac{\eta(d,c)}{2} \displaystyle\int_0^1 ((1 - \delta)^\alpha - \delta^\alpha) \xi'(c + (1 - \delta)\eta(d,c)) d\delta$

$$= \frac{\eta(d,c)}{2} \Big[\int_0^1 (1 - \delta)^\alpha \xi'(c + (1 - \delta)\eta(d,c)) d\delta$$

$$- \int_0^1 \delta^\alpha \xi'(c + (1 - \delta)\eta(d,c)) d\delta\Big]$$

$$= \frac{\xi(c) + \xi(c + \eta(d,c))}{2} - \frac{\alpha}{2} \Big[\int_0^1 \delta^{\alpha-1} \xi(c + (1 - \delta)\eta(d,c)) d\delta$$

$$+ \int_0^1 \delta^{\alpha-1} \xi(c + \delta\eta(d,c)) d\delta\Big]$$

$$= \frac{\xi(c) + \xi(c + \eta(d,c))}{2}$$

$$- \frac{\alpha}{2\eta^\alpha(d,c)} \Big[\int_{\psi^{-1}(c)}^{\psi^{-1}(c+\eta(d,c))} (c + \eta(d,c) - \psi(v))^{\alpha-1} (\xi \circ \psi)(v) \psi'(v) dv$$

$$+ \int_{\psi^{-1}(c)}^{\psi^{-1}(c+\eta(d,c))} (\psi(v) - c)^{\alpha-1} (\xi \circ \psi)(v) \psi'(v) dv\Big]$$

$$= \frac{\xi(c) + \xi(c + \eta(d,c))}{2} - \frac{\Gamma(\alpha+1)}{2\eta^\alpha(d,c)} \left[J^{\alpha:\psi}_{\psi^{-1}(c)+} (\xi \circ \psi)\psi^{-1}(c + \eta(d,c)) \right.$$

$$\left. + J^{\alpha:\psi}_{\psi^{-1}(c+\eta(d,c))-} (\xi \circ \psi)\psi^{-1}(c) \right]. \tag{4.12}$$

This completes the proof.

Remark 4.3.2. *When $\eta(d,c) = d - c$, then above lemma reduces to Lemma 3.1 of [98], i.e.*

$$\frac{\xi(c) + \xi(d)}{2} - \frac{\Gamma(\alpha+1)}{2(d-c)^\alpha} \left[J^{\alpha:\psi}_{\psi^{-1}(c)+} (\xi \circ \psi)(\psi^{-1}(d)) \right.$$

$$\left. + J^{\alpha:\psi}_{\psi^{-1}(d)-} (\xi \circ \psi)(\psi^{-1}(c)) \right]$$

$$= \frac{1}{2(d-c)^\alpha} \int_{\psi^{-1}(c)}^{\psi^{-1}(d)} [(\psi(v) - c)^\alpha - (d - \psi(v))^\alpha](\xi' \circ \psi)(v)\psi'(v)dv$$

$$= \frac{d-c}{2} \int_0^1 ((1-\delta)^\alpha - \delta^\alpha)\xi'(c + (1-\delta)(d-c))d\delta.$$

Lemma 4.3.2. *Let $K \subseteq \mathbb{R}$ be an open invex subset with respect to $\eta : K \times K \to \mathbb{R}$ and $c, d \in K$ with $c < c + \eta(d,c)$. Suppose that $\xi : K \to \mathbb{R}$ is a differentiable function. If ξ' is preinvex function on K and $\xi' \in L^1[c, c + \eta(d,c)]$, $\psi(x)$ is an increasing and positive monotone function on $(c, c + \eta(d,c))$, having a continuous derivative $\psi'(x)$ on $(c, c + \eta(d,c))$ and $\alpha \in (0,1)$. Then,*

$$\frac{\Gamma(\alpha+1)}{2\eta^\alpha(d,c)} \left[J^{\alpha:\psi}_{\psi^{-1}(c)+} (\xi \circ \psi)\psi^{-1}(c + \eta(d,c)) + J^{\alpha:\psi}_{\psi^{-1}(c+\eta(d,c))-} (\xi \circ \psi)\psi^{-1}(c) \right]$$

$$- \xi \left(c + \frac{1}{2}\eta(d,c) \right) = \int_{\psi^{-1}(c)}^{\psi^{-1}(c+\eta(d,c))} k(\xi' \circ \psi)(v)\psi'(v)dv$$

$$+ \frac{1}{2\eta^\alpha(d,c)} \int_{\psi^{-1}(c)}^{\psi^{-1}(c+\eta(d,c))} [(c + \eta(d,c) - \psi(v))^\alpha - (\psi(v) - c)^\alpha]$$

$$\times (\xi' \circ \psi)(v)\psi'(v)dv,$$

where

$$k = \begin{cases} \frac{1}{2}, & \psi^{-1}\left(c + \frac{1}{2}\eta(d,c)\right) \le v \le \psi^{-1}(c + \eta(d,c)), \\ -\frac{1}{2}, & \psi^{-1}(c) < v < \psi^{-1}\left(c + \frac{1}{2}\eta(d,c)\right). \end{cases}$$

Proof Let

$$N_1 = -\frac{1}{2} \int_{\psi^{-1}(c)}^{\psi^{-1}\left(c + \frac{1}{2}\eta(d,c)\right)} (\xi' \circ \psi)(v)\psi'(v)dv$$

$$= \frac{1}{2}\xi(c) - \frac{1}{2}\xi \left(c + \frac{1}{2}\eta(d,c) \right), \tag{4.13}$$

$$N_2 = \frac{1}{2} \int_{\psi^{-1}\left(c+\frac{1}{2}\eta(d,c)\right)}^{\psi^{-1}(c+\eta(d,c))} (\xi' \circ \psi)(v)\psi'(v)dv$$

$$= \frac{1}{2}\xi\left(c+\eta(d,c)\right) - \frac{1}{2}\xi\left(c+\frac{1}{2}\eta(d,c)\right), \tag{4.14}$$

$$N_3 = \frac{1}{2\eta^\alpha(d,c)} \int_{\psi^{-1}(c)}^{\psi^{-1}(c+\eta(d,c))} (c+\eta(d,c)-\psi(v))^\alpha (\xi' \circ \psi)(v)\psi'(v)dv$$

$$= -\frac{1}{2}\xi(c) + \frac{\alpha}{2\eta^\alpha(d,c)} \int_{\psi^{-1}(c)}^{\psi^{-1}(c+\eta(d,c))} (c+\eta(d,c)-\psi(v))^{\alpha-1} (\xi \circ \psi)(v)\psi'(v)dv$$

$$= -\frac{1}{2}\xi(c) + \frac{\Gamma(\alpha+1)}{2\eta^\alpha(d,c)} J^{\alpha:\psi}_{\psi^{-1}(c)+}(\xi \circ \psi)\psi^{-1}(c+\eta(d,c)), \tag{4.15}$$

and

$$N_4 = -\frac{1}{2\eta^\alpha(d,c)} \int_{\psi^{-1}(c)}^{\psi^{-1}(c+\eta(d,c))} (\psi(v)-c)^\alpha (\xi' \circ \psi)(v)\psi'(v)dv$$

$$= -\frac{1}{2}\xi(c+\eta(d,c)) + \frac{\alpha}{2\eta^\alpha(d,c)} \int_{\psi^{-1}(c)}^{\psi^{-1}(c+\eta(d,c))} (\psi(v)-c)^{\alpha-1} (\xi \circ \psi)(v)\psi'(v)dv$$

$$= -\frac{1}{2}\xi(c+\eta(d,c)) + \frac{\Gamma(\alpha+1)}{2\eta^\alpha(d,c)} J^{\alpha:\psi}_{\psi^{-1}(c+\eta(d,c))-}(\xi \circ \psi)\psi^{-1}(c). \tag{4.16}$$

Adding (4.13), (4.14), (4.15) and (4.16), we have

$$N_1 + N_2 + N_3 + N_4 = \frac{\Gamma(\alpha+1)}{2\eta^\alpha(d,c)} \left[J^{\alpha:\psi}_{\psi^{-1}(c)+}(\xi \circ \psi)\psi^{-1}(c+\eta(d,c)) \right.$$

$$\left. + J^{\alpha:\psi}_{\psi^{-1}(c+\eta(d,c))-}(\xi \circ \psi)\psi^{-1}(c) \right] - \xi\left(c+\frac{1}{2}\eta(d,c)\right).$$

This completes the proof.

Remark 4.3.3. *When $\eta(d,c) = d - c$, then above lemma reduces to Lemma 3.2 of [98], i.e.*

$$\frac{\Gamma(\alpha+1)}{2(d-c)^\alpha} \left[J^{\alpha:\psi}_{\psi^{-1}(c)+}(\xi \circ \psi)\psi^{-1}(d) + J^{\alpha:\psi}_{\psi^{-1}(d)-}(\xi \circ \psi)(\psi^{-1}(c)) \right]$$

$$-\xi\left(\frac{c+d}{2}\right) = \int_{\psi^{-1}(c)}^{\psi^{-1}(d)} k(\xi' \circ \psi)(v)\psi'(v)dv$$

$$+ \frac{1}{2(d-c)^\alpha} \int_{\psi^{-1}(c)}^{\psi^{-1}(d)} [(d-\psi(v))^\alpha - (\psi(v)-c)^\alpha](\xi' \circ \psi)(v)\psi'(v)dv.$$

Now, we derive some fractional Hermite–Hadamard inequalities using the above results.

Theorem 4.3.2. *Let $K \subseteq \mathbb{R}$ be an open invex subset with respect to η : $K \times K \to \mathbb{R}$ and $c, d \in K$ with $c < c + \eta(d, c)$ such that $\xi' \in L^1[c, c + \eta(d, c)]$. Suppose that $\xi : K \to \mathbb{R}$ is a differentiable function. If $|\xi'|$ is preinvex function on K, $\psi(x)$ is an increasing and positive monotone function on $(c, c + \eta(d, c))$, having a continuous derivative $\psi'(x)$ on $(c, c + \eta(d, c))$ and $\alpha \in (0, 1)$. Then,*

$$\left| \frac{\xi(c) + \xi(c + \eta(d, c))}{2} - \frac{\Gamma(\alpha + 1)}{2\eta^\alpha(d, c)} \left[J^{\alpha:\psi}_{\psi^{-1}(c)^+} (\xi \circ \psi) \psi^{-1}(c + \eta(d, c)) \right. \right.$$
$$\left. \left. + J^{\alpha:\psi}_{\psi^{-1}(c+\eta(d,c))^-} (\xi \circ \psi) \psi^{-1}(c) \right] \right| \le \frac{\eta(d, c)}{2(\alpha + 1)} \left(1 - \frac{1}{2^\alpha} \right) [|\xi'(c)| + |\xi'(d)|].$$

Proof From Lemma 4.3.1 and definition of preinvexity, we have

$$\left| \frac{\xi(c) + \xi(c + \eta(d, c))}{2} - \frac{\Gamma(\alpha + 1)}{2\eta^\alpha(d, c)} \left[J^{\alpha:\psi}_{\psi^{-1}(c)^+} (\xi \circ \psi) \psi^{-1}(c + \eta(d, c)) \right. \right.$$
$$\left. \left. + J^{\alpha:\psi}_{\psi^{-1}(c+\eta(d,c))^-} (\xi \circ \psi) \psi^{-1}(c) \right] \right|$$
$$= \left| \frac{1}{2\eta^\alpha(d, c)} \int_{\psi^{-1}(c)}^{\psi^{-1}(c+\eta(d,c))} [(\psi(v) - c)^\alpha - (c + \eta(d, c) - \psi(v))^\alpha] \right.$$
$$\left. \times (\xi' \circ \psi)(v) \psi'(v) dv \right|$$
$$\le \frac{\eta(d, c)}{2} \int_0^1 |(1 - \delta)^\alpha - \delta^\alpha| |\xi'(c + (1 - \delta)\eta(d, c))| d\delta$$
$$\le \frac{\eta(d, c)}{2} \int_0^1 |(1 - \delta)^\alpha - \delta^\alpha| [\delta |\xi'(c)| + (1 - \delta)|\xi'(d)|] d\delta$$
$$= \frac{\eta(d, c)}{2} \left[|\xi'(c)| \left(\int_0^{1/2} \delta(1 - \delta)^\alpha d\delta - \int_0^{1/2} \delta^{\alpha+1} d\delta + \int_{1/2}^1 \delta^{\alpha+1} d\delta \right. \right.$$
$$\left. - \int_{1/2}^1 \delta(1 - \delta)^\alpha d\delta \right) + |\xi'(d)| \left(\int_0^{1/2} (1 - \delta)^{\alpha+1} d\delta - \int_0^{1/2} (1 - \delta)\delta^\alpha d\delta \right.$$
$$\left. \left. + \int_{1/2}^1 (1 - \delta)\delta^\alpha d\delta - \int_{1/2}^1 (1 - \delta)^{\alpha+1} d\delta \right) \right]$$
$$= \frac{\eta(d, c)}{2(\alpha + 1)} \left(1 - \frac{1}{2^\alpha} \right) [|\xi'(c)| + |\xi'(d)|].$$

This completes the proof.

Remark 4.3.4. *When $\eta(d, c) = d - c$, then above theorem reduces to Theorem 3.4 of [98], i.e.*

$$\left| \frac{\xi(c) + \xi(d)}{2} - \frac{\Gamma(\alpha + 1)}{2(d - c)^\alpha} \left[J^{\alpha:\psi}_{\psi^{-1}(c)^+} (\xi \circ \psi) \psi^{-1}(d) \right. \right.$$
$$\left. \left. + J^{\alpha:\psi}_{\psi^{-1}(d)^-} (\xi \circ \psi) \psi^{-1}(c) \right] \right| \le \frac{d - c}{2(\alpha + 1)} \left(1 - \frac{1}{2^\alpha} \right) [|\xi'(c)| + |\xi'(d)|].$$

Example 4.3.1. *Let $c = 0, d = 2, \xi(x) = x, \psi(x) = x$ and $\eta(d,c) = \frac{d-2c}{2}$. Then all the assumptions of above theorem are satisfied. Clearly,*

$$\frac{\xi(c) + \xi(c + \eta(d,c))}{2} = \frac{\xi(c) + \xi(c + \frac{d-2c}{2})}{2} = \frac{1}{2}, \qquad (4.17)$$

and

$$\frac{\Gamma(\alpha+1)}{2\eta^\alpha(d,c)} \left[J^{\alpha:\psi}_{\psi^{-1}(c)+}(\xi \circ \psi)\psi^{-1}(c+\eta(d,c)) + J^{\alpha:\psi}_{\psi^{-1}(c+\eta(d,c))-}(\xi \circ \psi)\psi^{-1}(c) \right]$$
$$= \frac{\alpha}{2} \left[\int_0^1 (1-v)^{\alpha-1}vdv + \int_0^1 v^{\alpha-1}vdv \right] = \frac{\alpha}{2} \left[\frac{\Gamma\alpha\Gamma2}{\Gamma(\alpha+2)} + \frac{1}{\alpha+1} \right] = \frac{1}{2}. \qquad (4.18)$$

From (4.17) and (4.18), we get

$$\left| \frac{\xi(c) + \xi(c + \eta(d,c))}{2} - \frac{\Gamma(\alpha+1)}{2\eta^\alpha(d,c)} \left[J^{\alpha:\psi}_{\psi^{-1}(c)+}(\xi \circ \psi)\psi^{-1}(c+\eta(d,c)) \right. \right.$$
$$\left. \left. + J^{\alpha:\psi}_{\psi^{-1}(c+\eta(d,c))-}(\xi \circ \psi)\psi^{-1}(c) \right] \right| = 0$$

Next,

$$\frac{\eta(d,c)}{2(\alpha+1)} \left(1 - \frac{1}{2^\alpha} \right) [|\xi'(c)| + |\xi'(d)|] = \frac{d-2c}{2(\alpha+1)} \left(1 - \frac{1}{2^\alpha} \right)$$
$$= \frac{1}{(\alpha+1)} \left(1 - \frac{1}{2^\alpha} \right) > 0.$$

Theorem 4.3.3. *Let $K \subseteq \mathbb{R}$ be an open invex subset with respect to $\eta : K \times K \to \mathbb{R}$ and $c, d \in K$ with $c < c + \eta(d,c)$ such that $\xi' \in L^1[c, c+\eta(d,c)]$. Suppose that $\xi : K \to \mathbb{R}$ is a differentiable function. If $|\xi'|$ is preinvex function on K, $\psi(x)$ is an increasing and positive monotone function on $(c, c+\eta(d,c))$, having a continuous derivative $\psi'(x)$ on $(c, c+\eta(d,c))$ and $\alpha \in (0,1)$. Then,*

$$\left| \frac{\Gamma(\alpha+1)}{2\eta^\alpha(d,c)} \left[J^{\alpha:\psi}_{\psi^{-1}(c)+}(\xi \circ \psi)\psi^{-1}(c+\eta(d,c)) + J^{\alpha:\psi}_{\psi^{-1}(c+\eta(d,c))-}(\xi \circ \psi)\psi^{-1}(c) \right] \right.$$
$$\left. - \xi\left(c + \frac{1}{2}\eta(d,c) \right) \right| \leq \frac{|\xi(d) - \xi(c)|}{2} + \frac{\eta(d,c)}{2(\alpha+1)} \left(1 - \frac{1}{2^\alpha} \right) [|\xi'(c)| + |\xi'(d)|].$$

Proof Using Lemma 4.3.2, we have

$$\left| \frac{\Gamma(\alpha+1)}{2\eta^\alpha(d,c)} \left[J^{\alpha:\psi}_{\psi^{-1}(c)+}(\xi \circ \psi)\psi^{-1}(c+\eta(d,c)) + J^{\alpha:\psi}_{\psi^{-1}(c+\eta(d,c))-}(\xi \circ \psi)\psi^{-1}(c) \right] \right.$$
$$\left. - \xi\left(c + \frac{1}{2}\eta(d,c) \right) \right| \leq \left| \int_{\psi^{-1}(c)}^{\psi^{-1}(c+\eta(d,c))} k(\xi' \circ \psi)(v)\psi'(v)dv \right|$$
$$+ \left| \frac{1}{2\eta^\alpha(d,c)} \int_{\psi^{-1}(c)}^{\psi^{-1}(c+\eta(d,c))} [(c+\eta(d,c) - \psi(v))^\alpha - (\psi(v) - c)^\alpha] \right.$$
$$\times (\xi' \circ \psi)(v)\psi'(v)dv \Bigg|.$$

Using previous theorem and definition of preinvexity, we have

$$\left| \frac{\Gamma(\alpha+1)}{2\eta^\alpha(d,c)} \left[J_{\psi^{-1}(c)^+}^{\alpha:\psi}(\xi \circ \psi)\psi^{-1}(c+\eta(d,c)) + J_{\psi^{-1}(c+\eta(d,c))^-}^{\alpha:\psi}(\xi \circ \psi)\psi^{-1}(c) \right] \right.$$
$$\left. - \xi\left(c+\frac{1}{2}\eta(d,c)\right) \right| \leq \frac{|\xi(d)-\xi(c)|}{2} + \frac{\eta(d,c)}{2(\alpha+1)}\left(1-\frac{1}{2^\alpha}\right)[|\xi'(c)|+|\xi'(d)|].$$

This completes the proof.

Remark 4.3.5. *When $\eta(d,c)=d-c$, then above theorem reduces to Theorem 3.5 of [98], i.e.*

$$\left| \frac{\Gamma(\alpha+1)}{2(d-c)^\alpha} \left[J_{\psi^{-1}(c)^+}^{\alpha:\psi}(\xi \circ \psi)(\psi^{-1}(d)) + J_{\psi^{-1}(d)^-}^{\alpha:\psi}(\xi \circ \psi)(\psi^{-1}(c)) \right] \right.$$
$$\left. - \xi\left(\frac{c+d}{2}\right) \right| \leq \frac{|\xi(d)-\xi(c)|}{2} + \frac{(d-c)}{2(\alpha+1)}\left(1-\frac{1}{2^\alpha}\right)[|\xi'(c)|+|\xi'(d)|].$$

Example 4.3.2. *Let $c=0, d=1, \xi(x)=x/2, \psi(x)=x$ and $\eta(d,c)=d-3c$. Then all the assumptions of above theorem are satisfied. Clearly,*

$$\frac{\Gamma(\alpha+1)}{2\eta^\alpha(d,c)} \left[J_{\psi^{-1}(c)^+}^{\alpha:\psi}(\xi \circ \psi)\psi^{-1}(c+\eta(d,c)) + J_{\psi^{-1}(c+\eta(d,c))^-}^{\alpha:\psi}(\xi \circ \psi)\psi^{-1}(c) \right]$$
$$= \frac{\alpha}{4}\left[\int_0^1 (1-v)^{\alpha-1} v\, dv + \int_0^1 v^{\alpha-1} v\, dv \right] = \frac{\alpha}{4}\left[\frac{\Gamma\alpha\Gamma 2}{\Gamma(\alpha+2)} + \frac{1}{\alpha+1} \right] = \frac{1}{4},$$
$$(4.19)$$

and

$$\xi\left(c+\frac{1}{2}\eta(d,c)\right) = \xi\left(c+\frac{1}{2}(d-3c)\right) = \frac{1}{4}. \qquad (4.20)$$

From (4.19) and (4.20), we get

$$\left| \frac{\Gamma(\alpha+1)}{2\eta^\alpha(d,c)} \left[J_{\psi^{-1}(c)^+}^{\alpha:\psi}(\xi \circ \psi)\psi^{-1}(c+\eta(d,c)) + J_{\psi^{-1}(c+\eta(d,c))^-}^{\alpha:\psi}(\xi \circ \psi)\psi^{-1}(c) \right] \right.$$
$$\left. - \xi\left(c+\frac{1}{2}\eta(d,c)\right) \right| = 0. \qquad (4.21)$$

Next,

$$\frac{|\xi(d)-\xi(c)|}{2} + \frac{\eta(d,c)}{2(\alpha+1)}\left(1-\frac{1}{2^\alpha}\right)[|\xi'(c)|+|\xi'(d)|]$$
$$= \frac{1}{4} + \frac{1}{2(\alpha+1)}\left(1-\frac{1}{2^\alpha}\right) > 0.$$

Theorem 4.3.4. *Let $K \subseteq \mathbb{R}$ be an open invex subset with respect to $\eta : K \times K \to \mathbb{R}$ and $c, d \in K$ with $c < c + \eta(d,c)$ such that $\xi' \in L^1[c, c+\eta(d,c)]$.*

Suppose that $\xi : K \to \mathbb{R}$ is a differentiable function. If $|\xi'|^q$ is preinvex function on K for some fixed $q > 1$, $\psi(x)$ is an increasing and positive monotone function on $(c, c + \eta(d, c))$, having a continuous derivative $\psi'(x)$ on $(c, c + \eta(d, c))$ and $\alpha \in (0, 1)$. Then,

$$\left| \frac{\xi(c) + \xi(c + \eta(d, c))}{2} - \frac{\Gamma(\alpha + 1)}{2\eta^\alpha(d, c)} \left[J^{\alpha:\psi}_{\psi^{-1}(c)^+} (\xi \circ \psi)\psi^{-1}(c + \eta(d, c)) \right. \right.$$

$$\left. \left. + J^{\alpha:\psi}_{\psi^{-1}(c+\eta(d,c))^-} (\xi \circ \psi)\psi^{-1}(c) \right] \right| \leq \frac{\eta(d, c)}{2(\alpha p + 1)^{\frac{1}{p}}} \left(\frac{|\xi'(c)|^q + |\xi'(d)|^q}{2} \right)^{\frac{1}{q}},$$

where $\frac{1}{p} + \frac{1}{q} = 1$.

Proof From Lemma 4.3.1, we have

$$\left| \frac{\xi(c) + \xi(c + \eta(d, c))}{2} - \frac{\Gamma(\alpha + 1)}{2\eta^\alpha(d, c)} \left[J^{\alpha:\psi}_{\psi^{-1}(c)^+} (\xi \circ \psi)\psi^{-1}(c + \eta(d, c)) \right. \right.$$

$$\left. \left. + J^{\alpha:\psi}_{\psi^{-1}(c+\eta(d,c))^-} (\xi \circ \psi)\psi^{-1}(c) \right] \right|$$

$$\leq \frac{\eta(d, c)}{2} \int_0^1 |\delta^\alpha - (1 - \delta)^\alpha| |\xi'(c + \delta\eta(d, c))| d\delta.$$

Using Hölder's inequality and definition of preinvexity, we have

$$\left| \frac{\xi(c) + \xi(c + \eta(d, c))}{2} - \frac{\Gamma(\alpha + 1)}{2\eta^\alpha(d, c)} \left[J^{\alpha:\psi}_{\psi^{-1}(c)^+} (\xi \circ \psi)\psi^{-1}(c + \eta(d, c)) \right. \right.$$

$$\left. \left. + J^{\alpha:\psi}_{\psi^{-1}(c+\eta(d,c))^-} (\xi \circ \psi)\psi^{-1}(c) \right] \right|$$

$$\leq \frac{\eta(d, c)}{2} \left(\int_0^1 |\delta^\alpha - (1 - \delta)^\alpha|^p d\delta \right)^{\frac{1}{p}} \left(\int_0^1 |\xi'(c + \delta\eta(d, c))|^q d\delta \right)^{\frac{1}{q}}$$

$$\leq \frac{\eta(d, c)}{2} \left(\int_0^1 |1 - 2\delta|^{\alpha p} d\delta \right)^{\frac{1}{p}} \left(\int_0^1 ((1 - \delta)|\xi'(c)|^q + \delta|\xi'(d)|^q) \right)^{\frac{1}{q}}$$

$$= \frac{\eta(d, c)}{2(\alpha p + 1)^{\frac{1}{p}}} \left(\frac{|\xi'(c)|^q + |\xi'(d)|^q}{2} \right)^{\frac{1}{q}}.$$

This completes the proof.

Corollary 4.3.1. *If $\eta(d, c) = d - c$, then we have the following result:*

$$\left| \frac{\xi(c) + \xi(d)}{2} - \frac{\Gamma(\alpha + 1)}{2(d - c)^\alpha} \left[J^{\alpha:\psi}_{\psi^{-1}(c)^+} (\xi \circ \psi)\psi^{-1}(d) \right. \right.$$

$$\left. \left. + J^{\alpha:\psi}_{\psi^{-1}(d)^-} (\xi \circ \psi)\psi^{-1}(c) \right] \right| \leq \frac{(d - c)}{2(\alpha p + 1)^{\frac{1}{p}}} \left(\frac{|\xi'(c)|^q + |\xi'(d)|^q}{2} \right)^{\frac{1}{q}}.$$

Theorem 4.3.5. *Let $K \subseteq \mathbb{R}$ be an open invex subset with respect to η : $K \times K \to \mathbb{R}$ and $c, d \in K$ with $c < c + \eta(d, c)$ such that $\xi' \in L^1[c, c + \eta(d, c)]$. Suppose that $\xi : K \to \mathbb{R}$ is a differentiable function. If $|\xi'|^q$ is preinvex function on K for some fixed $q > 1$, $\psi(x)$ is an increasing and positive monotone function on $(c, c + \eta(d, c))$, having a continuous derivative $\psi'(x)$ on $(c, c + \eta(d, c))$ and $\alpha \in (0, 1)$. Then,*

Proof Using Lemma 4.3.1, Hölder's inequality and definition of preinvexity, we have

$$\left| \frac{\xi(c) + \xi(c + \eta(d, c))}{2} - \frac{\Gamma(\alpha + 1)}{2\eta^\alpha(d, c)} \left[J^{\alpha:\psi}_{\psi^{-1}(c)+} (\xi \circ \psi)\psi^{-1}(c + \eta(d, c)) \right. \right.$$

$$\left. \left. + J^{\alpha:\psi}_{\psi^{-1}(c+\eta(d,c))-} (\xi \circ \psi)\psi^{-1}(c) \right] \right| \le \frac{\eta(d, c)}{2} \left(\int_0^1 |\delta^\alpha - (1 - \delta)^\alpha| d\delta \right)^{\frac{1}{p}}$$

$$\times \left(\int_0^1 |\delta^\alpha - (1 - \delta)^\alpha| |\xi'(c + \delta\eta(d, c))|^q d\delta \right)^{\frac{1}{q}}$$

$$= \frac{\eta(d, c)}{2} \left(\int_0^{1/2} ((1 - \delta)^\alpha - \delta^\alpha) d\delta + \int_{1/2}^1 (\delta^\alpha - (1 - \delta)^\alpha) d\delta \right)^{\frac{1}{p}}$$

$$\times \left(\int_0^{1/2} ((1 - \delta)^\alpha - \delta^\alpha)((1 - \delta)|\xi'(c)|^q + \delta|\xi'(d)|^q) d\delta \right.$$

$$\left. + \int_{1/2}^1 (\delta^\alpha - (1 - \delta)^\alpha)((1 - \delta)|\xi'(c)|^q + \delta|\xi'(d)|^q) d\delta \right)^{\frac{1}{q}}$$

$$= \frac{\eta(d, c)}{(\alpha + 1)} \left(1 - \frac{1}{2^\alpha} \right) \left[\frac{|\xi'(c)|^q + |\xi'(d)|^q}{2} \right]^{\frac{1}{q}}.$$

Corollary 4.3.2. *If $\eta(d, c) = d - c$, then we have the following result:*

$$\left| \frac{\xi(c) + \xi(d)}{2} - \frac{\Gamma(\alpha + 1)}{2(d - c)^\alpha} \left[J^{\alpha:\psi}_{\psi^{-1}(c)+} (\xi \circ \psi)\psi^{-1}(d) \right. \right.$$

$$\left. \left. + J^{\alpha:\psi}_{\psi^{-1}(d)-} (\xi \circ \psi)\psi^{-1}(c) \right] \right| \le \frac{d - c}{(\alpha + 1)} \left(1 - \frac{1}{2^\alpha} \right) \left[\frac{|\xi'(c)|^q + |\xi'(d)|^q}{2} \right]^{\frac{1}{q}}.$$

4.4　Application to Special Means

We shall start with the following proposition [158]. Here, $A(c, d), L_n(c, d)$ and $H(c, d)$ are Arithmetic mean, Generalized logarithmic mean and Harmonic mean, respectively, which are defined in Section 2.3.1 of Chapter 2.

Proposition 4.4.1. *Let* $c, c + \eta(d,c) \in \mathbb{R}^+, c < c + \eta(d,c)$. *Then*

$$|A(c^n, (c + \eta(d,c))^n) - L_n^n(c, c + \eta(d,c))| \le \frac{n\eta(d,c)}{2(p+1)^{\frac{1}{p}}} \left(\frac{c^{(n-1)q} + d^{(n-1)q}}{2} \right)^{\frac{1}{q}}.$$

Proof Applying Theorem 4.3.4 with $\xi(x) = x^n, \psi(x) = x, \alpha = 1$. Then we compute the result easily.

Proposition 4.4.2. *Let* $c, c + \eta(d,c) \in \mathbb{R}^+, c < c + \eta(d,c)$. *Then*

$$|A(e^c, e^{c+\eta(d,c)}) - L(e^c, e^{c+\eta(d,c)})| \le \frac{\eta(d,c)}{2(p+1)^{\frac{1}{p}}} \left(\frac{e^{cq} + e^{dq}}{2} \right)^{\frac{1}{q}}.$$

Proof Applying Theorem 4.3.4 with $\xi(x) = e^x, \psi(x) = x, \alpha = 1$. Then we compute the result easily.

Proposition 4.4.3. *Let* $c, c + \eta(d,c) \in \mathbb{R}^+, c < c + \eta(d,c)$. *Then*

$$|H^{-1}(c, c + \eta(d,c)) - L^{-1}(c, c + \eta(d,c))| \le \frac{\eta(d,c)}{2(p+1)^{\frac{1}{p}}} \left[\frac{1}{2} \left(\frac{1}{c^{2q}} + \frac{1}{d^{2q}} \right) \right]^{\frac{1}{q}}.$$

Proof Applying Theorem 4.3.4 with $\xi(x) = \frac{1}{x}, \psi(x) = x, \alpha = 1$. Then we compute the result easily.

Proposition 4.4.4. *Let* $c, c + \eta(d,c) \in \mathbb{R}^+, c < c + \eta(d,c)$. *Then*

$$|A(c^n, (c + \eta(d,c))^n) - L_n^n(c, c + \eta(d,c))| \le \frac{n\eta(d,c)}{4} \left(\frac{c^{(n-1)q} + d^{(n-1)q}}{2} \right)^{\frac{1}{q}}.$$

Proof Applying Theorem 4.3.5 with $\xi(x) = x^n, \psi(x) = x, \alpha = 1$. Then we compute the result easily.

Proposition 4.4.5. *Let* $c, c + \eta(d,c) \in \mathbb{R}^+, c < c + \eta(d,c)$. *Then*

$$|A(e^c, e^{c+\eta(d,c)}) - L(e^c, e^{c+\eta(d,c)})| \le \frac{\eta(d,c)}{4} \left(\frac{e^{cq} + e^{dq}}{2} \right)^{\frac{1}{q}}.$$

Proof Applying Theorem 4.3.5 with $\xi(x) = e^x, \psi(x) = x, \alpha = 1$. Then we compute the result easily.

Proposition 4.4.6. *Let* $c, c + \eta(d,c) \in \mathbb{R}^+, c < c + \eta(d,c)$. *Then*

$$|H^{-1}(c, c + \eta(d,c)) - L^{-1}(c, c + \eta(d,c))| \le \frac{\eta(d,c)}{4} \left[\frac{1}{2} \left(\frac{1}{c^{2q}} + \frac{1}{d^{2q}} \right) \right]^{\frac{1}{q}}.$$

Proof Applying Theorem 4.3.5 with $\xi(x) = \frac{1}{x}, \psi(x) = x, \alpha = 1$. Then we compute the result easily.

4.5   Generalized (m,h)-Preunivex Mappings via k-Fractional Integrals

We present the definition of (m, h)-preunivex function [157].

Definition 4.5.1. *Let $K \subseteq \mathbb{R}$ be an open m-invex subset with respect to $\eta : K \times K \times (0, 1] \to \mathbb{R}$ and let $h : [0, 1] \to \mathbb{R}\backslash\{0\}$. A function $\xi : K \to \mathbb{R}$ is said to be generalized (m, h)-preunivex with respect to η, ϕ and b if*

$$\xi(mx + \delta\eta(y, x, m)) \leq m\xi(x) + h(\delta)b(y, x, h(\delta))\phi[\xi(y) - m\xi(x)].$$

Remark 4.5.1. *Now, we will discuss some special cases of Definition 4.5.1, we have*

(1) choosing $h(\delta) = 1$, we obtain the definition of generalized (m, p)-preunivex functions;

(2) choosing $h(\delta) = \delta^s$ for $s \in (0, 1]$, we obtain the definition of generalized (m, s)-Breckner-preunivex functions;

(3) choosing $h(\delta) = \delta^{-s}$ for $s \in (0, 1]$, we obtain the definition of generalized (m, s)-Godunova-Levin-Dragomir-preunivex functions;

(4) choosing $h(\delta) = \delta(1-\delta)$, we obtain the definition of generalized (m, tgs)-preunivex functions;

(5) choosing $h(\delta) = \dfrac{\sqrt{\delta}}{2\sqrt{1-\delta}}$, we obtain the definition of generalized m-MT-preunivex functions.

By using Lemma 4.2.1, we prove the following theorem [157].

Theorem 4.5.1. *Let $K \subseteq \mathbb{R}$ be an open m-invex subset with respect to $\eta : K \times K \times (0, 1] \to \mathbb{R}\backslash\{0\}$ for some fixed $m \in (0, 1]$, and let $c, d \in K$, $c < d$ with $\eta(d, c, m) > 0$. Assume that $\xi : K \to \mathbb{R}$ is a twice differentiable function on K such that ξ'' is integrable on $[mc, mc + \eta(d, c, m)]$. If $|\xi''|^q$ for $q \geq 1$ is a generalized (m, h)-preunivex function with respect to η, ϕ and b and $h : [0, 1] \to \mathbb{R}\backslash\{0\}$, then the following inequality for k-fractional integrals with $x \in [c, d]$, $\lambda \in [0, 1]$, $\alpha > 0$, $k > 0$ exists:*

$$|I_{\xi,\eta}(\alpha, k; x, \lambda, m, c, d)| \leq A_0^{1-\frac{1}{q}}(k, \alpha, \lambda) \left[\left| \frac{\eta^{\frac{\alpha}{k}+2}(x, c, m)}{(\frac{\alpha}{k}+1)\eta(d, c, m)} \right| \right.$$

$$\times \left(mA_0(k, \alpha, \lambda)|\xi''(c)|^q + \phi(|\xi''(x)|^q - m|\xi''(c)|^q)A_1(k, \alpha, \lambda, c, x; h) \right)^{\frac{1}{q}}$$

$$+ \left| \frac{(-1)^{\frac{\alpha}{k}+2}\eta^{\frac{\alpha}{k}+2}(x, d, m)}{(\frac{\alpha}{k}+1)\eta(d, c, m)} \right|$$

$$\left. \times \left(mA_0(k, \alpha, \lambda)|\xi''(d)|^q + \phi(|\xi''(x)|^q - m|\xi''(d)|^q)A_2(k, \alpha, \lambda, d, x; h) \right)^{\frac{1}{q}} \right],$$

where $A_0(k, \alpha, \lambda) = \int_0^1 \delta \left| \left(\frac{\alpha}{k} + 1 \right) \lambda - \delta^{\frac{\alpha}{k}} \right| d\delta$

$$= \begin{cases} \frac{\frac{\alpha}{k} [(\frac{\alpha}{k}+1)\lambda]^{1+\frac{2k}{\alpha}}}{\frac{\alpha}{k}+2} - \frac{(\frac{\alpha}{k}+1)\lambda}{2} + \frac{1}{\frac{\alpha}{k}+2}, & 0 \leq \lambda \leq \frac{1}{\frac{\alpha}{k}+1}, \\ \frac{(\frac{\alpha}{k}+1)\lambda}{2} - \frac{1}{\frac{\alpha}{k}+2}, & \frac{1}{\frac{\alpha}{k}+1} < \lambda \leq 1, \end{cases}$$

$$A_1(k, \alpha, \lambda, c, x; h) = \int_0^1 \delta \left| \left(\frac{\alpha}{k} + 1 \right) \lambda - \delta^{\frac{\alpha}{k}} \right| h(\delta) b(x, c, h(\delta)) d\delta, \quad 0 \leq \lambda \leq 1,$$

and

$$A_2(k, \alpha, \lambda, d, x; h) = \int_0^1 \delta \left| \left(\frac{\alpha}{k} + 1 \right) \lambda - \delta^{\frac{\alpha}{k}} \right| h(\delta) b(x, d, h(\delta)) d\delta, \quad 0 \leq \lambda \leq 1.$$

Proof For the proof of this theorem, we will apply Lemma 4.2.1 and the Power mean inequality, we have

$$|I_{\xi,\eta}(\alpha, k; x, \lambda, m, c, d)| \leq \left| \frac{\eta^{\frac{\alpha}{k}+2}(x, c, m)}{(\frac{\alpha}{k}+1)\eta(d, c, m)} \right| \left(\int_0^1 \delta \left| \left(\frac{\alpha}{k} + 1 \right) \lambda - \delta^{\frac{\alpha}{k}} \right| d\delta \right)^{1-\frac{1}{q}}$$

$$\times \left(\int_0^1 \delta \left| \left(\frac{\alpha}{k} + 1 \right) \lambda - \delta^{\frac{\alpha}{k}} \right| |\xi''(mc + \delta\eta(x, c, m))|^q d\delta \right)^{\frac{1}{q}}$$

$$+ \left| \frac{(-1)^{\frac{\alpha}{k}+2}\eta^{\frac{\alpha}{k}+2}(x, d, m)}{(\frac{\alpha}{k}+1)\eta(d, c, m)} \right| \left(\int_0^1 \delta \left| \left(\frac{\alpha}{k} + 1 \right) \lambda - \delta^{\frac{\alpha}{k}} \right| d\delta \right)^{1-\frac{1}{q}}$$

$$\times \left(\int_0^1 \delta \left| \left(\frac{\alpha}{k} + 1 \right) \lambda - \delta^{\frac{\alpha}{k}} \right| |\xi''(md + \delta\eta(x, d, m))|^q d\delta \right)^{\frac{1}{q}}$$

$$= A_0^{1-\frac{1}{q}}(k, \alpha, \lambda)$$

$$\times \left[\left| \frac{\eta^{\frac{\alpha}{k}+2}(x, c, m)}{(\frac{\alpha}{k}+1)\eta(d, c, m)} \right| \left(\int_0^1 \delta \left| \left(\frac{\alpha}{k} + 1 \right) \lambda - \delta^{\frac{\alpha}{k}} \right| |\xi''(mc + \delta\eta(x, c, m))|^q d\delta \right)^{\frac{1}{q}} \right.$$

$$+ \left| \frac{(-1)^{\frac{\alpha}{k}+2}\eta^{\frac{\alpha}{k}+2}(x, d, m)}{(\frac{\alpha}{k}+1)\eta(d, c, m)} \right|$$

$$\left. \times \left(\int_0^1 \delta \left| \left(\frac{\alpha}{k} + 1 \right) \lambda - \delta^{\frac{\alpha}{k}} \right| |\xi''(md + \delta\eta(x, d, m))|^q d\delta \right)^{\frac{1}{q}} \right]. \tag{4.22}$$

Since $|\xi''(x)|^q$ is generalized (m, h)-preunivex on $[mc, mc + \eta(d, c, m)]$, we get

$$\int_0^1 \delta \left| \left(\frac{\alpha}{k} + 1 \right) \lambda - \delta^{\frac{\alpha}{k}} \right| |\xi''(mc + \delta\eta(x, c, m))|^q d\delta$$

$$\leq \int_0^1 \left[\delta \left| \left(\frac{\alpha}{k} + 1 \right) \lambda - \delta^{\frac{\alpha}{k}} \right| (m|\xi''(c)|^q + h(\delta)b(x, c, h(\delta)) \right.$$

$$\left. \times \phi(|\xi''(x)|^q - m|\xi''(c)|^q)) \right] dt$$

$$= mA_0(k, \alpha, \lambda)|\xi''(c)|^q + \phi(|\xi''(x)|^q - m|\xi''(c)|^q)A_1(k, \alpha, \lambda, c, x; h) \tag{4.23}$$

and

$$\int_0^1 \delta \left| \left(\frac{\alpha}{k} + 1 \right) \lambda - \delta^{\frac{\alpha}{k}} \right| |\xi''(md + \delta\eta(x,d,m))|^q d\delta$$

$$\leq \int_0^1 \left[\delta \left| \left(\frac{\alpha}{k} + 1 \right) \lambda - \delta^{\frac{\alpha}{k}} \right| (m|\xi''(d)|^q + h(\delta)b(x,d,h(\delta)) \right.$$

$$\left. \times \phi(|\xi''(x)|^q - m|\xi''(d)|^q)) \right] d\delta$$

$$= mA_0(k,\alpha,\lambda)|\xi''(d)|^q + \phi(|\xi''(x)|^q - m|\xi''(d)|^q)A_2(k,\alpha,\lambda,d,x;h).$$

$$(4.24)$$

Hence, if we use (4.23) and (4.24) in (4.22), then we have the desired result. This completes the proof.

Now, we will discuss some special cases of Theorem 4.5.1 [157].

Corollary 4.5.1. *In Theorem 4.5.1, if $|\xi''|^q$ for $q \geq 1$ is generalized (m,s)-Breckner-preunivex functions, then for $s \in (0,1]$ and $m \in (0,1]$, we have*

$$|I_{\xi,\eta}(\alpha,k;x,\lambda,m,c,d)| \leq A_0^{1-\frac{1}{q}}(k,\alpha,\lambda) \left[\left| \frac{\eta^{\frac{\alpha}{k}+2}(x,c,m)}{(\frac{\alpha}{k}+1)\eta(d,c,m)} \right| \right.$$

$$\times (mA_0(k,\alpha,\lambda)|\xi''(c)|^q + \phi(|\xi''(x)|^q - m|\xi''(c)|^q)B_1(k,\alpha;\lambda,s,c,x))^{\frac{1}{q}}$$

$$+ \left| \frac{(-1)^{\frac{\alpha}{k}+2}\eta^{\frac{\alpha}{k}+2}(x,d,m)}{(\frac{\alpha}{k}+1)\eta(d,c,m)} \right|$$

$$\left. \times (mA_0(k,\alpha,\lambda)|\xi''(d)|^q + \phi(|\xi''(x)|^q - m|\xi''(d)|^q)B_2(k,\alpha;\lambda,s,d,x))^{\frac{1}{q}} \right],$$

where

$$B_1(k,\alpha;\lambda,s,c,x) = \int_0^1 \delta^{s+1} \left| \left(\frac{\alpha}{k} + 1 \right) \lambda - \delta^{\frac{\alpha}{k}} \right| b(x,c,\delta^s) d\delta$$

and

$$B_2(k,\alpha;\lambda,s,d,x) = \int_0^1 \delta^{s+1} \left| \left(\frac{\alpha}{k} + 1 \right) \lambda - \delta^{\frac{\alpha}{k}} \right| b(x,d,\delta^s) d\delta.$$

Corollary 4.5.2. *In Corollary 4.5.1, if $\phi : \mathbb{R} \to \mathbb{R}$ is defined by $\phi(c) = c$ and $b(.,.,.) \equiv 1$, then we have*

$$|I_{\xi,\eta}(\alpha,k;x,\lambda,m,c,d)| \leq A_0^{1-\frac{1}{q}}(k,\alpha,\lambda) \left[\left| \frac{\eta^{\frac{\alpha}{k}+2}(x,c,m)}{(\frac{\alpha}{k}+1)\eta(d,c,m)} \right| \right.$$

$$\times \left(mA_0(k,\alpha,\lambda)|\xi''(c)|^q + (|\xi''(x)|^q - m|\xi''(c)|^q)H_1(k,\alpha,\lambda;s) \right)^{\frac{1}{q}}$$

$$+ \left| \frac{(-1)^{\frac{\alpha}{k}+2}\eta^{\frac{\alpha}{k}+2}(x,d,m)}{(\frac{\alpha}{k}+1)\eta(d,c,m)} \right|$$

$$\left. \times \left(mA_0(k,\alpha,\lambda)|\xi''(d)|^q + (|\xi''(x)|^q - m|\xi''(d)|^q)H_1(k,\alpha,\lambda;s) \right)^{\frac{1}{q}} \right],$$

where

$$H_1(k,\alpha,\lambda;s) = \int_0^1 \delta^{s+1} \left| \left(\tfrac{\alpha}{k}+1\right)\lambda - \delta^{\frac{\alpha}{k}} \right| d\delta$$

$$= \begin{cases} \dfrac{\frac{2\alpha}{k}[(\frac{\alpha}{k}+1)\lambda]^{1+\frac{k(s+2)}{\alpha}}}{(s+2)(\frac{\alpha}{k}+s+2)} - \dfrac{(\frac{\alpha}{k}+1)\lambda}{s+2} + \dfrac{1}{\frac{\alpha}{k}+s+2}, & 0 \le \lambda \le \frac{1}{\frac{\alpha}{k}+1}, \\ \dfrac{(\frac{\alpha}{k}+1)\lambda}{s+2} - \dfrac{1}{\frac{\alpha}{k}+s+2}, & \frac{1}{\frac{\alpha}{k}+1} < \lambda \le 1. \end{cases}$$

Corollary 4.5.3. *In Corollary 4.5.2, if the mapping $\eta(d,c,m) = d-mc$ along with $m=1$, taking $x = \frac{c+d}{2}$, then for $s \in (0,1]$, we have*

$$\left| \frac{2^{\frac{\alpha}{k}-1}}{(d-c)^{\frac{\alpha}{k}-1}} I_\xi \left(\alpha,k;\frac{c+d}{2},\lambda,1,c,d\right) \right| \le \frac{(d-c)^2}{8(\frac{\alpha}{k}+1)} A_0^{1-\frac{1}{q}}(k,\alpha,\lambda)$$

$$\times \left[\left(A_0(k,\alpha,\lambda)|\xi''(c)|^q + \left(\left|\xi''\left(\frac{c+d}{2}\right)\right|^q - |\xi''(c)|^q \right) H_1(k,\alpha,\lambda;s) \right)^{\frac{1}{q}} \right.$$

$$\left. + \left(A_0(k,\alpha,\lambda)|\xi''(d)|^q + \left(\left|\xi''\left(\frac{c+d}{2}\right)\right|^q - |\xi''(d)|^q \right) H_1(k,\alpha,\lambda;s) \right)^{\frac{1}{q}} \right].$$

Remark 4.5.2. *In Corollary 4.5.3,*

(1) if $\lambda = 0$, then we have

$$\left| \xi\left(\frac{c+d}{2}\right) - \frac{2^{\frac{\alpha}{k}-1}\Gamma_k(\alpha+k)}{(d-c)^{\frac{\alpha}{k}}} [_k J^\alpha_{(\frac{c+d}{2})-}\xi(c) + {}_k J^\alpha_{(\frac{c+d}{2})+}\xi(d)] \right|$$

$$\le \frac{(d-c)^2}{8(\frac{\alpha}{k}+1)} \left(\frac{1}{\frac{\alpha}{k}+2}\right)^{1-\frac{1}{q}}$$

$$\times \left[\left(\frac{1}{\frac{\alpha}{k}+2}|\xi''(c)|^q + \frac{1}{\frac{\alpha}{k}+s+2} \left(\left|\xi''\left(\frac{c+d}{2}\right)\right|^q - |\xi''(c)|^q \right) \right)^{\frac{1}{q}} \right.$$

$$\left. + \left(\frac{1}{\frac{\alpha}{k}+2}|\xi''(d)|^q + \frac{1}{\frac{\alpha}{k}+s+2} \left(\left|\xi''\left(\frac{c+d}{2}\right)\right|^q - |\xi''(d)|^q \right) \right)^{\frac{1}{q}} \right],$$

(2) if we choose $k = \alpha = 1$ and $\lambda = 0$, then we get

$$\left| \xi\left(\frac{c+d}{2}\right) - \frac{1}{d-c}\int_c^d \xi(x)dx \right|$$

$$\le \frac{(d-c)^2}{16} \left(\frac{1}{3}\right)^{1-\frac{1}{q}} \left[\left(\frac{1}{3}|\xi''(c)|^q + \frac{1}{s+3} \left(\left|\xi''\left(\frac{c+d}{2}\right)\right|^q - |\xi''(c)|^q \right) \right)^{\frac{1}{q}} \right.$$

$$\left. + \left(\frac{1}{3}|\xi''(d)|^q + \frac{1}{s+3} \left(\left|\xi''\left(\frac{c+d}{2}\right)\right|^q - |\xi''(d)|^q \right) \right)^{\frac{1}{q}} \right],$$

(3) if $\lambda = 1$, then we have

$$\left| \frac{\xi(c) + \xi(d)}{2} - \frac{2^{\frac{\alpha}{k}-1}\Gamma_k(\alpha+k)}{(d-c)^{\frac{\alpha}{k}}} \left[{}_kJ^{\alpha}_{(\frac{c+d}{2})-}\xi(c) + {}_kJ^{\alpha}_{(\frac{c+d}{2})+}\xi(d) \right] \right|$$

$$\leq \frac{(d-c)^2}{8(\frac{\alpha}{k}+1)} \left(\frac{\frac{\alpha}{k}(\frac{\alpha}{k}+3)}{2(\frac{\alpha}{k}+2)} \right)^{1-\frac{1}{q}} \left[\left(\left(\frac{\frac{\alpha}{k}(\frac{\alpha}{k}+3)}{2(\frac{\alpha}{k}+2)} \right) |\xi''(c)|^q \right. \right.$$

$$+ \frac{\frac{\alpha}{k}(\frac{\alpha}{k}+s+3)}{(\frac{\alpha}{k}+s+2)(s+2)} \left(\left| \xi''\left(\frac{c+d}{2}\right) \right|^q - |\xi''(c)|^q \right) \right)^{\frac{1}{q}}$$

$$+ \left(\left(\frac{\frac{\alpha}{k}(\frac{\alpha}{k}+3)}{2(\frac{\alpha}{k}+2)} \right) |\xi''(d)|^q \right.$$

$$\left. + \frac{\frac{\alpha}{k}(\frac{\alpha}{k}+s+3)}{(\frac{\alpha}{k}+s+2)(s+2)} \left(\left| \xi''\left(\frac{c+d}{2}\right) \right|^q - |\xi''(d)|^q \right) \right)^{\frac{1}{q}} \right],$$

(4) if we choose $k = \alpha = \lambda = 1$, then we get

$$\left| \frac{\xi(c) + \xi(d)}{2} - \frac{1}{d-c} \int_c^d \xi(x)dx \right| \leq \frac{(d-c)^2}{16} \left(\frac{2}{3} \right)^{1-\frac{1}{q}}$$

$$\times \left[\left(\frac{2}{3}|\xi''(c)|^q + \frac{(s+4)}{(s+2)(s+3)} \left(\left| \xi''\left(\frac{c+d}{2}\right) \right|^q - |\xi''(c)|^q \right) \right)^{\frac{1}{q}} \right.$$

$$\left. + \left(\frac{2}{3}|\xi''(d)|^q + \frac{(s+4)}{(s+2)(s+3)} \left(\left| \xi''\left(\frac{c+d}{2}\right) \right|^q - |\xi''(d)|^q \right) \right)^{\frac{1}{q}} \right],$$

(5) if $\lambda = \frac{1}{3}$, then we have

$$\left| \frac{1}{6} \left[\xi(c) + 4\xi\left(\frac{c+d}{2}\right) + \xi(d) \right] \right.$$

$$\left. - \frac{2^{\frac{\alpha}{k}-1}\Gamma_k(\alpha+k)}{(d-c)^{\frac{\alpha}{k}}} \left[{}_kJ^{\alpha}_{(\frac{c+d}{2})-}\xi(c) + {}_kJ^{\alpha}_{(\frac{c+d}{2})+}\xi(d) \right] \right|$$

$$\leq \frac{(d-c)^2}{8(\frac{\alpha}{k}+1)} A_0^{1-\frac{1}{q}} \left(k, \alpha, \frac{1}{3} \right) \left[\left(A_0 \left(k, \alpha, \frac{1}{3} \right) |\xi''(c)|^q \right. \right.$$

$$+ \left(\left| \xi''\left(\frac{c+d}{2}\right) \right|^q - |\xi''(c)|^q \right) H_1 \left(k, \alpha, \frac{1}{3}, s \right) \right)^{\frac{1}{q}}$$

$$+ \left(A_0 \left(k, \alpha, \frac{1}{3} \right) |\xi''(d)|^q \right.$$

$$\left. + \left(\left| \xi''\left(\frac{c+d}{2}\right) \right|^q - |\xi''(d)|^q \right) H_1 \left(k, \alpha, \frac{1}{3}, s \right) \right)^{\frac{1}{q}} \right],$$

(6) if we choose $k = \alpha = 1$ and $\lambda = \frac{1}{3}$, then we get

$$\left| \frac{1}{6} \left[\xi(c) + 4\xi\left(\frac{c+d}{2}\right) + \xi(d) \right] - \frac{1}{d-c} \int_c^d \xi(x)dx \right| \leq \frac{(d-c)^2}{16} \left(\frac{8}{81}\right)^{1-\frac{1}{q}}$$

$$\times \left[\left(\frac{8}{81} |\xi''(c)|^q + \left(\left| \xi''\left(\frac{c+d}{2}\right) \right|^q - |\xi''(c)|^q \right) H_1\left(1, 1, \frac{1}{3}, s\right) \right)^{\frac{1}{q}} \right.$$

$$\left. + \left(\frac{8}{81} |\xi''(d)|^q + \left(\left| \xi''\left(\frac{c+d}{2}\right) \right|^q - |\xi''(d)|^q \right) H_1\left(1, 1, \frac{1}{3}, s\right) \right)^{\frac{1}{q}} \right],$$

(7) if $\lambda = \frac{1}{2}$, then we have

$$\left| \frac{1}{4} \left[\xi(c) + 2\xi\left(\frac{c+d}{2}\right) + \xi(d) \right] \right.$$

$$\left. - \frac{2^{\frac{\alpha}{k}-1}\Gamma_k(\alpha+k)}{(d-c)^{\frac{\alpha}{k}}} [_k J^\alpha_{(\frac{c+d}{2})-}\xi(c) + {}_k J^\alpha_{(\frac{c+d}{2})+}\xi(d)] \right|$$

$$\leq \frac{(d-c)^2}{8(\frac{\alpha}{k}+1)} A_0^{1-\frac{1}{q}} \left(k, \alpha, \frac{1}{2}\right)$$

$$\times \left[\left(A_0\left(k, \alpha, \frac{1}{2}\right) |\xi''(c)|^q + \left(\left| \xi''\left(\frac{c+d}{2}\right) \right|^q - |\xi''(c)|^q \right) H_1\left(k, \alpha, \frac{1}{2}, s\right) \right)^{\frac{1}{q}} \right.$$

$$\left. + \left(A_0\left(k, \alpha, \frac{1}{2}\right) |\xi''(d)|^q + \left(\left| \xi''\left(\frac{c+d}{2}\right) \right|^q - |\xi''(d)|^q \right) H_1\left(k, \alpha, \frac{1}{2}, s\right) \right)^{\frac{1}{q}} \right],$$

(8) if we choose $k = \alpha = 1$ and $\lambda = \frac{1}{2}$, then we get

$$\left| \frac{1}{4} \left[\xi(c) + 2\xi\left(\frac{c+d}{2}\right) + \xi(d) \right] - \frac{1}{d-c} \int_c^d \xi(x)dx \right| \leq \frac{(d-c)^2}{16} \left(\frac{1}{6}\right)^{1-\frac{1}{q}}$$

$$\times \left[\left(\frac{1}{6} |\xi''(c)|^q + \left(\left| \xi''\left(\frac{c+d}{2}\right) \right|^q - |\xi''(c)|^q \right) H_1\left(1, 1, \frac{1}{2}, s\right) \right)^{\frac{1}{q}} \right.$$

$$\left. + \left(\frac{1}{6} |\xi''(d)|^q + \left(\left| \xi''\left(\frac{c+d}{2}\right) \right|^q - |\xi''(d)|^q \right) H_1\left(1, 1, \frac{1}{2}, s\right) \right)^{\frac{1}{q}} \right].$$

Corollary 4.5.4. *In Theorem 4.5.1, if $|\xi''|^q$ for $q \geq 1$ is generalized (m, tgs)-preunivex functions, then for $m \in (0, 1]$, we have*

$$|I_{\xi,\eta}(\alpha, k; x, \lambda, m, c, d)| \leq A_0^{1-\frac{1}{q}}(k, \alpha, \lambda) \left[\left| \frac{\eta^{\frac{\alpha}{k}+2}(x, c, m)}{(\frac{\alpha}{k}+1)\eta(d, c, m)} \right| \right.$$

$$\times \left(m A_0(k, \alpha, \lambda) |\xi''(c)|^q + \phi(|\xi''(x)|^q - m|\xi''(c)|^q) E_1(k, \alpha, \lambda, \delta, c, x) \right)^{\frac{1}{q}}$$

$$+ \left| \frac{(-1)^{\frac{\alpha}{k}+2} \eta^{\frac{\alpha}{k}+2}(x, d, m)}{(\frac{\alpha}{k}+1)\eta(d, c, m)} \right|$$

$$\left. \times \left(m A_0(k, \alpha, \lambda) |\xi''(d)|^q + \phi(|\xi''(x)|^q - m|\xi''(d)|^q) E_2(k, \alpha, \lambda, \delta, d, x) \right)^{\frac{1}{q}} \right],$$

where

$$E_1(k, \alpha, \lambda, \delta, c, x) = \int_0^1 \delta^2(1-\delta) \left| \left(\frac{\alpha}{k}+1\right)\lambda - \delta^{\frac{\alpha}{k}} \right| b(x, c, \delta(1-\delta)) d\delta$$

and

$$E_2(k, \alpha, \lambda, \delta, d, x) = \int_0^1 \delta^2(1-\delta) \left| \left(\frac{\alpha}{k}+1\right)\lambda - \delta^{\frac{\alpha}{k}} \right| b(x, d, \delta(1-\delta)) d\delta.$$

Corollary 4.5.5. *In Corollary 4.5.4, if $\phi : \mathbb{R} \to \mathbb{R}$ is defined by $\phi(c) = c$ and $b(., ., .) \equiv 1$, then we have*

$$|I_{\xi,\eta}(\alpha, k; x, \lambda, m, c, d)| \leq A_0^{1-\frac{1}{q}}(k, \alpha, \lambda) \left[\left| \frac{\eta^{\frac{\alpha}{k}+2}(x, c, m)}{(\frac{\alpha}{k}+1)\eta(d, c, m)} \right| \right.$$

$$\times \left(m A_0(k, \alpha, \lambda) |\xi''(c)|^q + (|\xi''(x)|^q - m|\xi''(c)|^q) N_1(k, \alpha, \lambda) \right)^{\frac{1}{q}}$$

$$+ \left| \frac{(-1)^{\frac{\alpha}{k}+2} \eta^{\frac{\alpha}{k}+2}(x, d, m)}{(\frac{\alpha}{k}+1)\eta(d, c, m)} \right|$$

$$\left. \times \left(m A_0(k, \alpha, \lambda) |\xi''(d)|^q + (|\xi''(x)|^q - m|\xi''(d)|^q) N_1(k, \alpha, \lambda) \right)^{\frac{1}{q}} \right],$$

where
$N_1(k, \alpha, \lambda) = \int_0^1 \delta^2(1-\delta) \left| \left(\frac{\alpha}{k}+1\right)\lambda - \delta^{\frac{\alpha}{k}} \right| d\delta$

$$= \begin{cases} \frac{\frac{2\alpha}{k}[(\frac{\alpha}{k}+1)\lambda]^{1+\frac{3k}{\alpha}}}{3(\frac{\alpha}{k}+3)} - \frac{\frac{\alpha}{k}[(\frac{\alpha}{k}+1)\lambda]^{1+\frac{4k}{\alpha}}}{2(\frac{\alpha}{k}+4)} - \frac{(\frac{\alpha}{k}+1)\lambda}{12} + \frac{1}{(\frac{\alpha}{k}+3)(\frac{\alpha}{k}+4)}, & 0 \leq \lambda \leq \frac{1}{\frac{\alpha}{k}+1}, \\ \frac{(\frac{\alpha}{k}+1)\lambda}{12} - \frac{1}{(\frac{\alpha}{k}+3)(\frac{\alpha}{k}+4)}, & \frac{1}{\frac{\alpha}{k}+1} < \lambda \leq 1. \end{cases}$$

Corollary 4.5.6. *In Corollary 4.5.5, if the mapping $\eta(d, c, m) = d - mc$ along with $m = 1$, taking $x = \frac{c+d}{2}$, then we have*

$$\left| \frac{2^{\frac{\alpha}{k}-1}}{(d-c)^{\frac{\alpha}{k}-1}} I_\xi \left(\alpha, k; \frac{c+d}{2}, \lambda, 1, c, d \right) \right| \leq \frac{(d-c)^2}{8(\frac{\alpha}{k}+1)} A_0^{1-\frac{1}{q}}(k, \alpha, \lambda)$$

$$\times \left[\left(A_0(k, \alpha, \lambda) |\xi''(c)|^q + \left(\left| \xi''\left(\frac{c+d}{2} \right) \right|^q - |\xi''(c)|^q \right) N_1(k, \alpha, \lambda) \right)^{\frac{1}{q}} \right.$$

$$\left. + \left(A_0(k, \alpha, \lambda) |\xi''(d)|^q + \left(\left| \xi''\left(\frac{c+d}{2} \right) \right|^q - |\xi''(d)|^q \right) N_1(k, \alpha, \lambda) \right)^{\frac{1}{q}} \right].$$

Remark 4.5.3. *In Corollary 4.5.6,*

(1) if $\lambda = 0$, then we have

$$\left| \xi\left(\frac{c+d}{2} \right) - \frac{2^{\frac{\alpha}{k}-1}\Gamma_k(\alpha+k)}{(d-c)^{\frac{\alpha}{k}}} [{}_k J^\alpha_{(\frac{c+d}{2})-}\xi(c) + {}_k J^\alpha_{(\frac{c+d}{2})+}\xi(d)] \right|$$

$$\leq \frac{(d-c)^2}{8(\frac{\alpha}{k}+1)} \left(\frac{1}{\frac{\alpha}{k}+2} \right)^{1-\frac{1}{q}}$$

$$\times \left[\left(\frac{1}{\frac{\alpha}{k}+2} |\xi''(c)|^q + \frac{1}{(\frac{\alpha}{k}+3)(\frac{\alpha}{k}+4)} \left(\left| \xi''\left(\frac{c+d}{2} \right) \right|^q - |\xi''(c)|^q \right) \right)^{\frac{1}{q}} \right.$$

$$\left. + \left(\frac{1}{\frac{\alpha}{k}+2} |\xi''(d)|^q + \frac{1}{(\frac{\alpha}{k}+3)(\frac{\alpha}{k}+4)} \left(\left| \xi''\left(\frac{c+d}{2} \right) \right|^q - |\xi''(d)|^q \right) \right)^{\frac{1}{q}} \right],$$

(2) if we choose $k = \alpha = 1$ and $\lambda = 0$, then we get

$$\left| \xi\left(\frac{c+d}{2} \right) - \frac{1}{d-c} \int_c^d \xi(x)dx \right| \leq \frac{(d-c)^2}{16} \left(\frac{1}{3} \right)^{1-\frac{1}{q}}$$

$$\times \left[\left(\frac{1}{3} |\xi''(c)|^q + \frac{1}{20} \left(\left| \xi''\left(\frac{c+d}{2} \right) \right|^q - |\xi''(c)|^q \right) \right)^{\frac{1}{q}} \right.$$

$$\left. + \left(\frac{1}{3} |\xi''(d)|^q + \frac{1}{20} \left(\left| \xi''\left(\frac{c+d}{2} \right) \right|^q - |\xi''(d)|^q \right) \right)^{\frac{1}{q}} \right],$$

(3) if $\lambda = 1$, then we have

$$\left| \frac{\xi(c) + \xi(d)}{2} - \frac{2^{\frac{\alpha}{k}-1}\Gamma_k(\alpha+k)}{(d-c)^{\frac{\alpha}{k}}} \left[{}_kJ^{\alpha}_{(\frac{c+d}{2})^-}\xi(c) + {}_kJ^{\alpha}_{(\frac{c+d}{2})^+}\xi(d) \right] \right|$$

$$\leq \frac{(d-c)^2}{8(\frac{\alpha}{k}+1)} \left(\frac{\frac{\alpha}{k}(\frac{\alpha}{k}+3)}{2(\frac{\alpha}{k}+2)} \right)^{1-\frac{1}{q}} \left[\left(\left(\frac{\frac{\alpha}{k}(\frac{\alpha}{k}+3)}{2(\frac{\alpha}{k}+2)} \right) |\xi''(c)|^q \right. \right.$$

$$+ \frac{\frac{\alpha}{k}(\frac{\alpha^2}{k^2}+8\frac{\alpha}{k}+19)}{12(\frac{\alpha}{k}+3)(\frac{\alpha}{k}+4)} \left(\left| \xi'' \left(\frac{c+d}{2} \right) \right|^q - |\xi''(c)|^q \right) \right)^{\frac{1}{q}}$$

$$+ \left(\left(\frac{\frac{\alpha}{k}(\frac{\alpha}{k}+3)}{2(\frac{\alpha}{k}+2)} \right) |\xi''(d)|^q \right.$$

$$\left. + \frac{\frac{\alpha}{k}(\frac{\alpha^2}{k^2}+8\frac{\alpha}{k}+19)}{12(\frac{\alpha}{k}+3)(\frac{\alpha}{k}+4)} \left(\left| \xi'' \left(\frac{c+d}{2} \right) \right|^q - |\xi''(d)|^q \right) \right)^{\frac{1}{q}} \right],$$

(4) if we choose $k = \alpha = \lambda = 1$, then we get

$$\left| \frac{\xi(c) + \xi(d)}{2} - \frac{1}{d-c}\int_c^d \xi(x)dx \right| \leq \frac{(d-c)^2}{16} \left(\frac{2}{3} \right)^{1-\frac{1}{q}}$$

$$\times \left[\left(\frac{2}{3}|\xi''(c)|^q + \frac{7}{60} \left(\left| \xi'' \left(\frac{c+d}{2} \right) \right|^q - |\xi''(c)|^q \right) \right)^{\frac{1}{q}} \right.$$

$$\left. + \left(\frac{2}{3}|\xi''(d)|^q + \frac{7}{60} \left(\left| \xi'' \left(\frac{c+d}{2} \right) \right|^q - |\xi''(d)|^q \right) \right)^{\frac{1}{q}} \right],$$

(5) if $\lambda = \frac{1}{3}$, then we have

$$\left| \frac{1}{6} \left[\xi(c) + 4\xi \left(\frac{c+d}{2} \right) + \xi(d) \right] \right.$$

$$\left. - \frac{2^{\frac{\alpha}{k}-1}\Gamma_k(\alpha+k)}{(d-c)^{\frac{\alpha}{k}}} \left[{}_kJ^{\alpha}_{(\frac{c+d}{2})^-}\xi(c) + {}_kJ^{\alpha}_{(\frac{c+d}{2})^+}\xi(d) \right] \right|$$

$$\leq \frac{(d-c)^2}{8(\frac{\alpha}{k}+1)} A_0^{1-\frac{1}{q}} \left(k, \alpha, \frac{1}{3} \right)$$

$$\times \left[\left(A_0 \left(k, \alpha, \frac{1}{3} \right) |\xi''(c)|^q + \left(\left| \xi'' \left(\frac{c+d}{2} \right) \right|^q - |\xi''(c)|^q \right) N_1 \left(k, \alpha, \frac{1}{3} \right) \right)^{\frac{1}{q}} \right.$$

$$\left. + \left(A_0 \left(k, \alpha, \frac{1}{3} \right) |\xi''(d)|^q + \left(\left| \xi'' \left(\frac{c+d}{2} \right) \right|^q - |\xi''(d)|^q \right) N_1 \left(k, \alpha, \frac{1}{3} \right) \right)^{\frac{1}{q}} \right],$$

(6) if we choose $k = \alpha = \lambda = 1$ and $\lambda = \frac{1}{3}$, then we get

$$\left| \frac{1}{6} \left[\xi(c) + 4\xi \left(\frac{c+d}{2} \right) + \xi(d) \right] - \frac{1}{d-c} \int_c^d \xi(x) dx \right| \le \frac{(d-c)^2}{16} \left(\frac{8}{81} \right)^{1-\frac{1}{q}}$$

$$\times \left[\left(\frac{8}{81} |\xi''(c)|^q + \left(\left| \xi'' \left(\frac{c+d}{2} \right) \right|^q - |\xi''(c)|^q \right) N_1 \left(1, 1, \frac{1}{3} \right) \right)^{\frac{1}{q}} \right.$$

$$\left. + \left(\frac{8}{81} |\xi''(d)|^q + \left(\left| \xi'' \left(\frac{c+d}{2} \right) \right|^q - |\xi''(d)|^q \right) N_1 \left(1, 1, \frac{1}{3} \right) \right)^{\frac{1}{q}} \right],$$

(7) if $\lambda = \frac{1}{2}$, then we have

$$\left| \frac{1}{4} \left[\xi(c) + 2\xi \left(\frac{c+d}{2} \right) + \xi(d) \right] \right.$$

$$\left. - \frac{2^{\frac{\alpha}{k}-1} \Gamma_k(\alpha+k)}{(d-c)^{\frac{\alpha}{k}}} \left[{}_k J^\alpha_{\left(\frac{c+d}{2} \right)^-} \xi(c) + {}_k J^\alpha_{\left(\frac{c+d}{2} \right)^+} \xi(d) \right] \right|$$

$$\le \frac{(d-c)^2}{8(\frac{\alpha}{k}+1)} A_0^{1-\frac{1}{q}} \left(k, \alpha, \frac{1}{2} \right)$$

$$\times \left[\left(A_0 \left(k, \alpha, \frac{1}{2} \right) |\xi''(c)|^q + \left(\left| \xi'' \left(\frac{c+d}{2} \right) \right|^q - |\xi''(c)|^q \right) N_1 \left(k, \alpha, \frac{1}{2} \right) \right)^{\frac{1}{q}} \right.$$

$$\left. + \left(A_0 \left(k, \alpha, \frac{1}{2} \right) |\xi''(d)|^q + \left(\left| \xi'' \left(\frac{c+d}{2} \right) \right|^q - |\xi''(d)|^q \right) N_1 \left(k, \alpha, \frac{1}{2} \right) \right)^{\frac{1}{q}} \right],$$

(8) if we choose $k = \alpha = 1$ and $\lambda = \frac{1}{2}$, then we get

$$\left| \frac{1}{4} \left[\xi(c) + 2\xi \left(\frac{c+d}{2} \right) + \xi(d) \right] - \frac{1}{d-c} \int_c^d \xi(x) dx \right| \le \frac{(d-c)^2}{16} \left(\frac{1}{6} \right)^{1-\frac{1}{q}}$$

$$\times \left[\left(\frac{1}{6} |\xi''(c)|^q + \left(\left| \xi'' \left(\frac{c+d}{2} \right) \right|^q - |\xi''(c)|^q \right) N_1 \left(1, 1, \frac{1}{2} \right) \right)^{\frac{1}{q}} \right.$$

$$\left. + \left(\frac{1}{6} |\xi''(d)|^q + \left(\left| \xi'' \left(\frac{c+d}{2} \right) \right|^q - |\xi''(d)|^q \right) N_1 \left(1, 1, \frac{1}{2} \right) \right)^{\frac{1}{q}} \right].$$

Corollary 4.5.7. *In Theorem 4.5.1, if $|\xi''|^q$ for $q \geq 1$ is generalized m-MT-preunivex functions, then for $m \in (0,1]$, we have*

$$
|I_{\xi,\eta}(\alpha, k; x, \lambda, m, c, d)| \leq A_0^{1-\frac{1}{q}}(k, \alpha, \lambda) \left[\left| \frac{\eta^{\frac{\alpha}{k}+2}(x, c, m)}{(\frac{\alpha}{k}+1)\eta(d, c, m)} \right| \right.
$$

$$
\times \left(mA_0(k, \alpha, \lambda)|\xi''(c)|^q + \phi(|\xi''(x)|^q - m|\xi''(c)|^q)G_1(k, \alpha, \lambda, \delta, c, x) \right)^{\frac{1}{q}}
$$

$$
+ \left| \frac{(-1)^{\frac{\alpha}{k}+2}\eta^{\frac{\alpha}{k}+2}(x, d, m)}{(\frac{\alpha}{k}+1)\eta(d, c, m)} \right|
$$

$$
\left. \times \left(mA_0(k, \alpha, \lambda)|\xi''(d)|^q + \phi(|\xi''(x)|^q - m|\xi''(d)|^q)G_2(k, \alpha, \lambda, \delta, d, x) \right)^{\frac{1}{q}} \right],
$$

where

$$
G_1(k, \alpha, \lambda, \delta, c, x) = \int_0^1 \frac{\delta\sqrt{\delta}}{2\sqrt{1-\delta}} \left| \left(\frac{\alpha}{k}+1 \right)\lambda - \delta^{\frac{\alpha}{k}} \right| b\left(x, c, \frac{\sqrt{\delta}}{2\sqrt{1-\delta}} \right) d\delta
$$

and

$$
G_2(k, \alpha, \lambda, \delta, d, x) = \int_0^1 \frac{\delta\sqrt{\delta}}{2\sqrt{1-\delta}} \left| \left(\frac{\alpha}{k}+1 \right)\lambda - \delta^{\frac{\alpha}{k}} \right| b\left(x, d, \frac{\sqrt{\delta}}{2\sqrt{1-\delta}} \right) d\delta.
$$

Corollary 4.5.8. *In Corollary 4.5.7, if $\phi : \mathbb{R} \to \mathbb{R}$ is defined by $\phi(c) = c$ and $b(.,.,.) \equiv 1$, then we have*

$$
|I_{\xi,\eta}(\alpha, k; x, \lambda, m, c, d)| \leq A_0^{1-\frac{1}{q}}(k, \alpha, \lambda) \left[\left| \frac{\eta^{\frac{\alpha}{k}+2}(x, c, m)}{(\frac{\alpha}{k}+1)\eta(d, c, m)} \right| \right.
$$

$$
\times \left(mA_0(k, \alpha, \lambda)|\xi''(c)|^q + (|\xi''(x)|^q - m|\xi''(c)|^q)R_1(k, \alpha, \lambda) \right)^{\frac{1}{q}}
$$

$$
+ \left| \frac{(-1)^{\frac{\alpha}{k}+2}\eta^{\frac{\alpha}{k}+2}(x, d, m)}{(\frac{\alpha}{k}+1)\eta(d, c, m)} \right|
$$

$$
\left. \times \left(mA_0(k, \alpha, \lambda)|\xi''(d)|^q + (|\xi''(x)|^q - m|\xi''(d)|^q)R_1(k, \alpha, \lambda) \right)^{\frac{1}{q}} \right],
$$

where
$R_1(k, \alpha, \lambda) = \int_0^1 \frac{\delta\sqrt{\delta}}{2\sqrt{1-\delta}} \left| \left(\frac{\alpha}{k}+1 \right)\lambda - \delta^{\frac{\alpha}{k}} \right| d\delta$

$$
= \begin{cases} \left[\begin{array}{l} \frac{1}{2}\beta(\frac{\alpha}{k}+\frac{5}{2}, \frac{1}{2}) - \frac{3\lambda\pi(\frac{\alpha}{k}+1)}{16} \\ +(\frac{\alpha}{k}+1)\lambda\beta([(\frac{\alpha}{k}+1)\lambda]^{\frac{k}{\alpha}}; \frac{5}{2}, \frac{1}{2}) \\ -\beta([(\frac{\alpha}{k}+1)\lambda]^{\frac{k}{\alpha}}; \frac{\alpha}{k}+\frac{5}{2}, \frac{1}{2}) \end{array} \right], & 0 \leq \lambda \leq \frac{1}{\frac{\alpha}{k}+1}, \\ \frac{3\lambda\pi(\frac{\alpha}{k}+1)}{16} - \frac{1}{2}\beta(\frac{\alpha}{k}+\frac{5}{2}, \frac{1}{2}), & \frac{1}{\frac{\alpha}{k}+1} < \lambda \leq 1. \end{cases}
$$

Corollary 4.5.9. *In Corollary 4.5.8, if the mapping $\eta(d,c,m) = d - mc$ along with $m = 1$, taking $x = \frac{c+d}{2}$, then we have*

$$\left| \frac{2^{\frac{\alpha}{k}-1}}{(d-c)^{\frac{\alpha}{k}-1}} I_\xi\left(\alpha,k;\frac{c+d}{2},\lambda,1,c,d\right) \right| \leq \frac{(d-c)^2}{8(\frac{\alpha}{k}+1)} A_0^{1-\frac{1}{q}}(k,\alpha,\lambda)$$

$$\times \left[\left(A_0(k,\alpha,\lambda)|\xi''(c)|^q + \left(\left|\xi''\left(\frac{c+d}{2}\right)\right|^q - |\xi''(c)|^q \right) R_1(k,\alpha,\lambda) \right)^{\frac{1}{q}} \right.$$

$$\left. + \left(A_0(k,\alpha,\lambda)|\xi''(d)|^q + \left(\left|\xi''\left(\frac{c+d}{2}\right)\right|^q - |\xi''(d)|^q \right) R_1(k,\alpha,\lambda) \right)^{\frac{1}{q}} \right].$$

Remark 4.5.4. *In Corollary 4.5.9,*

(1) if $\lambda = 0$, then we have

$$\left| \xi\left(\frac{c+d}{2}\right) - \frac{2^{\frac{\alpha}{k}-1}\Gamma_k(\alpha+k)}{(d-c)^{\frac{\alpha}{k}}}[{}_kJ^\alpha_{(\frac{c+d}{2})-}\xi(c) + {}_kJ^\alpha_{(\frac{c+d}{2})+}\xi(d)] \right|$$

$$\leq \frac{(d-c)^2}{8(\frac{\alpha}{k}+1)}\left(\frac{1}{\frac{\alpha}{k}+2}\right)^{1-\frac{1}{q}}$$

$$\times \left[\left(\frac{1}{\frac{\alpha}{k}+2}|\xi''(c)|^q + \frac{1}{2}\beta\left(\frac{\alpha}{k}+\frac{5}{2},\frac{1}{2}\right)\left(\left|\xi''\left(\frac{c+d}{2}\right)\right|^q - |\xi''(c)|^q\right) \right)^{\frac{1}{q}} \right.$$

$$\left. + \left(\frac{1}{\frac{\alpha}{k}+2}|\xi''(d)|^q + \frac{1}{2}\beta\left(\frac{\alpha}{k}+\frac{5}{2},\frac{1}{2}\right)\left(\left|\xi''\left(\frac{c+d}{2}\right)\right|^q - |\xi''(d)|^q\right) \right)^{\frac{1}{q}} \right],$$

(2) if we choose $k = \alpha = 1$ and $\lambda = 0$, then we get

$$\left| \xi\left(\frac{c+d}{2}\right) - \frac{1}{d-c}\int_c^d \xi(x)dx \right|$$

$$\leq \frac{(d-c)^2}{16}\left(\frac{1}{3}\right)^{1-\frac{1}{q}}\left[\left(\frac{1}{3}|\xi''(c)|^q + \frac{5\pi}{32}\left(\left|\xi''\left(\frac{c+d}{2}\right)\right|^q - |\xi''(c)|^q\right) \right)^{\frac{1}{q}} \right.$$

$$\left. + \left(\frac{1}{3}|\xi''(d)|^q + \frac{5\pi}{32}\left(\left|\xi''\left(\frac{c+d}{2}\right)\right|^q - |\xi''(d)|^q\right) \right)^{\frac{1}{q}} \right],$$

(3) if $\lambda = 1$, then we have

$$\left| \frac{\xi(c) + \xi(d)}{2} - \frac{2^{\frac{\alpha}{k}-1}\Gamma_k(\alpha+k)}{(d-c)^{\frac{\alpha}{k}}} [{}_kJ^{\alpha}_{(\frac{c+d}{2})^-}\xi(c) + {}_kJ^{\alpha}_{(\frac{c+d}{2})^+}\xi(d)] \right|$$

$$\leq \frac{(d-c)^2}{8(\frac{\alpha}{k}+1)} \left(\frac{\frac{\alpha}{k}(\frac{\alpha}{k}+3)}{2(\frac{\alpha}{k}+2)} \right)^{1-\frac{1}{q}} \left[\left(\left(\frac{\frac{\alpha}{k}(\frac{\alpha}{k}+3)}{2(\frac{\alpha}{k}+2)} \right) |\xi''(c)|^q \right. \right.$$

$$+ \left(\frac{3\pi(\frac{\alpha}{k}+1)}{16} - \frac{1}{2}\beta\left(\frac{\alpha}{k}+\frac{5}{2},\frac{1}{2}\right) \right) \left(\left| \xi''\left(\frac{c+d}{2}\right) \right|^q - |\xi''(c)|^q \right) \right)^{\frac{1}{q}}$$

$$+ \left(\left(\frac{\frac{\alpha}{k}(\frac{\alpha}{k}+3)}{2(\frac{\alpha}{k}+2)} \right) |\xi''(d)|^q \right.$$

$$\left. + \left(\frac{3\pi(\frac{\alpha}{k}+1)}{16} - \frac{1}{2}\beta\left(\frac{\alpha}{k}+\frac{5}{2},\frac{1}{2}\right) \right) \left(\left| \xi''\left(\frac{c+d}{2}\right) \right|^q - |\xi''(d)|^q \right) \right)^{\frac{1}{q}} \right],$$

(4) if we choose $k = \alpha = \lambda = 1$, then we get

$$\left| \frac{\xi(c) + \xi(d)}{2} - \frac{1}{d-c}\int_c^d \xi(x)dx \right|$$

$$\leq \frac{(d-c)^2}{16} \left(\frac{2}{3}\right)^{1-\frac{1}{q}} \left[\left(\frac{2}{3}|\xi''(c)|^q + \frac{7\pi}{32}\left(\left| \xi''\left(\frac{c+d}{2}\right) \right|^q - |\xi''(c)|^q \right) \right)^{\frac{1}{q}} \right.$$

$$\left. + \left(\frac{2}{3}|\xi''(d)|^q + \frac{7\pi}{32}\left(\left| \xi''\left(\frac{c+d}{2}\right) \right|^q - |\xi''(d)|^q \right) \right)^{\frac{1}{q}} \right],$$

(5) if $\lambda = \frac{1}{3}$, then we have

$$\left| \frac{1}{6}\left[\xi(c) + 4\xi\left(\frac{c+d}{2}\right) + \xi(d) \right] \right.$$

$$\left. - \frac{2^{\frac{\alpha}{k}-1}\Gamma_k(\alpha+k)}{(d-c)^{\frac{\alpha}{k}}} [{}_kJ^{\alpha}_{(\frac{c+d}{2})^-}\xi(c) + {}_kJ^{\alpha}_{(\frac{c+d}{2})^+}\xi(d)] \right|$$

$$\leq \frac{(d-c)^2}{8(\frac{\alpha}{k}+1)} A_0^{1-\frac{1}{q}}\left(k,\alpha,\frac{1}{3}\right)$$

$$\times \left[\left(A_0\left(k,\alpha,\frac{1}{3}\right)|\xi''(c)|^q + \left(\left| \xi''\left(\frac{c+d}{2}\right) \right|^q - |\xi''(c)|^q \right) R_1\left(k,\alpha,\frac{1}{3}\right) \right)^{\frac{1}{q}} \right.$$

$$\left. + \left(A_0\left(k,\alpha,\frac{1}{3}\right)|\xi''(d)|^q + \left(\left| \xi''\left(\frac{c+d}{2}\right) \right|^q - |\xi''(d)|^q \right) R_1\left(k,\alpha,\frac{1}{3}\right) \right)^{\frac{1}{q}} \right],$$

(6) if we choose $k = \alpha = 1$ and $\lambda = \frac{1}{3}$, then we get

$$\left| \frac{1}{6} \left[\xi(c) + 4\xi \left(\frac{c+d}{2} \right) + \xi(d) \right] - \frac{1}{d-c} \int_c^d \xi(x)dx \right| \leq \frac{(d-c)^2}{16} \left(\frac{8}{81} \right)^{1-\frac{1}{q}}$$

$$\times \left[\left(\frac{8}{81} |\xi''(c)|^q + \left(\left| \xi'' \left(\frac{c+d}{2} \right) \right|^q - |\xi''(c)|^q \right) R_1 \left(1,1,\frac{1}{3} \right) \right)^{\frac{1}{q}} \right.$$

$$\left. + \left(\frac{8}{81} |\xi''(d)|^q + \left(\left| \xi'' \left(\frac{c+d}{2} \right) \right|^q - |\xi''(d)|^q \right) R_1 \left(1,1,\frac{1}{3} \right) \right)^{\frac{1}{q}} \right],$$

(7) if $\lambda = \frac{1}{2}$, then we have

$$\left| \frac{1}{4} \left[\xi(c) + 2\xi \left(\frac{c+d}{2} \right) + \xi(d) \right] \right.$$

$$\left. - \frac{2^{\frac{\alpha}{k}-1}\Gamma_k(\alpha+k)}{(d-c)^{\frac{\alpha}{k}}} [{}_kJ^\alpha_{(\frac{c+d}{2})^-}\xi(c) + {}_kJ^\alpha_{(\frac{c+d}{2})^+}\xi(d)] \right|$$

$$\leq \frac{(d-c)^2}{8(\frac{\alpha}{k}+1)} A_0^{1-\frac{1}{q}} \left(k,\alpha,\frac{1}{2} \right)$$

$$\times \left[\left(A_0 \left(k,\alpha,\frac{1}{2} \right) |\xi''(c)|^q + \left(\left| \xi'' \left(\frac{c+d}{2} \right) \right|^q - |\xi''(c)|^q \right) R_1 \left(k,\alpha,\frac{1}{2} \right) \right)^{\frac{1}{q}} \right.$$

$$\left. + \left(A_0 \left(k,\alpha,\frac{1}{2} \right) |\xi''(d)|^q + \left(\left| \xi'' \left(\frac{c+d}{2} \right) \right|^q - |\xi''(d)|^q \right) R_1 \left(k,\alpha,\frac{1}{2} \right) \right)^{\frac{1}{q}} \right],$$

(8) if we choose $k = \alpha = 1$ and $\lambda = \frac{1}{2}$, then we get

$$\left| \frac{1}{4} \left[\xi(c) + 2\xi \left(\frac{c+d}{2} \right) + \xi(d) \right] - \frac{1}{d-c} \int_c^d \xi(x)dx \right| \leq \frac{(d-c)^2}{16} \left(\frac{1}{6} \right)^{1-\frac{1}{q}}$$

$$\times \left[\left(\frac{1}{6} |\xi''(c)|^q + \left(\left| \xi'' \left(\frac{c+d}{2} \right) \right|^q - |\xi''(c)|^q \right) R_1 \left(1,1,\frac{1}{2} \right) \right)^{\frac{1}{q}} \right.$$

$$\left. + \left(\frac{1}{6} |\xi''(d)|^q + \left(\left| \xi'' \left(\frac{c+d}{2} \right) \right|^q - |\xi''(d)|^q \right) R_1 \left(1,1,\frac{1}{2} \right) \right)^{\frac{1}{q}} \right].$$

We present the following theorem taken from [157].

Theorem 4.5.2. *Let $K \subseteq \mathbb{R}$ be an open m-invex subset with respect to η : $K \times K \times (0,1] \to \mathbb{R}\backslash\{0\}$ for some fixed $m \in (0,1]$, and let $c,d \in K$, $c < d$ with $\eta(d,c,m) > 0$. Assume that $\xi : K \to \mathbb{R}$ is a twice differentiable function on K such that ξ'' is integrable on $[mc, mc + \eta(d,c,m)]$. If $|\xi''|^q$ for $q > 1$ is a generalized (m,h)-preunivex function with respect to η, ϕ and b and*

$h : [0, 1] \to \mathbb{R}\backslash\{0\}$, *then the following inequality for k-fractional integrals with* $x \in [c, d], \ \lambda \in [0, 1], \ \alpha > 0, \ k > 0$ *exists:*

$$|I_{\xi,\eta}(\alpha, k; x, \lambda, m, c, d)| \le A_3(k, \alpha, \lambda, p)^{\frac{1}{p}} \left[\left| \frac{\eta^{\frac{\alpha}{k}+2}(x, c, m)}{(\frac{\alpha}{k}+1)\eta(d, c, m)} \right| \right.$$

$$\times \left(m|\xi''(c)|^q + \phi(|\xi''(x)|^q - m|\xi''(c)|^q) \int_0^1 h(\delta)b(x, c, h(\delta))d\delta \right)^{\frac{1}{q}}$$

$$+ \left| \frac{(-1)^{\frac{\alpha}{k}+2}\eta^{\frac{\alpha}{k}+2}(x, d, m)}{(\frac{\alpha}{k}+1)\eta(d, c, m)} \right|$$

$$\times \left. \left(m|\xi''(d)|^q + \phi(|\xi''(x)|^q - m|\xi''(d)|^q) \int_0^1 h(\delta)b(x, d, h(\delta))d\delta \right)^{\frac{1}{q}} \right],$$

where $p = \frac{q}{q-1}$ *and*

$$A_3(k, \alpha, \lambda, p) = \int_0^1 \delta^p \left| \left(\frac{\alpha}{k} + 1 \right) \lambda - \delta^{\frac{\alpha}{k}} \right|^p d\delta$$

$$= \begin{cases} \frac{1}{p(\frac{\alpha}{k}+1)+1}, & \lambda = 0, \\[2mm] \left[\begin{array}{l} \frac{k[(\frac{\alpha}{k}+1)\lambda]^{\frac{k+kp(\frac{\alpha}{k}+1)}{\alpha}}}{\alpha}\beta(\frac{k(1+p)}{\alpha}, 1+p) + \frac{k[1-(\frac{\alpha}{k}+1)\lambda]^{p+1}}{\alpha(p+1)} \\[2mm] \times \ _2F_1(1 - \frac{k(1+p)}{\alpha}, 1; p+2; 1-(\frac{\alpha}{k}+1)\lambda) \end{array} \right], & 0 < \lambda \le \frac{1}{\frac{\alpha}{k}+1}, \\[4mm] \frac{k[(\frac{\alpha}{k}+1)\lambda]^{\frac{k+kp(\frac{\alpha}{k}+1)}{\alpha}}}{\alpha}\beta(\frac{1}{(\frac{\alpha}{k}+1)\lambda}; \frac{k(1+p)}{\alpha}, 1+p), & \frac{1}{\frac{\alpha}{k}+1} < \lambda \le 1. \end{cases}$$

Proof Using Lemma 4.2.1 and Hölder's inequality, we have

$$|I_{\xi,\eta}(\alpha, k; x, \lambda, m, c, d)| \le \left| \frac{\eta^{\frac{\alpha}{k}+2}(x, c, m)}{(\frac{\alpha}{k}+1)\eta(d, c, m)} \right|$$

$$\times \left[\int_0^1 \delta^p \left| \left(\frac{\alpha}{k} + 1 \right) \lambda - \delta^{\frac{\alpha}{k}} \right|^p d\delta \right]^{\frac{1}{p}} \times \left[\int_0^1 |\xi''(mc + \delta\eta(x, c, m))|^q d\delta \right]^{\frac{1}{q}}$$

$$+ \left| \frac{(-1)^{\frac{\alpha}{k}+2}\eta^{\frac{\alpha}{k}+2}(x, d, m)}{(\frac{\alpha}{k}+1)\eta(d, c, m)} \right| \left[\int_0^1 \delta^p \left| \left(\frac{\alpha}{k} + 1 \right) \lambda - \delta^{\frac{\alpha}{k}} \right|^p d\delta \right]^{\frac{1}{p}}$$

$$\times \left[\int_0^1 |\xi''(md + \delta\eta(x, d, m))|^q d\delta \right]^{\frac{1}{q}}. \tag{4.25}$$

Since $|\xi''|^q$ is generalized (m, h)-preunivex on $[mc, mc + \eta(d, c, m)]$, we get

$$\int_0^1 |\xi''(mc + \delta\eta(x, c, m))|^q d\delta$$

$$\le \int_0^1 [m|\xi''(c)|^q + h(\delta)b(x, c, h(\delta))\phi(|\xi''(x)|^q - m|\xi''(c)|^q)]d\delta$$

$$= m|\xi''(c)|^q + \phi(|\xi''(x)|^q - m|\xi''(c)|^q) \int_0^1 h(\delta)b(x, c, h(\delta))d\delta \tag{4.26}$$

and

$$\int_0^1 |\xi''(md + \delta\eta(x,d,m))|^q d\delta$$

$$\leq \int_0^1 [m|\xi''(d)|^q + h(\delta)b(x,d,h(\delta))\phi(|\xi''(x)|^q - m|\xi''(d)|^q)]d\delta$$

$$= m|\xi''(d)|^q + \phi(|\xi''(x)|^q - m|\xi''(d)|^q)\int_0^1 h(\delta)b(x,d,h(\delta))d\delta. \quad (4.27)$$

Using (4.26) and (4.27) in (4.25), we have

$$|I_{\xi,\eta}(\alpha,k;x,\lambda,m,c,d)| \leq A_3(k,\alpha,\lambda,p)^{\frac{1}{p}} \left[\left| \frac{\eta^{\frac{\alpha}{k}+2}(x,c,m)}{(\frac{\alpha}{k}+1)\eta(d,c,m)} \right| \right.$$

$$\times \left(m|\xi''(c)|^q + \phi(|\xi''(x)|^q - m|\xi''(c)|^q)\int_0^1 h(\delta)b(x,c,h(\delta))d\delta \right)^{\frac{1}{q}}$$

$$+ \left| \frac{(-1)^{\frac{\alpha}{k}+2}\eta^{\frac{\alpha}{k}+2}(x,d,m)}{(\frac{\alpha}{k}+1)\eta(d,c,m)} \right|$$

$$\times \left. \left(m|\xi''(d)|^q + \phi(|\xi''(x)|^q - m|\xi''(d)|^q)\int_0^1 h(\delta)b(x,d,h(\delta))d\delta \right)^{\frac{1}{q}} \right].$$

This completes the proof.

Now, we will discuss some special cases of Theorem 4.5.2 [157].

Corollary 4.5.10. *In Theorem 4.5.2, if we use the generalized (m,s)-Breckner-preunivexity of $|\xi''|^q$ along with $q > 1$ and $p = \frac{q}{q-1}$, then for $s \in (0,1]$ and $m \in (0,1]$, we have the following inequality:*

$$|I_{\xi,\eta}(\alpha,k;x,\lambda,m,c,d)| \leq A_3(k,\alpha,\lambda,p)^{\frac{1}{p}} \left[\left| \frac{\eta^{\frac{\alpha}{k}+2}(x,c,m)}{(\frac{\alpha}{k}+1)\eta(d,c,m)} \right| \right.$$

$$\times \left(m|\xi''(c)|^q + \phi(|\xi''(x)|^q - m|\xi''(c)|^q)\int_0^1 \delta^s b(x,c,\delta^s)d\delta \right)^{\frac{1}{q}}$$

$$+ \left| \frac{(-1)^{\frac{\alpha}{k}+2}\eta^{\frac{\alpha}{k}+2}(x,d,m)}{(\frac{\alpha}{k}+1)\eta(d,c,m)} \right|$$

$$\times \left. \left(m|\xi''(d)|^q + \phi(|\xi''(x)|^q - m|\xi''(d)|^q)\int_0^1 \delta^s b(x,d,\delta^s)d\delta \right)^{\frac{1}{q}} \right].$$

Corollary 4.5.11. *In Corollary 4.5.10, if* $\phi : \mathbb{R} \to \mathbb{R}$ *is defined by* $\phi(c) = c$ *and* $b(.,.,.) \equiv 1$*, then we have*

$$|I_{\xi,\eta}(\alpha, k; x, \lambda, m, c, d)| \leq A_3(k, \alpha, \lambda, p)^{\frac{1}{p}}$$

$$\times \left[\left| \frac{\eta^{\frac{\alpha}{k}+2}(x, c, m)}{(\frac{\alpha}{k}+1)\eta(d, c, m)} \right| \left(m|\xi''(c)|^q + \frac{(|\xi''(x)|^q - m|\xi''(c)|^q)}{(s+1)} \right)^{\frac{1}{q}} \right.$$

$$\left. + \left| \frac{(-1)^{\frac{\alpha}{k}+2}\eta^{\frac{\alpha}{k}+2}(x, d, m)}{(\frac{\alpha}{k}+1)\eta(d, c, m)} \right| \left(m|\xi''(d)|^q + \frac{(|\xi''(x)|^q - m|\xi''(d)|^q)}{(s+1)} \right)^{\frac{1}{q}} \right].$$

Corollary 4.5.12. *In Corollary 4.5.11, if the mapping* $\eta(d, c, m) = d - mc$ *together with* $m = 1$*, choosing* $x = \frac{c+d}{2}$*, for* $s \in (0, 1]$*, we have the following inequality:*

$$\left| \frac{2^{\frac{\alpha}{k}-1}}{(d-c)^{\frac{\alpha}{k}-1}} I_\xi \left(\alpha, k; \frac{c+d}{2}, \lambda, 1, c, d \right) \right|$$

$$\leq \frac{(d-c)^2}{8(\frac{\alpha}{k}+1)} A_3(k, \alpha, \lambda, p)^{\frac{1}{p}} \left[\left(|\xi''(c)|^q + \frac{(|\xi''(\frac{c+d}{2})|^q - |\xi''(c)|^q)}{(s+1)} \right)^{\frac{1}{q}} \right.$$

$$\left. + \left(|\xi''(d)|^q + \frac{(|\xi''(\frac{c+d}{2})|^q - |\xi''(d)|^q)}{(s+1)} \right)^{\frac{1}{q}} \right].$$

Remark 4.5.5. *In Corollary 4.5.12,*

(1) *if* $\lambda = 0$*, then we have*

$$\left| \xi\left(\frac{c+d}{2}\right) - \frac{2^{\frac{\alpha}{k}-1}\Gamma_k(\alpha+k)}{(d-c)^{\frac{\alpha}{k}}} [{}_kJ^\alpha_{(\frac{c+d}{2})^-}\xi(c) + {}_kJ^\alpha_{(\frac{c+d}{2})^+}\xi(d)] \right|$$

$$\leq \frac{(d-c)^2}{8(\frac{\alpha}{k}+1)} \left[\frac{1}{p(\frac{\alpha}{k}+1)+1} \right]^{\frac{1}{p}} \left[\left(|\xi''(c)|^q + \frac{(|\xi''(\frac{c+d}{2})|^q - |\xi''(c)|^q)}{s+1} \right)^{\frac{1}{q}} \right.$$

$$\left. + \left(|\xi''(d)|^q + \frac{(|\xi''(\frac{c+d}{2})|^q - |\xi''(d)|^q)}{s+1} \right)^{\frac{1}{q}} \right],$$

(2) *if we choose* $k = \alpha = 1$ *and* $\lambda = 0$*, then we get*

$$\left| \xi\left(\frac{c+d}{2}\right) - \frac{1}{d-c} \int_c^d \xi(x)dx \right|$$

$$\leq \frac{(d-c)^2}{16} \left(\frac{1}{2p+1} \right)^{\frac{1}{p}} \left[\left(|\xi''(c)|^q + \frac{(|\xi''(\frac{c+d}{2})|^q - |\xi''(c)|^q)}{s+1} \right)^{\frac{1}{q}} \right.$$

$$\left. + \left(|\xi''(d)|^q + \frac{(|\xi''(\frac{c+d}{2})|^q - |\xi''(d)|^q)}{s+1} \right)^{\frac{1}{q}} \right],$$

(3) if $\lambda = 1$, then we have

$$\left| \frac{\xi(c) + \xi(d)}{2} - \frac{2^{\frac{\alpha}{k}-1}\Gamma_k(\alpha+k)}{(d-c)^{\frac{\alpha}{k}}} [_k J^\alpha_{(\frac{c+d}{2})^-} \xi(c) + {}_k J^\alpha_{(\frac{c+d}{2})^+} \xi(d)] \right|$$

$$\leq \frac{(d-c)^2}{8(\frac{\alpha}{k}+1)} A_3(k,\alpha,1,p)^{\frac{1}{p}} \left[\left(|\xi''(c)|^q + \frac{(|\xi''(\frac{c+d}{2})|^q - |\xi''(c)|^q)}{s+1} \right)^{\frac{1}{q}} \right.$$

$$\left. + \left(|\xi''(d)|^q + \frac{(|\xi''(\frac{c+d}{2})|^q - |\xi''(d)|^q)}{s+1} \right)^{\frac{1}{q}} \right],$$

(4) if we choose $k = \alpha = \lambda = 1$, then we get

$$\left| \frac{\xi(c) + \xi(d)}{2} - \frac{1}{d-c} \int_c^d \xi(x)dx \right|$$

$$\leq \frac{(d-c)^2}{16} \left[2^{1+2p} \beta \left(\frac{1}{2}; 1+p, 1+p \right) \right]^{\frac{1}{p}}$$

$$\times \left[\left(|\xi''(c)|^q + \frac{(|\xi''(\frac{c+d}{2})|^q - |\xi''(c)|^q)}{s+1} \right)^{\frac{1}{q}} \right.$$

$$\left. + \left(|\xi''(d)|^q + \frac{(|\xi''(\frac{c+d}{2})|^q - |\xi''(d)|^q)}{s+1} \right)^{\frac{1}{q}} \right],$$

(5) if $\lambda = \frac{1}{3}$, then we have

$$\left| \frac{1}{6} \left[\xi(c) + 4\xi \left(\frac{c+d}{2} \right) + \xi(d) \right] \right.$$

$$\left. - \frac{2^{\frac{\alpha}{k}-1}\Gamma_k(\alpha+k)}{(d-c)^{\frac{\alpha}{k}}} [_k J^\alpha_{(\frac{c+d}{2})^-} \xi(c) + {}_k J^\alpha_{(\frac{c+d}{2})^+} \xi(d)] \right|$$

$$\leq \frac{(d-c)^2}{8(\frac{\alpha}{k}+1)} A_3 \left(k,\alpha,\frac{1}{3},p \right)^{\frac{1}{p}} \left[\left(|\xi''(c)|^q + \frac{(|\xi''(\frac{c+d}{2})|^q - |\xi''(c)|^q)}{s+1} \right)^{\frac{1}{q}} \right.$$

$$\left. + \left(|\xi''(d)|^q + \frac{(|\xi''(\frac{c+d}{2})|^q - |\xi''(d)|^q)}{s+1} \right)^{\frac{1}{q}} \right],$$

(6) if we choose $k = \alpha = 1$ and $\lambda = \frac{1}{3}$, then we get

$$\left| \frac{1}{6} \left[\xi(c) + 4\xi \left(\frac{c+d}{2} \right) + \xi(d) \right] - \frac{1}{d-c} \int_c^d \xi(x)dx \right|$$

$$\leq \frac{(d-c)^2}{16} A_3 \left(1,1,\frac{1}{3},p\right)^{\frac{1}{p}} \left[\left(|\xi''(c)|^q + \frac{(|\xi''(\frac{c+d}{2})|^q - |\xi''(c)|^q)}{s+1}\right)^{\frac{1}{q}}\right.$$

$$\left. + \left(|\xi''(d)|^q + \frac{(|\xi''(\frac{c+d}{2})|^q - |\xi''(d)|^q)}{s+1}\right)^{\frac{1}{q}}\right],$$

(7) if $\lambda = \frac{1}{2}$, then we have

$$\left|\frac{1}{4}\left[\xi(c) + 2\xi\left(\frac{c+d}{2}\right) + \xi(d)\right]\right.$$

$$\left. - \frac{2^{\frac{\alpha}{k}-1}\Gamma_k(\alpha+k)}{(d-c)^{\frac{\alpha}{k}}}[_kJ^\alpha_{(\frac{c+d}{2})-}\xi(c) + \,_kJ^\alpha_{(\frac{c+d}{2})+}\xi(d)]\right|$$

$$\leq \frac{(d-c)^2}{8(\frac{\alpha}{k}+1)} A_3 \left(k,\alpha,\frac{1}{2},p\right)^{\frac{1}{p}} \left[\left(|\xi''(c)|^q + \frac{(|\xi''(\frac{c+d}{2})|^q - |\xi''(c)|^q)}{s+1}\right)^{\frac{1}{q}}\right.$$

$$\left. + \left(|\xi''(d)|^q + \frac{(|\xi''(\frac{c+d}{2})|^q - |\xi''(d)|^q)}{s+1}\right)^{\frac{1}{q}}\right],$$

(8) if we choose $k = \alpha = 1$ and $\lambda = \frac{1}{2}$, then we get

$$\left|\frac{1}{4}\left[\xi(c) + 2\xi\left(\frac{c+d}{2}\right) + \xi(d)\right] - \frac{1}{d-c}\int_c^d \xi(x)dx\right|$$

$$\leq \frac{(d-c)^2}{16}\beta^{\frac{1}{p}}(1+p,1+p)\left[\left(|\xi''(c)|^q + \frac{(|\xi''(\frac{c+d}{2})|^q - |\xi''(c)|^q)}{s+1}\right)^{\frac{1}{q}}\right.$$

$$\left. + \left(|\xi''(d)|^q + \frac{(|\xi''(\frac{c+d}{2})|^q - |\xi''(d)|^q)}{s+1}\right)^{\frac{1}{q}}\right].$$

Corollary 4.5.13. *In Theorem 4.5.2, if we use the generalized (m,tgs)-preunivexity of $|\xi''|^q$ along with $q > 1$ and $p = \frac{q}{q-1}$, then for $m \in (0,1]$, we have the following inequality:*

$$|I_{\xi,\eta}(\alpha,k;x,\lambda,m,c,d)| \leq A_3(k,\alpha,\lambda,p)^{\frac{1}{p}}\left[\left|\frac{\eta^{\frac{\alpha}{k}+2}(x,c,m)}{(\frac{\alpha}{k}+1)\eta(d,c,m)}\right|\right.$$

$$\times \left(m|\xi''(c)|^q + \phi(|\xi''(x)|^q - m|\xi''(c)|^q)\int_0^1 \delta(1-\delta)b(x,c,\delta(1-\delta))d\delta\right)^{\frac{1}{q}}$$

$$+ \left| \frac{(-1)^{\frac{\alpha}{k}+2} \eta^{\frac{\alpha}{k}+2}(x,d,m)}{(\frac{\alpha}{k}+1)\eta(d,c,m)} \right|$$

$$\times \left(m|\xi''(d)|^q + \phi(|\xi''(x)|^q - m|\xi''(d)|^q) \int_0^1 \delta(1-\delta) b(x,d,\delta(1-\delta)) d\delta \right)^{\frac{1}{q}} \right].$$

Corollary 4.5.14. *In Corollary 4.5.13, if $\phi : \mathbb{R} \to \mathbb{R}$ is defined by $\phi(c) = c$ and $b(.,.,.) \equiv 1$, then we have*

$$|I_{\xi,\eta}(\alpha,k;x,\lambda,m,c,d)| \le A_3(k,\alpha,\lambda,p)^{\frac{1}{p}}$$

$$\times \left[\left| \frac{\eta^{\frac{\alpha}{k}+2}(x,c,m)}{(\frac{\alpha}{k}+1)\eta(d,c,m)} \right| \left(m|\xi''(c)|^q + \frac{(|\xi''(x)|^q - m|\xi''(c)|^q)}{6} \right)^{\frac{1}{q}} \right.$$

$$\left. + \left| \frac{(-1)^{\frac{\alpha}{k}+2} \eta^{\frac{\alpha}{k}+2}(x,d,m)}{(\frac{\alpha}{k}+1)\eta(d,c,m)} \right| \left(m|\xi''(d)|^q + \frac{(|\xi''(x)|^q - m|\xi''(d)|^q)}{6} \right)^{\frac{1}{q}} \right].$$

Corollary 4.5.15. *In Corollary 4.5.14, if the mapping $\eta(d,c,m) = d - mc$ together with $m = 1$, choosing $x = \frac{c+d}{2}$, we have the following inequality:*

$$\left| \frac{2^{\frac{\alpha}{k}-1}}{(d-c)^{\frac{\alpha}{k}-1}} I_{\xi} \left(\alpha, k; \frac{c+d}{2}, \lambda, 1, c, d \right) \right|$$

$$\le \frac{(d-c)^2}{8(\frac{\alpha}{k}+1)} A_3(k,\alpha,\lambda,p)^{\frac{1}{p}} \left[\left(|\xi''(c)|^q + \frac{(|\xi''(\frac{c+d}{2})|^q - |\xi''(c)|^q)}{6} \right)^{\frac{1}{q}} \right.$$

$$\left. + \left(|\xi''(d)|^q + \frac{(|\xi''(\frac{c+d}{2})|^q - |\xi''(d)|^q)}{6} \right)^{\frac{1}{q}} \right].$$

Remark 4.5.6. *In Corollary 4.5.15,*

(1) if $\lambda = 0$, then we have

$$\left| \xi \left(\frac{c+d}{2} \right) - \frac{2^{\frac{\alpha}{k}-1} \Gamma_k(\alpha+k)}{(d-c)^{\frac{\alpha}{k}}} [{}_k J^{\alpha}_{(\frac{c+d}{2})-} \xi(c) + {}_k J^{\alpha}_{(\frac{c+d}{2})+} \xi(d)] \right|$$

$$\le \frac{(d-c)^2}{8(\frac{\alpha}{k}+1)} \left[\frac{1}{p(\frac{\alpha}{k}+1)+1} \right]^{\frac{1}{p}} \left[\left(|\xi''(c)|^q + \frac{(|\xi''(\frac{c+d}{2})|^q - |\xi''(c)|^q)}{6} \right)^{\frac{1}{q}} \right.$$

$$\left. + \left(|\xi''(d)|^q + \frac{(|\xi''(\frac{c+d}{2})|^q - |\xi''(d)|^q)}{6} \right)^{\frac{1}{q}} \right],$$

(2) if we choose $k = \alpha = 1$ and $\lambda = 0$, then we get

$$\left| \xi\left(\frac{c+d}{2}\right) - \frac{1}{d-c}\int_c^d \xi(x)dx \right|$$

$$\leq \frac{(d-c)^2}{16}\left(\frac{1}{2p+1}\right)^{\frac{1}{p}}\left[\left(|\xi''(c)|^q + \frac{(|\xi''(\frac{c+d}{2})|^q - |\xi''(c)|^q)}{6}\right)^{\frac{1}{q}}\right.$$

$$\left. + \left(|\xi''(d)|^q + \frac{(|\xi''(\frac{c+d}{2})|^q - |\xi''(d)|^q)}{6}\right)^{\frac{1}{q}}\right],$$

(3) if $\lambda = 1$, then we have

$$\left| \frac{\xi(c)+\xi(d)}{2} - \frac{2^{\frac{\alpha}{k}-1}\Gamma_k(\alpha+k)}{(d-c)^{\frac{\alpha}{k}}}[{}_kJ^\alpha_{(\frac{c+d}{2})-}\xi(c) + {}_kJ^\alpha_{(\frac{c+d}{2})+}\xi(d)] \right|$$

$$\leq \frac{(d-c)^2}{8(\frac{\alpha}{k}+1)}A_3(k,\alpha,1,p)^{\frac{1}{p}}\left[\left(|\xi''(c)|^q + \frac{(|\xi''(\frac{c+d}{2})|^q - |\xi''(c)|^q)}{6}\right)^{\frac{1}{q}}\right.$$

$$\left. + \left(|\xi''(d)|^q + \frac{(|\xi''(\frac{c+d}{2})|^q - |\xi''(d)|^q)}{6}\right)^{\frac{1}{q}}\right],$$

(4) if we choose $k = \alpha = \lambda = 1$, then we get

$$\left| \frac{\xi(c)+\xi(d)}{2} - \frac{1}{d-c}\int_c^d \xi(x)dx \right| \leq \frac{(d-c)^2}{16}\left[2^{1+2p}\beta\left(\frac{1}{2};1+p,1+p\right)\right]^{\frac{1}{p}}$$

$$\times\left[\left(|\xi''(c)|^q + \frac{(|\xi''(\frac{c+d}{2})|^q - |\xi''(c)|^q)}{6}\right)^{\frac{1}{q}}\right.$$

$$\left. + \left(|\xi''(d)|^q + \frac{(|\xi''(\frac{c+d}{2})|^q - |\xi''(d)|^q)}{6}\right)^{\frac{1}{q}}\right],$$

(5) if $\lambda = \frac{1}{3}$, then we have

$$\left| \frac{1}{6}\left[\xi(c) + 4\xi\left(\frac{c+d}{2}\right) + \xi(d)\right] \right.$$

$$\left. - \frac{2^{\frac{\alpha}{k}-1}\Gamma_k(\alpha+k)}{(d-c)^{\frac{\alpha}{k}}}[{}_kJ^\alpha_{(\frac{c+d}{2})-}\xi(c) + {}_kJ^\alpha_{(\frac{c+d}{2})+}\xi(d)] \right|$$

$$\leq \frac{(d-c)^2}{8(\frac{\alpha}{k}+1)} A_3\left(k,\alpha,\frac{1}{3},p\right)^{\frac{1}{p}}\left[\left(|\xi''(c)|^q + \frac{(|\xi''(\frac{c+d}{2})|^q - |\xi''(c)|^q)}{6}\right)^{\frac{1}{q}}\right.$$

$$\left. + \left(|\xi''(d)|^q + \frac{(|\xi''(\frac{c+d}{2})|^q - |\xi''(d)|^q)}{6}\right)^{\frac{1}{q}}\right],$$

(6) if we choose $k = \alpha = 1$ and $\lambda = \frac{1}{3}$, then we get

$$\left| \frac{1}{6}\left[\xi(c) + 4\xi\left(\frac{c+d}{2}\right) + \xi(d)\right] - \frac{1}{d-c}\int_c^d \xi(x)dx \right|$$

$$\leq \frac{(d-c)^2}{16} A_3\left(1,1,\frac{1}{3},p\right)^{\frac{1}{p}}\left[\left(|\xi''(c)|^q + \frac{(|\xi''(\frac{c+d}{2})|^q - |\xi''(c)|^q)}{6}\right)^{\frac{1}{q}}\right.$$

$$\left. + \left(|\xi''(d)|^q + \frac{(|\xi''(\frac{c+d}{2})|^q - |\xi''(d)|^q)}{6}\right)^{\frac{1}{q}}\right],$$

(7) if $\lambda = \frac{1}{2}$, then we have

$$\left| \frac{1}{4}\left[\xi(c) + 2\xi\left(\frac{c+d}{2}\right) + \xi(d)\right] \right.$$

$$\left. - \frac{2^{\frac{\alpha}{k}-1}\Gamma_k(\alpha+k)}{(d-c)^{\frac{\alpha}{k}}}\left[{}_kJ^\alpha_{(\frac{c+d}{2})^-}\xi(c) + {}_kJ^\alpha_{(\frac{c+d}{2})^+}\xi(d)\right] \right|$$

$$\leq \frac{(d-c)^2}{8(\frac{\alpha}{k}+1)} A_3\left(k,\alpha,\frac{1}{2},p\right)^{\frac{1}{p}}\left[\left(|\xi''(c)|^q + \frac{(|\xi''(\frac{c+d}{2})|^q - |\xi''(c)|^q)}{6}\right)^{\frac{1}{q}}\right.$$

$$\left. + \left(|\xi''(d)|^q + \frac{(|\xi''(\frac{c+d}{2})|^q - |\xi''(d)|^q)}{6}\right)^{\frac{1}{q}}\right],$$

(8) if we choose $k = \alpha = 1$ and $\lambda = \frac{1}{2}$, then we get

$$\left| \frac{1}{4}\left[\xi(c) + 2\xi\left(\frac{c+d}{2}\right) + \xi(d)\right] - \frac{1}{d-c}\int_c^d \xi(x)dx \right|$$

$$\leq \frac{(d-c)^2}{16}\beta^{\frac{1}{p}}(1+p,1+p)\left[\left(|\xi''(c)|^q + \frac{(|\xi''(\frac{c+d}{2})|^q - |\xi''(c)|^q)}{6}\right)^{\frac{1}{q}}\right.$$

$$\left. + \left(|\xi''(d)|^q + \frac{(|\xi''(\frac{c+d}{2})|^q - |\xi''(d)|^q)}{6}\right)^{\frac{1}{q}}\right].$$

Corollary 4.5.16. *In Theorem 4.5.2, if we use the generalized $m - MT$-preunivexity of $|\xi''|^q$ along with $q > 1$ and $p = \frac{q}{q-1}$, then for $m \in (0,1]$, we have the following inequality:*

$$|I_{\xi,\eta}(\alpha,k;x,\lambda,m,c,d)| \le A_3(k,\alpha,\lambda,p)^{\frac{1}{p}} \left[\left| \frac{\eta^{\frac{\alpha}{k}+2}(x,c,m)}{(\frac{\alpha}{k}+1)\eta(d,c,m)} \right| (m|\xi''(c)|^q \right.$$

$$+ \phi(|\xi''(x)|^q - m|\xi''(c)|^q) \int_0^1 \frac{\sqrt{\delta}}{2\sqrt{(1-\delta)}} b\left(x,c,\frac{\sqrt{\delta}}{2\sqrt{(1-\delta)}}\right) d\delta \bigg)^{\frac{1}{q}}$$

$$+ \left| \frac{(-1)^{\frac{\alpha}{k}+2}\eta^{\frac{\alpha}{k}+2}(x,d,m)}{(\frac{\alpha}{k}+1)\eta(d,c,m)} \right| (m|\xi''(d)|^q$$

$$\left. + \phi(|\xi''(x)|^q - m|\xi''(d)|^q) \int_0^1 \frac{\sqrt{\delta}}{2\sqrt{(1-\delta)}} b\left(x,d,\frac{\sqrt{\delta}}{2\sqrt{(1-\delta)}}\right) d\delta \bigg)^{\frac{1}{q}} \right].$$

Corollary 4.5.17. *In Corollary 4.5.16, if $\phi : \mathbb{R} \to \mathbb{R}$ is defined by $\phi(c) = c$ and $b(.,.,.) \equiv 1$, then we have*

$$|I_{\xi,\eta}(\alpha,k;x,\lambda,m,c,d)| \le A_3(k,\alpha,\lambda,p)^{\frac{1}{p}}$$

$$\times \left[\left| \frac{\eta^{\frac{\alpha}{k}+2}(x,c,m)}{(\frac{\alpha}{k}+1)\eta(d,c,m)} \right| \left(m|\xi''(c)|^q + \frac{\pi}{4}(|\xi''(x)|^q - m|\xi''(c)|^q) \right)^{\frac{1}{q}} \right.$$

$$\left. + \left| \frac{(-1)^{\frac{\alpha}{k}+2}\eta^{\frac{\alpha}{k}+2}(x,d,m)}{(\frac{\alpha}{k}+1)\eta(d,c,m)} \right| \left(m|\xi''(d)|^q + \frac{\pi}{4}(|\xi''(x)|^q - m|\xi''(d)|^q) \right)^{\frac{1}{q}} \right].$$

Corollary 4.5.18. *In Corollary 4.5.17, if the mapping $\eta(d,c,m) = d - mc$ together with $m = 1$, choosing $x = \frac{c+d}{2}$, we have the following inequality:*

$$\left| \frac{2^{\frac{\alpha}{k}-1}}{(d-c)^{\frac{\alpha}{k}-1}} I_\xi\left(\alpha,k;\frac{c+d}{2},\lambda,1,c,d\right) \right|$$

$$\le \frac{(d-c)^2}{8(\frac{\alpha}{k}+1)} A_3(k,\alpha,\lambda,p)^{\frac{1}{p}} \left[\left(|\xi''(c)|^q + \frac{\pi}{4}\left(\left|\xi''\left(\frac{c+d}{2}\right)\right|^q - |\xi''(c)|^q \right) \right)^{\frac{1}{q}} \right.$$

$$\left. + \left(|\xi''(d)|^q + \frac{\pi}{4}\left(\left|\xi''\left(\frac{c+d}{2}\right)\right|^q - |\xi''(d)|^q \right) \right)^{\frac{1}{q}} \right].$$

Remark 4.5.7. *In Corollary 4.5.18,*

(1) if $\lambda = 0$, then we have

$$\left| \xi\left(\frac{c+d}{2}\right) - \frac{2^{\frac{\alpha}{k}-1}\Gamma_k(\alpha+k)}{(d-c)^{\frac{\alpha}{k}}} [_k J^\alpha_{(\frac{c+d}{2})^-}\xi(c) + {_k J^\alpha_{(\frac{c+d}{2})^+}}\xi(d)] \right|$$

$$\le \frac{(d-c)^2}{8(\frac{\alpha}{k}+1)} \left[\frac{1}{p(\frac{\alpha}{k}+1)+1} \right]^{\frac{1}{p}} \left[\left(|\xi''(c)|^q + \frac{\pi}{4}\left(\left|\xi''\left(\frac{c+d}{2}\right)\right|^q - |\xi''(c)|^q \right) \right)^{\frac{1}{q}} \right.$$

$$\left. + \left(|\xi''(d)|^q + \frac{\pi}{4}\left(\left|\xi''\left(\frac{c+d}{2}\right)\right|^q - |\xi''(d)|^q \right) \right)^{\frac{1}{q}} \right],$$

(2) if we choose $k = \alpha = 1$ and $\lambda = 0$, then we get

$$\left| \xi\left(\frac{c+d}{2}\right) - \frac{1}{d-c}\int_c^d \xi(x)dx \right|$$

$$\leq \frac{(d-c)^2}{16}\left(\frac{1}{2p+1}\right)^{\frac{1}{p}}\left[\left(|\xi''(c)|^q + \frac{\pi}{4}\left(\left|\xi''\left(\frac{c+d}{2}\right)\right|^q - |\xi''(c)|^q\right)\right)^{\frac{1}{q}}\right.$$

$$\left. + \left(|\xi''(d)|^q + \frac{\pi}{4}\left(\left|\xi''\left(\frac{c+d}{2}\right)\right|^q - |\xi''(d)|^q\right)\right)^{\frac{1}{q}}\right],$$

(3) if $\lambda = 1$, then we have

$$\left| \frac{\xi(c)+\xi(d)}{2} - \frac{2^{\frac{\alpha}{k}-1}\Gamma_k(\alpha+k)}{(d-c)^{\frac{\alpha}{k}}}[_kJ^\alpha_{(\frac{c+d}{2})^-}\xi(c) + \,_kJ^\alpha_{(\frac{c+d}{2})^+}\xi(d)] \right|$$

$$\leq \frac{(d-c)^2}{8(\frac{\alpha}{k}+1)}A_3(k,\alpha,1,p)^{\frac{1}{p}}\left[\left(|\xi''(c)|^q + \frac{\pi}{4}\left(\left|\xi''\left(\frac{c+d}{2}\right)\right|^q - |\xi''(c)|^q\right)\right)^{\frac{1}{q}}\right.$$

$$\left. + \left(|\xi''(d)|^q + \frac{\pi}{4}\left(\left|\xi''\left(\frac{c+d}{2}\right)\right|^q - |\xi''(d)|^q\right)\right)^{\frac{1}{q}}\right],$$

(4) if we choose $k = \alpha = \lambda = 1$, then we get

$$\left| \frac{\xi(c)+\xi(d)}{2} - \frac{1}{d-c}\int_c^d \xi(x)dx \right| \leq \frac{(d-c)^2}{16}\left[2^{1+2p}\beta\left(\frac{1}{2};1+p,1+p\right)\right]^{\frac{1}{p}}$$

$$\times\left[\left(|\xi''(c)|^q + \frac{\pi}{4}\left(\left|\xi''\left(\frac{c+d}{2}\right)\right|^q - |\xi''(c)|^q\right)\right)^{\frac{1}{q}}\right.$$

$$\left. + \left(|\xi''(d)|^q + \frac{\pi}{4}\left(\left|\xi''\left(\frac{c+d}{2}\right)\right|^q - |\xi''(d)|^q\right)\right)^{\frac{1}{q}}\right],$$

(5) if $\lambda = \frac{1}{3}$, then we have

$$\left| \frac{1}{6}\left[\xi(c) + 4\xi\left(\frac{c+d}{2}\right) + \xi(d)\right]\right.$$

$$\left. - \frac{2^{\frac{\alpha}{k}-1}\Gamma_k(\alpha+k)}{(d-c)^{\frac{\alpha}{k}}}[_kJ^\alpha_{(\frac{c+d}{2})^-}\xi(c) + \,_kJ^\alpha_{(\frac{c+d}{2})^+}\xi(d)] \right|$$

$$\leq \frac{(d-c)^2}{8(\frac{\alpha}{k}+1)}A_3\left(k,\alpha,\frac{1}{3},p\right)^{\frac{1}{p}}\left[\left(|\xi''(c)|^q + \frac{\pi}{4}\left(\left|\xi''\left(\frac{c+d}{2}\right)\right|^q - |\xi''(c)|^q\right)\right)^{\frac{1}{q}}\right.$$

$$\left. + \left(|\xi''(d)|^q + \frac{\pi}{4}\left(\left|\xi''\left(\frac{c+d}{2}\right)\right|^q - |\xi''(d)|^q\right)\right)^{\frac{1}{q}}\right],$$

(6) if we choose $k = \alpha = 1$ and $\lambda = \frac{1}{3}$, then we get

$$\left| \frac{1}{6} \left[\xi(c) + 4\xi\left(\frac{c+d}{2}\right) + \xi(d) \right] - \frac{1}{d-c} \int_c^d \xi(x)dx \right|$$

$$\leq \frac{(d-c)^2}{16} A_3\left(1,1,\frac{1}{3},p\right)^{\frac{1}{p}} \left[\left(|\xi''(c)|^q + \frac{\pi}{4}\left(\left|\xi''\left(\frac{c+d}{2}\right)\right|^q - |\xi''(c)|^q\right) \right)^{\frac{1}{q}} \right.$$

$$\left. + \left(|\xi''(d)|^q + \frac{\pi}{4}\left(\left|\xi''\left(\frac{c+d}{2}\right)\right|^q - |\xi''(d)|^q\right) \right)^{\frac{1}{q}} \right],$$

(7) if $\lambda = \frac{1}{2}$, then we have

$$\left| \frac{1}{4}\left[\xi(c) + 2\xi\left(\frac{c+d}{2}\right) + \xi(d) \right] \right.$$

$$\left. - \frac{2^{\frac{\alpha}{k}-1}\Gamma_k(\alpha+k)}{(d-c)^{\frac{\alpha}{k}}} \left[{}_kJ^{\alpha}_{\left(\frac{c+d}{2}\right)^-}\xi(c) + {}_kJ^{\alpha}_{\left(\frac{c+d}{2}\right)^+}\xi(d) \right] \right|$$

$$\leq \frac{(d-c)^2}{8(\frac{\alpha}{k}+1)} A_3\left(k,\alpha,\frac{1}{2},p\right)^{\frac{1}{p}} \left[\left(|\xi''(c)|^q + \frac{\pi}{4}\left(\left|\xi''\left(\frac{c+d}{2}\right)\right|^q - |\xi''(c)|^q\right) \right)^{\frac{1}{q}} \right.$$

$$\left. + \left(|\xi''(d)|^q + \frac{\pi}{4}\left(\left|\xi''\left(\frac{c+d}{2}\right)\right|^q - |\xi''(d)|^q\right) \right)^{\frac{1}{q}} \right],$$

(8) if we choose $k = \alpha = 1$ and $\lambda = \frac{1}{2}$, then we get

$$\left| \frac{1}{4}\left[\xi(c) + 2\xi\left(\frac{c+d}{2}\right) + \xi(d) \right] - \frac{1}{d-c}\int_c^d \xi(x)dx \right|$$

$$\leq \frac{(d-c)^2}{16} \beta^{\frac{1}{p}}(1+p,1+p) \left[\left(|\xi''(c)|^q + \frac{\pi}{4}\left(\left|\xi''\left(\frac{c+d}{2}\right)\right|^q - |\xi''(c)|^q\right) \right)^{\frac{1}{q}} \right.$$

$$\left. + \left(|\xi''(d)|^q + \frac{\pi}{4}\left(\left|\xi''\left(\frac{c+d}{2}\right)\right|^q - |\xi''(d)|^q\right) \right)^{\frac{1}{q}} \right].$$

Chapter 5

Some Majorization Integral Inequalities for Functions Defined on Rectangles via Strong Convexity

5.1 Introduction

Hardy *et al.* [58] introduced the notion of majorization and showed that a necessary and sufficient condition that $u \prec v$ is that there exists a doubly stochastic matrix $\tilde{P}$ such that $u = v\tilde{P}$. Schur [150] proved that the eigen values majorize the diagonal elements of a positive semidefinite Hermitian matrix. Majorization theory has interesting applications in different fields of mathematics, such as linear algebra, geometry, probability, statistics, group theory, optimization, etc. Niezgoda and Pečarić [115] extended the Hardy–Littlewood–Polya theorem on majorization from convex functions to invex ones and considered some variants for pseudo invex and quasi invex functions. For more details, we can refer to [41, 77, 79–82, 95, 103, 133, 150, 179, 180].

Khan *et al.* [84] introduced a new class of functions known as coordinate strongly convex function. It is well known that a twice differentiable function $\xi : \Delta = [a,b] \times [c,d] \to \mathbb{R}$ is coordinate strongly convex with respect to $\mu_1, \mu_2 > 0$ on Δ if and only if the partial mappings $\xi_y : [a,b] \to \mathbb{R}$ defined by $\xi_y(u) = \xi(u,y)$ and $\xi_x : [c,d] \to \mathbb{R}$ defined by $\xi_x(v) = \xi(x,v)$, satisfied $\xi_y''(u) \geq 2\mu_1$ and $\xi_x''(v) \geq 2\mu_2$ for all $u \in [a,b]$ and $v \in [c,d]$ (see, [84]). Further, Wu *et al.* [174] established some defined versions of majorization inequality involving twice differentiable convex functions by using Taylor's theorem with mean value form of the remainder and also given some interesting applications. In particular, many important inequalities can be found in the literature [1, 30, 31, 74–76, 78, 83].

Recently, Zaheer Ullah *et al.* [179] established a monotonicity property for the function involving the strongly convex function and proved the classical majorization theorem by using strongly convex functions for majorized n-tuples. Zaheer Ullah *et al.* [180] obtained integral majorization type and generalized Favard's inequalities for the class of strongly convex functions. Further, Wu *et al.* [173] established some majorization integral inequalities and Favard type inequalities for functions defined on rectangles.

DOI: 10.1201/9781003408284-5

The organization of this chapter is as follows: In Section 5.2, we recall some basic results that are necessary for our main results. In Section 5.3, we extend several integral majorization type and generalized Favard's inequalities from functions defined on intervals to functions defined on rectangles via strong convexity. The results obtained in this paper are the generalizations of the results given in [173, 180].

5.2 Preliminaries

In this section, we recall some basic results which are useful for our main results.

Maligranda *et al.* [102] proved the following lemma of weighted version of majorization.

Lemma 5.2.1. *Let ψ be a weight function on $[a, b]$.*

(a) If l is decreasing function on $[a, b]$, then

$$\int_a^b l(u)\psi(u)du \int_a^x \psi(u)du \leq \int_a^x l(u)\psi(u)du \int_a^b \psi(u)du \quad for\ all\ x \in [a, b].$$

(b) If l is increasing function on $[a, b]$, then

$$\int_a^x l(u)\psi(u)du \int_a^b \psi(u)du \leq \int_a^b l(u)\psi(u)du \int_a^x \psi(u)du \quad for\ all\ x \in [a, b].$$

Zaheer Ullah *et al.* [180] obtained the following results for strongly convex functions.

Lemma 5.2.2. *Let g be a real-valued function defined on $[a, b]$. Then the following statements are true.*

(a) If g be a strongly concave function with modulus μ, then

(i) the function $P_1(u) = g(u)/(u - a) - \mu u$ is decreasing on $(a, b]$;

(ii) the function $Q_1(u) = g(u)/(b - u) + \mu u$ is increasing on $[a, b)$.

(b) If g be a strongly convex function with modulus μ, then

(i) the function $P_1(u) = g(u)/(u - a) - \mu u$ is increasing on $(a, b]$, if $g(a) = 0$;

(ii) the function $Q_1(u) = g(u)/(b - u) + \mu u$ is decreasing on $[a, b)$, if $g(b) = 0$.

Theorem 5.2.1. *Let $\mu > 0, \xi : [0, \infty) \to \mathbb{R}$ be a continuous strongly convex function with modulus μ, and let f, g and w be three positive and integrable functions defined on $[a, b]$ such that*

$$\int_a^x g(u)w(u)du \leq \int_a^x f(u)w(u)du \ \text{ for all } x \in [a, b) \tag{5.1}$$

and

$$\int_a^b g(u)w(u)du = \int_a^b f(u)w(u)du. \tag{5.2}$$

Then the following statements are true.

(a) If g is decreasing on $[a, b]$, then we have

$$\int_a^b \xi\{f(u)\}w(u)du \geq \int_a^b \xi\{g(u)\}w(u)du + \mu \int_a^b \{g(u) - f(u)\}^2 w(u)du. \tag{5.3}$$

(b) If f is increasing on $[a, b]$, then we have

$$\int_a^b \xi\{g(u)\}w(u)du \geq \int_a^b \xi\{f(u)\}w(u)du + \mu \int_a^b \{g(u) - f(u)\}^2 w(u)du. \tag{5.4}$$

The following interesting inequality is given by Favard [49].

Theorem 5.2.2. *Let f be a nonnegative continuous concave function on $[a, b]$, not identically zero, and let ξ be a convex function on $[0, 2\bar{f}]$, where*

$$\bar{f} = \frac{2}{b-a} \int_a^b f(u)du.$$

Then,

$$\frac{1}{b-a} \int_a^b \xi\{f(u)\}du \leq \int_0^1 \xi(s\bar{f}).$$

5.3 Majorization Integral Inequalities for Strong Convexity

In this section, first, we prove some majorization integral inequalities for functions defined on rectangles via strong convexity [154].

Theorem 5.3.1. *Let w and ρ be positive continuous functions on $[a,b]$ and $[c,d]$, respectively, and let f, g and h, k be positive differentiable functions on $[a,b]$ and $[c,d]$, respectively. Suppose that $\xi : [0,\infty) \times [0,\infty) \to \mathbb{R}$ is a strongly convex function with modulus $\mu > 0$ and that*

$$\int_a^x g(u)w(u)du \leq \int_a^x f(u)w(u)du \quad \text{for all } x \in [a,b),$$

$$\int_c^y k(v)\rho(v)dv \leq \int_c^y h(v)\rho(v)dv \quad \text{for all } y \in [c,d),$$

$$\int_a^b g(u)w(u)du = \int_a^b f(u)w(u)du$$

and

$$\int_c^d k(v)\rho(v)dv = \int_c^d h(v)\rho(v)dv.$$

(a) If g and k are decreasing functions on $[a,b]$ and $[c,d]$, respectively, then

$$\int_a^b \int_c^d \xi(g(u),k(v))w(u)\rho(v)dudv \leq \int_a^b \int_c^d \xi(f(u),h(v))w(u)\rho(v)dudv$$
$$- \mu \int_a^b \int_c^d \|(g(u),k(v)) - (f(u),h(v))\|^2 w(u)\rho(v)dudv.$$

(b) If f and h are increasing functions on $[a,b]$ and $[c,d]$, respectively, then

$$\int_a^b \int_c^d \xi(f(u),h(v))w(u)\rho(v)dudv \leq \int_a^b \int_c^d \xi(g(u),k(v))w(u)\rho(v)dudv$$
$$- \mu \int_a^b \int_c^d \|(g(u),k(v)) - (f(u),h(v))\|^2 w(u)\rho(v)dudv.$$

Proof

(a) By the definition of strong convexity, we have

$$\xi(x,y) - \xi(w,z) \geq \langle \nabla_+ \xi(w,z), (x-w,y-z) \rangle + \mu \|(w,z)-(x,y)\|^2$$
$$\text{for all } (x,y),(w,z) \in [0,\infty) \times [0,\infty),$$

that is,

$$\xi(x,y) - \xi(w,z) \geq \frac{\partial \xi_+(w,z)}{\partial w}(x-w) + \frac{\partial \xi_+(w,z)}{\partial z}(y-z)$$
$$+ \mu \|(w,z)-(x,y)\|^2, \quad \text{for all } (x,y),(w,z) \in [0,\infty) \times [0,\infty). \quad (5.5)$$

Putting $x = f(u), y = h(v), w = g(u)$ and $z = k(v)$ in (5.5), we get

$$\xi(f(u), h(v)) - \xi(g(u), k(v))$$
$$\geq \frac{\partial \xi_+(g(u), k(v))}{\partial g(u)}(f(u) - g(u)) + \frac{\partial \xi_+(g(u), k(v))}{\partial k(v)}(h(v) - k(v))$$
$$+ \mu \|((g(u), k(v)) - (f(u), h(v))\|^2.$$

Suppose $\Theta_v^1(u) = \frac{\partial \xi_+(\alpha, \beta)}{\partial \alpha}\big|_{\alpha = g(u), \beta = k(v)}$, $\Theta_v^2(u) = \frac{\partial \xi_+(\alpha, \beta)}{\partial \beta}\big|_{\alpha = g(u), \beta = k(v)}$,
$\Theta_v^3(u) = \frac{\partial^2 \xi_+(\alpha, \beta)}{\partial^2 \alpha}\big|_{\alpha = g(u), \beta = k(v)}$, $\Theta_v^4(u) = \frac{\partial^2 \xi_+(\alpha, \beta)}{\partial^2 \beta}\big|_{\alpha = g(u), \beta = k(v)}$.
Then, we get

$$\xi(f(u), h(v)) - \xi(g(u), k(v))$$
$$\geq \Theta_v^1(u)(f(u) - g(u)) + \Theta_v^2(u)(h(v) - k(v))$$
$$+ \mu \|((g(u), k(v)) - (f(u), h(v))\|^2. \tag{5.6}$$

Assume that,

$$F(x) = \int_a^x (f(u) - g(u))w(u)du \text{ for all } x \in [a, b]$$

and

$$G(y) = \int_c^y (h(v) - k(v))\rho(v)dv \text{ for all } y \in [c, d].$$

From the assumptions of Theorem 5.3.1, we have
$F(x) \geq 0$, $G(y) \geq 0$ for all $x \in [a, b]$, $y \in [c, d]$ and $F(a) = F(b) = G(c) = G(d) = 0$.
Multiplying both sides by $w(u)\rho(v)$ in (5.6), we have

$$[\xi(f(u), h(v)) - \xi(g(u), k(v))]w(u)\rho(v)$$
$$\geq \Theta_v^1(u)[(f(u) - g(u))]w(u)\rho(v) + \Theta_v^2(u)[(h(v) - k(v))]w(u)\rho(v)$$
$$+ \mu \|(g(u), k(v)) - (f(u), h(v))\|^2 w(u)\rho(v).$$

Integrating both sides above inequality, we have

$$\int_a^b \int_c^d [\xi(f(u), h(v)) - \xi(g(u), k(v))]w(u)\rho(v)dudv$$
$$\geq \int_a^b \int_c^d \Theta_v^1(u)[(f(u) - g(u))]w(u)\rho(v)dudv$$
$$+ \int_a^b \int_c^d \Theta_v^2(u)[(h(v) - k(v))]w(u)\rho(v)dudv$$
$$+ \mu \int_a^b \int_c^d \|(g(u), k(v)) - (f(u), h(v))\|^2 w(u)\rho(v)dudv. \tag{5.7}$$

Using Fubini's theorem in above inequality, we get

$$\int_a^b \int_c^d [\xi(f(u), h(v)) - \xi(g(u), k(v))] w(u)\rho(v) du dv$$

$$\geq \int_c^d \rho(v) \left[\int_a^b \Theta_v^1(u) dF(u) \right] dv$$

$$+ \int_a^b w(u) \left[\int_c^d \Theta_v^2(u) dG(v) \right] du$$

$$+ \mu \int_a^b \int_c^d \|(g(u), k(v)) - (f(u), h(v))\|^2 w(u)\rho(v) du dv. \qquad (5.8)$$

This implies,

$$\int_a^b \int_c^d [\xi(f(u), h(v)) - \xi(g(u), k(v))] w(u)\rho(v) du dv$$

$$\geq \int_c^d \rho(v) \left[\Theta_v^1(u) F(u) \Big|_a^b - \int_a^b \Theta_v^3(u) g'(u) F(u) du \right] dv$$

$$+ \int_a^b w(u) \left[\Theta_v^2(u) G(v) \Big|_c^d - \int_c^d \Theta_v^4(u) k'(v) G(v) dv \right] du$$

$$+ \mu \int_a^b \int_c^d \|(g(u), k(v)) - (f(u), h(v))\|^2 w(u)\rho(v) du dv, \qquad (5.9)$$

which yields,

$$\int_a^b \int_c^d [\xi(f(u), h(v)) - \xi(g(u), k(v))] w(u)\rho(v) du dv$$

$$\geq - \int_c^d \int_a^b \Theta_v^3(u) g'(u) F(u)\rho(v) du dv$$

$$- \int_a^b \int_c^d \Theta_v^4(u) k'(v) G(v) w(u) dv du$$

$$+ \mu \int_a^b \int_c^d \|(g(u), k(v)) - (f(u), h(v))\|^2 w(u)\rho(v) du dv. \qquad (5.10)$$

From Lemma 1.1.2, ξ is coordinate strongly convex function, therefore $\Theta_v^3(u) \geq 0$, $\Theta_v^4(u) \geq 0$.
Since, g and k are decreasing functions, therefore $g'(u) \leq 0$ and $k'(v) \leq 0$.
Thus, it follows that

$$- \int_c^d \int_a^b \Theta_v^3(u) g'(u) F(u)\rho(v) du dv \geq 0 \qquad (5.11)$$

and

$$-\int_a^b \int_c^d \Theta_v^4(u)k'(v)G(v)w(u)dvdu \geq 0. \qquad (5.12)$$

From (5.10), (5.11) and (5.12), we get

$$\int_a^b \int_c^d [\xi(f(u),h(v)) - \xi(g(u),k(v))]w(u)\rho(v)dudv$$

$$-\mu \int_a^b \int_c^d \|(g(u),k(v)) - (f(u),h(v))\|^2 w(u)\rho(v)dudv \geq 0. \qquad (5.13)$$

This implies,

$$\int_a^b \int_c^d \xi(g(u),k(v))w(u)\rho(v)dudv \leq \int_a^b \int_c^d \xi(f(u),h(v))w(u)\rho(v)dudv$$

$$-\mu \int_a^b \int_c^d \|(g(u),k(v)) - (f(u),h(v))\|^2 w(u)\rho(v)dudv.$$

(b) The proof of Theorem 5.3.1(b) is similar to Theorem 5.3.1(a).

Remark 5.3.1. *If $\mu = 0$, then above theorem reduces to the Theorem 3.1 of [173], i.e.,*

(a) If g and k are decreasing functions on $[a,b]$ and $[c,d]$, respectively, then

$$\int_a^b \int_c^d \xi(g(u),k(v))w(u)\rho(v)dudv \leq \int_a^b \int_c^d \xi(f(u),h(v))w(u)\rho(v)dudv.$$

(b) If f and h are increasing functions on $[a,b]$ and $[c,d]$, respectively, then

$$\int_a^b \int_c^d \xi(f(u),h(v))w(u)\rho(v)dudv \leq \int_a^b \int_c^d \xi(g(u),k(v))w(u)\rho(v)dudv.$$

Theorem 5.3.2. *Let w and ρ be positive continuous functions on $[a,b]$ and $[c,d]$, respectively, and let f, g and h, k be positive differentiable functions on $[a,b]$ and $[c,d]$, respectively. Suppose that $\xi : [0,\infty) \times [0,\infty) \to \mathbb{R}$ is a strongly convex function with modulus μ.*

(a) Let $\frac{f}{g}$ and $\frac{h}{k}$ be decreasing functions on $[a,b]$ and $[c,d]$, respectively. If f and h are increasing functions on $[a,b]$ and $[c,d]$, respectively, then

$$\int_a^b \int_c^d \xi\left(\frac{f(u)}{\int_a^b f(u)w(u)du}, \frac{h(v)}{\int_c^d h(v)\rho(v)dv}\right) w(u)\rho(v)dudv$$

$$\leq \int_a^b \int_c^d \xi\left(\frac{g(u)}{\int_a^b g(u)w(u)du}, \frac{k(v)}{\int_c^d k(v)\rho(v)dv}\right) w(u)\rho(v)dudv$$

$$
- \mu \int_a^b \int_c^d \left\| \left(\frac{g(u)}{\int_a^b g(u)w(u)du}, \frac{k(v)}{\int_c^d k(v)\rho(v)dv} \right) \right.
$$
$$
\left. - \left(\frac{f(u)}{\int_a^b f(u)w(u)du}, \frac{h(v)}{\int_c^d h(v)\rho(v)dv} \right) \right\|^2
$$
$$
\times w(u)\rho(v)dudv. \tag{5.14}
$$

(b) Let $\frac{f}{g}$ and $\frac{h}{k}$ be increasing functions on $[a,b]$ and $[c,d]$, respectively. If g and k are increasing functions on $[a,b]$ and $[c,d]$, respectively, then

$$
\int_a^b \int_c^d \xi \left(\frac{g(u)}{\int_a^b g(u)w(u)du}, \frac{k(v)}{\int_c^d k(v)\rho(v)dv} \right) w(u)\rho(v)dudv
$$
$$
\leq \int_a^b \int_c^d \xi \left(\frac{f(u)}{\int_a^b f(u)w(u)du}, \frac{h(v)}{\int_c^d h(v)\rho(v)dv} \right) w(u)\rho(v)dudv
$$
$$
- \mu \int_a^b \int_c^d \left\| \left(\frac{g(u)}{\int_a^b g(u)w(u)du}, \frac{k(v)}{\int_c^d k(v)\rho(v)dv} \right) \right.
$$
$$
\left. - \left(\frac{f(u)}{\int_a^b f(u)w(u)du}, \frac{h(v)}{\int_c^d h(v)\rho(v)dv} \right) \right\|^2
$$
$$
\times w(u)\rho(v)dudv. \tag{5.15}
$$

Proof

(a) Applying Lemma 5.2.1(a) with substitution $\psi(u) = g(u)w(u)$ and $l(u) = \frac{f(u)}{g(u)}$, we have

$$
\int_a^b f(u)w(u)du \int_a^x g(u)w(u)du
$$
$$
\leq \int_a^x f(u)w(u)du \int_a^b g(u)w(u)du \quad \text{for all } x \in [a,b],
$$

which yields,

$$
\int_a^x \left(\frac{g(u)}{\int_a^b g(u)w(u)du} \right) w(u)du
$$
$$
\leq \int_a^x \left(\frac{f(u)}{\int_a^b f(u)w(u)du} \right) w(u)du \quad \text{for all } x \in [a,b]. \tag{5.16}
$$

Also, substituting $\psi(v) = k(v)\rho(v)$ and $l(v) = \frac{h(v)}{k(v)}$ in Lemma 5.2.1(a), we have

$$\int_c^d h(v)\rho(v)dv \int_c^y k(v)\rho(v)dv$$

$$\leq \int_c^y h(v)\rho(v)dv \int_c^d k(v)\rho(v)dv \quad \text{for all } y \in [c,d],$$

which yields,

$$\int_c^y \left(\frac{k(v)}{\int_c^d k(v)\rho(v)dv} \right) \rho(v)dv$$

$$\leq \int_c^y \left(\frac{h(v)}{\int_c^d h(v)\rho(v)dv} \right) \rho(v)dv \quad \text{for all } y \in [c,d]. \tag{5.17}$$

Additionally, it is easy to observe that from (5.16) and (5.17), we have

$$\int_a^b \left(\frac{g(u)}{\int_a^b g(u)w(u)du} \right) w(u)du = \int_a^b \left(\frac{f(u)}{\int_a^b f(u)w(u)du} \right) w(u)du, \tag{5.18}$$

$$\int_c^d \left(\frac{k(v)}{\int_c^d k(v)\rho(v)dv} \right) \rho(v)dv = \int_c^d \left(\frac{h(v)}{\int_c^d h(v)\rho(v)dv} \right) \rho(v)dv. \tag{5.19}$$

Applying (5.16), (5.17), (5.18) and (5.19) in Theorem 5.3.1(b), we have

$$\int_a^b \int_c^d \xi \left(\frac{f(u)}{\int_a^b f(u)w(u)du}, \frac{h(v)}{\int_c^d h(v)\rho(v)dv} \right) w(u)\rho(v)dudv$$

$$\leq \int_a^b \int_c^d \xi \left(\frac{g(u)}{\int_a^b g(u)w(u)du}, \frac{k(v)}{\int_c^d k(v)\rho(v)dv} \right) w(u)\rho(v)dudv$$

$$- \mu \int_a^b \int_c^d \left\| \left(\frac{g(u)}{\int_a^b g(u)w(u)du}, \frac{k(v)}{\int_c^d k(v)\rho(v)dv} \right) \right.$$

$$\left. - \left(\frac{f(u)}{\int_a^b f(u)w(u)du}, \frac{h(v)}{\int_c^d h(v)\rho(v)dv} \right) \right\|^2$$

$$\times w(u)\rho(v)dudv. \tag{5.20}$$

(b) The proof of Theorem 5.3.2(b) is similar to Theorem 5.3.2(a).

Remark 5.3.2. *If $\mu = 0$, then above theorem reduces to the Theorem 3.2 of [173], i.e.,*

(a) *Let $\frac{f}{g}$ and $\frac{h}{k}$ be decreasing functions on $[a,b]$ and $[c,d]$, respectively. If f and h are increasing functions on $[a,b]$ and $[c,d]$, respectively, then*

$$\int_a^b \int_c^d \xi \left(\frac{f(u)}{\int_a^b f(u)w(u)du}, \frac{h(v)}{\int_c^d h(v)\rho(v)dv} \right) w(u)\rho(v)dudv$$

$$\leq \int_a^b \int_c^d \xi \left(\frac{g(u)}{\int_a^b g(u)w(u)du}, \frac{k(v)}{\int_c^d k(v)\rho(v)dv} \right) w(u)\rho(v)dudv.$$

(b) *Let $\frac{f}{g}$ and $\frac{h}{k}$ be increasing functions on $[a,b]$ and $[c,d]$, respectively. If g and k are increasing functions on $[a,b]$ and $[c,d]$, respectively, then*

$$\int_a^b \int_c^d \xi \left(\frac{g(u)}{\int_a^b g(u)w(u)du}, \frac{k(v)}{\int_c^d k(v)\rho(v)dv} \right) w(u)\rho(v)dudv$$

$$\leq \int_a^b \int_c^d \xi \left(\frac{f(u)}{\int_a^b f(u)w(u)du}, \frac{h(v)}{\int_c^d h(v)\rho(v)dv} \right) w(u)\rho(v)dudv.$$

Next, we establish some Favard's type inequalities for functions defined on rectangles via strong convexity [154].

Theorem 5.3.3. (a) *Let g and k be strongly concave functions with modulus μ_1 and μ_2 on $[a,b]$ and $[c,d]$, respectively, such that $f(u) = g(u) - \mu_1 u(u-a)$ and $h(v) = k(v) - \mu_2 v(v-c)$ are positive increasing functions. Also suppose ξ be a strongly convex function with modulus μ on $[0, 2\bar{g}_1] \times [0, 2\bar{k}_1]$, $\bar{z}_1 = a(1-\zeta)+b\zeta$, $\bar{z}_2 = c(1-\tau)+d\tau$, $\bar{g}_1 = \frac{(b-a)\int_a^b f(u)w(u)du}{2\int_a^b(u-a)w(u)du}$ and $\bar{k}_1 = \frac{(d-c)\int_c^d h(v)\rho(v)dv}{2\int_c^d(v-c)\rho(v)dv}$. Then*

$$\frac{1}{(b-a)(d-c)} \int_a^b \int_c^d \xi(f(u),h(v))w(u)\rho(v)dudv$$

$$\leq \int_0^1 \int_0^1 \xi(2\bar{g}_1\zeta, 2\bar{k}_1\tau)w(\bar{z}_1)\rho(\bar{z}_2)d\zeta d\tau$$

$$- \mu \int_0^1 \int_0^1 \|(2\bar{g}_1\zeta, 2\bar{k}_1\tau) - (g(\bar{z}_1) - \mu_1\bar{z}_1(b-a)\zeta, k(\bar{z}_2)$$

$$- \mu_2\bar{z}_2(d-c)\tau)\|^2 w(\bar{z}_1)\rho(\bar{z}_2)d\zeta d\tau. \tag{5.21}$$

If g and k be strongly convex functions with modulus μ_1 and μ_2 on $[a,b]$ and $[c,d]$, respectively, such that $f(u) = g(u) - \mu_1 u(u-a)$ and $h(v) = k(v) - \mu_2 v(v-c)$ are positive increasing functions and $g(a) = k(c) = 0$, then the reverse inequality in (5.21) holds.

(b) *Let g and k be strongly concave functions with modulus μ_1 and μ_2 on $[a,b]$ and $[c,d]$, respectively, such that $f(u) = g(u) + \mu_1 u(b-u)$ and $h(v) = k(v) + \mu_2 v(d-v)$ are positive decreasing functions. Also suppose ξ be a strongly convex function with modulus μ on $[0, 2\bar{g}_2] \times [0, 2\bar{k}_2]$, $\bar{\omega}_1 = a\zeta + b(1-\zeta)$, $\bar{\omega}_2 = c\tau + d(1-\tau)$, $\bar{g}_2 = \frac{(b-a)\int_a^b f(u)w(u)du}{2\int_a^b (b-u)w(u)du}$ and $\bar{k}_2 = \frac{(d-c)\int_c^d h(v)\rho(v)dv}{2\int_c^d (d-v)\rho(v)dv}$.
Then*

$$\frac{1}{(b-a)(d-c)} \int_a^b \int_c^d \xi(f(u), h(v))w(u)\rho(v)dudv$$

$$\leq \int_0^1 \int_0^1 \xi(2\bar{g}_2\zeta, 2\bar{k}_2\tau)w(\bar{\omega}_1)\rho(\bar{\omega}_2)d\zeta d\tau$$

$$- \mu \int_0^1 \int_0^1 \|(2\bar{g}_2\zeta, 2\bar{k}_2\tau) - (g(\bar{\omega}_1) - \mu_1\bar{\omega}_1(b-a)\zeta, k(\bar{\omega}_2)$$

$$- \mu_2\bar{\omega}_2(d-c)\tau)\|^2 w(\bar{\omega}_1)\rho(\bar{\omega}_2)d\zeta d\tau. \tag{5.22}$$

If g and k be strongly convex functions with modulus μ_1 and μ_2 on $[a,b]$ and $[c,d]$, respectively, such that $f(u) = g(u) + \mu_1 u(b-u)$ and $h(v) = k(v) + \mu_2 v(d-v)$ are positive increasing functions and $g(a) = k(c) = 0$, then the reverse inequality in (5.22) holds.

Proof

(a) From Lemma 5.2.2, we know that the function $P_1(u) = \frac{g(u)}{u-a} - \mu_1 u$ and $P_2(v) = \frac{k(v)}{v-c} - \mu_2 v$ is decreasing then substituting $\psi(u) = (u-a)w(u)$ and $l(u) = \frac{g(u)}{u-a} - \mu_1 u$ in Lemma 5.2.1(a), we get

$$\int_a^b f(u)w(u)du \int_a^x (u-a)w(u)du$$

$$\leq \int_a^x f(u)w(u)du \int_a^b (u-a)w(u)du \quad \text{for all } x \in [a,b],$$

that is,

$$\int_a^x \frac{(u-a)}{(b-a)} 2\bar{g}_1 w(u)du \leq \int_a^x f(u)w(u)du \quad \text{for all } x \in [a,b]. \tag{5.23}$$

Also, substitute $\psi(v) = (v-c)\rho(v)$ and $l(v) = \frac{k(v)}{v-c} - \mu_2 v$ in Lemma 5.2.1(a), we obtain

$$\int_c^d h(v)\rho(v)dv \int_c^y (v-c)\rho(v)dv$$

$$\leq \int_c^y h(v)\rho(v)dv \int_c^d (v-c)\rho(v)dv \quad \text{for all } y \in [c,d],$$

that is,

$$\int_c^y \frac{(v-c)}{(d-c)} 2\bar{k}_1 \rho(v)dv \le \int_c^y h(v)\rho(v)dv \quad \text{for all } y \in [c,d]. \quad (5.24)$$

Since f and h are increasing functions, therefore from (5.23), (5.24) and Theorem 5.3.1(b), we obtain

$$\int_a^b \int_c^d \xi(f(u), h(v))w(u)\rho(v)dudv$$
$$\le \int_a^b \int_c^d \xi\left(\frac{u-a}{b-a}2\bar{g}_1, \frac{v-c}{d-c}2\bar{k}_1\right)w(u)\rho(v)dudv$$
$$- \mu \int_a^b \int_c^d \left\|\left(\frac{u-a}{b-a}2\bar{g}_1, \frac{v-c}{d-c}2\bar{k}_1\right) - (f(u), h(v))\right\|^2 w(u)\rho(v)dudv.$$
$$(5.25)$$

Multiplying on both sides by $\frac{1}{(b-a)(d-c)}$ in (5.25), we get

$$\frac{1}{(b-a)(d-c)} \int_a^b \int_c^d \xi(f(u), h(v))w(u)\rho(v)dudv$$
$$\le \frac{1}{(b-a)(d-c)} \int_a^b \int_c^d \xi\left(\frac{u-a}{b-a}2\bar{g}_1, \frac{v-c}{d-c}2\bar{k}_1\right)w(u)\rho(v)dudv$$
$$- \frac{\mu}{(b-a)(d-c)} \int_a^b \int_c^d \left\|\left(\frac{u-a}{b-a}2\bar{g}_1, \frac{v-c}{d-c}2\bar{k}_1\right) - (f(u), h(v))\right\|^2$$
$$\times w(u)\rho(v)dudv.$$

Applying change of variable technique in above inequality, we have

$$\frac{1}{(b-a)(d-c)} \int_a^b \int_c^d \xi(f(u), h(v))w(u)\rho(v)dudv$$
$$\le \frac{1}{4\bar{g}_1\bar{k}_1} \int_0^{2\bar{g}_1} \int_0^{2\bar{k}_1} \xi(x,y)w\left(a+\frac{(b-a)x}{2\bar{g}_1}\right)\rho\left(c+\frac{(d-c)y}{2\bar{k}_1}\right)dxdy$$
$$- \frac{\mu}{4\bar{g}_1\bar{k}_1} \int_0^{2\bar{g}_1} \int_0^{2\bar{k}_1} \left\|(x,y) - \left(f\left(a+\frac{(b-a)x}{2\bar{g}_1}\right), h\left(c+\frac{(d-c)y}{2\bar{k}_1}\right)\right)\right\|^2$$
$$\times w\left(a+\frac{(b-a)x}{2\bar{g}_1}\right)\rho\left(c+\frac{(d-c)y}{2\bar{k}_1}\right)dxdy. \quad (5.26)$$

Since,

$$f\left(a+\frac{(b-a)x}{2\bar{g}_1}\right) = g\left(a+\frac{(b-a)x}{2\bar{g}_1}\right) - \mu_1\left(a+\frac{(b-a)x}{2\bar{g}_1}\right)\frac{(b-a)x}{2\bar{g}_1} \quad (5.27)$$

and

$$h\left(c + \frac{(d-c)y}{2\bar{k}_1}\right) = k\left(c + \frac{(d-c)y}{2\bar{k}_1}\right) - \mu_2\left(c + \frac{(d-c)y}{2\bar{k}_1}\right)\frac{(d-c)y}{2\bar{k}_1}.$$
(5.28)

From (5.26), (5.27) and (5.28), we get

$$\frac{1}{(b-a)(d-c)}\int_a^b\int_c^d \xi(f(u),h(v))w(u)\rho(v)dudv$$

$$\leq \frac{1}{4\bar{g}_1\bar{k}_1}\int_0^{2\bar{g}_1}\int_0^{2\bar{k}_1} \xi(x,y)w\left(a+\frac{(b-a)x}{2\bar{g}_1}\right)\rho\left(c+\frac{(d-c)y}{2\bar{k}_1}\right)dxdy$$

$$-\frac{\mu}{4\bar{g}_1\bar{k}_1}\int_0^{2\bar{g}_1}\int_0^{2\bar{k}_1} \left\|(x,y)-\left(g\left(a+\frac{(b-a)x}{2\bar{g}_1}\right)\right.\right.$$

$$-\mu_1\left(a+\frac{(b-a)x}{2\bar{g}_1}\right)\frac{(b-a)x}{2\bar{g}_1}, k\left(c+\frac{(d-c)y}{2\bar{k}_1}\right)$$

$$\left.\left.-\mu_2\left(c+\frac{(d-c)y}{2\bar{k}_1}\right)\frac{(d-c)y}{2\bar{k}_1}\right)\right\|^2 w\left(a+\frac{(b-a)x}{2\bar{g}_1}\right)$$

$$\times \rho\left(c+\frac{(d-c)y}{2\bar{k}_1}\right)dxdy.$$
(5.29)

Applying change of variable technique, we obtain

$$\frac{1}{(b-a)(d-c)}\int_a^b\int_c^d \xi(f(u),h(v))w(u)\rho(v)dudv$$

$$\leq \int_0^1\int_0^1 \xi(2\bar{g}_1\zeta, 2\bar{k}_1\tau)w(a+(b-a)\zeta)\rho(c+(d-c)\tau)d\zeta d\tau$$

$$-\mu\int_0^1\int_0^1 \|(2\bar{g}_1\zeta, 2\bar{k}_1\tau)-(g(a+(b-a)\zeta)$$

$$-\mu_1(a+(b-a)\zeta)(b-a)\zeta, k(c+(d-c)\tau)$$

$$-\mu_2(c+(d-c)\tau)(d-c)\tau)\|^2 w(a+(b-a)\zeta)\rho(c+(d-c)\tau)d\zeta d\tau.$$
(5.30)

This implies,

$$\frac{1}{(b-a)(d-c)}\int_a^b\int_c^d \xi(f(u),h(v))w(u)\rho(v)dudv$$

$$\leq \int_0^1\int_0^1 \xi(2\bar{g}_1\zeta, 2\bar{k}_1\tau)w(\bar{z}_1)\rho(\bar{z}_2)d\zeta d\tau$$

$$-\mu\int_0^1\int_0^1 \|(2\bar{g}_1\zeta, 2\bar{k}_1\tau)-(g(\bar{z}_1)-\mu_1\bar{z}_1(b-a)\zeta, k(\bar{z}_2)$$

$$-\mu_2\bar{z}_2(d-c)\tau)\|^2 w(\bar{z}_1)\rho(\bar{z}_2)d\zeta d\tau.$$
(5.31)

Similarly, we can proved the reverse inequality of (5.21).

(b) The proof of Theorem 5.3.3(b) is similar to Theorem 5.3.3(a).

Corollary 5.3.1. *If $\mu = 0$, then we have the following new results:*

(a)

$$\frac{1}{(b-a)(d-c)} \int_a^b \int_c^d \xi(f(u), h(v)) w(u) \rho(v) du dv$$
$$\leq \int_0^1 \int_0^1 \xi(2\bar{g}_1 \zeta, 2\bar{k}_1 \tau) w(\bar{z}_1) \rho(\bar{z}_2) d\zeta d\tau.$$

(b)

$$\frac{1}{(b-a)(d-c)} \int_a^b \int_c^d \xi(f(u), h(v)) w(u) \rho(v) du dv$$
$$\leq \int_0^1 \int_0^1 \xi(2\bar{g}_2 \zeta, 2\bar{k}_2 \tau) w(\bar{\omega}_1) \rho(\bar{\omega}_2) d\zeta d\tau.$$

Theorem 5.3.4. *Let $f(u) = g(u) - \mu_1 u(u-a)$, $h(v) = k(v) - \mu_2 v(v-c)$ are increasing functions on $(0,1)$ and $\frac{f}{P}$, $\frac{h}{Q}$ be decreasing functions on $(0,1)$. Also suppose f, h, P, Q, w and ρ are positive functions on $(0,1)$, and $fw, h\rho, Pw$ and $Q\rho$ are integrable on $(0,1)$ such that*

$$\phi = \frac{\int_0^1 f(u)w(u)du}{\int_0^1 P(u)w(u)du} \geq 0 \text{ and } \varphi = \frac{\int_0^1 h(v)\rho(v)dv}{\int_0^1 Q(v)\rho(v)dv} \geq 0. \quad (5.32)$$

Suppose ξ be a strongly convex function with modulus μ. Then,

$$\int_0^1 \int_0^1 \xi(mf(u), nh(v)) w(u) \rho(v) du dv$$
$$\leq \int_0^1 \int_0^1 \xi(m\phi P(u), n\varphi Q(v)) w(u) \rho(v) du dv$$
$$- \mu \int_0^1 \int_0^1 \|(m\phi P(u), n\varphi Q(v)) - (mf(u), nh(v))\|^2 w(u)\rho(v) du dv$$
for all $m, n > 0$.

Proof From $P > 0$ and (5.32), substituting $\psi(u) = P(u)w(u)$ and $l(u) = f(u)/P(u)$, in Lemma 5.2.1(a), we get

$$\int_0^x m\phi P(u)w(u)du \leq \int_0^x mf(u)w(u)du. \quad (5.33)$$

Similarly, $Q > 0$ and (5.32), substituting $\psi(v) = Q(v)\rho(v)$ and $l(v) = h(v)/Q(v)$, in Lemma 5.2.1(a), we get

$$\int_0^y n\varphi Q(v)\rho(v)dv \leq \int_0^y nh(v)\rho(v)dv. \quad (5.34)$$

Since f and h are increasing functions, therefore by using Theorem 5.3.1(b), we get

$$\int_0^1 \int_0^1 \xi(mf(u), nh(v))w(u)\rho(v)dudv$$

$$\leq \int_0^1 \int_0^1 \xi(m\phi P(u), n\varphi Q(v))w(u)\rho(v)dudv$$

$$- \mu \int_0^1 \int_0^1 \|(m\phi P(u), n\varphi Q(v)) - (mf(u), nh(v))\|^2 w(u)\rho(v)dudv.$$

Corollary 5.3.2. *If $\mu = 0$, then we have the following new result:*

$$\int_0^1 \int_0^1 \xi(mf(u), nh(v))w(u)\rho(v)dudv$$

$$\leq \int_0^1 \int_0^1 \xi(m\phi P(u), n\varphi Q(v))w(u)\rho(v)dudv$$

$$- \mu \int_0^1 \int_0^1 \|(m\phi P(u), n\varphi Q(v)) - (mf(u), nh(v))\|^2 w(u)\rho(v)dudv$$

for all $m, n > 0$.

Theorem 5.3.5. *Let $\xi : [0, \infty) \to \mathbb{R}$ be a continuous strongly convex function with modulus μ, $f(u) = g(u) - \mu_1 u(u - a)$ and $h(v) = k(v) - \mu_2 v(v - c)$, P, w and Q, ρ are positive integrable functions on $[a,b]$ and $[c,d]$, respectively, $z_1(u) = \frac{f(u)}{\int_a^b f(u)w(u)du}$, $z_2(u) = \frac{P(u)}{\int_a^b P(u)w(u)du}$, $\omega_1(v) = \frac{h(v)}{\int_c^d h(v)\rho(v)dv}$, and $\omega_2(v) = \frac{Q(v)}{\int_c^d Q(v)\rho(v)dv}$. Then the following statements are true.*

(a) If f and h are increasing on $[a,b]$ and $[c,d]$, respectively, and f/P and h/Q are decreasing on $[a,b]$ and $[c,d]$, respectively, then

$$\int_a^b \int_c^d \xi(z_1(u), \omega_1(v))w(u)\rho(v)dudv$$

$$\leq \int_a^b \int_c^d \xi(z_2(u), \omega_2(v))w(u)\rho(v)dudv$$

$$- \mu \int_a^b \int_c^d \|(z_2(u), \omega_2(v)) - (z_1(u), \omega_1(v))\|^2 w(u)\rho(v)dudv. \quad (5.35)$$

(b) If P and Q are increasing on $[a,b]$ and $[c,d]$, respectively, and f/P and h/Q are increasing on $[a,b]$ and $[c,d]$, respectively, then

$$\int_a^b \int_c^d \xi(z_2(u), \omega_2(v))w(u)\rho(v)dudv$$

$$\leq \int_a^b \int_c^d \xi(z_1(u), \omega_1(v)) w(u) \rho(v) du dv$$

$$- \mu \int_a^b \int_c^d \|(z_2(u), \omega_2(v)) - (z_1(u), \omega_1(v))\|^2 w(u) \rho(v) du dv. \quad (5.36)$$

Proof

(a) Since $P > 0$, then substituting $\psi(u) = P(u)w(u)$ and $l(u) = f(u)/P(u)$, in Lemma 5.2.1(a), we get

$$\int_a^x z_2(u) w(u) du \leq \int_a^x z_1(u) w(u) du. \quad (5.37)$$

Similarly, $Q > 0$ and (5.32), substituting $\psi(v) = Q(v)\rho(v)$ and $l(v) = h(v)/Q(v)$, in Lemma 5.2.1(a), we get

$$\int_c^y \omega_2(v) \rho(v) dv \leq \int_c^y \omega_1(u) \rho(v) dv. \quad (5.38)$$

f and h are increasing functions, then applying Theorem 5.3.1(b), we have

$$\int_a^b \int_c^d \xi(z_1(u), \omega_1(v)) w(u) \rho(v) du dv$$

$$\leq \int_a^b \int_c^d \xi(z_2(u), \omega_2(v)) w(u) \rho(v) du dv$$

$$- \mu \int_a^b \int_c^d \|(z_2(u), \omega_2(v)) - (z_1(u), \omega_1(v))\|^2 w(u) \rho(v) du dv. \quad (5.39)$$

(b) The proof of Theorem 5.3.5(b) is similar to Theorem 5.3.5(a).

Corollary 5.3.3. *If $\mu = 0$, then we have the following new result:*

(a)

$$\int_a^b \int_c^d \xi(z_1(u), \omega_1(v)) w(u) \rho(v) du dv$$

$$\leq \int_a^b \int_c^d \xi(z_2(u), \omega_2(v)) w(u) \rho(v) du dv.$$

(b)

$$\int_a^b \int_c^d \xi(z_2(u), \omega_2(v)) w(u) \rho(v) du dv$$

$$\leq \int_a^b \int_c^d \xi(z_1(u), \omega_1(v)) w(u) \rho(v) du dv.$$

Chapter 6

Hermite–Hadamard Type Inclusions for Interval-Valued Generalized Preinvex Functions

---

6.1 Introduction

Interval analysis is a new and growing branch of applied mathematics. It is an approach to computing that treats an interval as a new kind of numbers. Moore [109] computed arbitrarily sharp upper and lower bounds on exact solutions of many problems in applied mathematics by using interval arithmetic, interval-valued functions and integrals of interval-valued functions. Moore [109] showed that if a real-valued function $\xi(x)$ satisfies an ordinary Lipschitz condition in K, $|\xi(x) - \xi(y)| \leqq L|x - y|$ for $x, y \in K$, then the united extension of ξ is a Lipschitz interval extension in K.

Bhurjee and Panda [18] defined the interval-valued function in the parametric form and developed a methodology to study the existence of the solution of a general interval optimization problem. Lupulescu [100] introduced the differentiability and integrability for the interval-valued functions on time scales by using the concept of the generalized Hukuhara difference. Chalco-Cano *et al.* [26] proposed a new Ostrowski type inequalities for $gH-$ differentiable interval-valued functions and obtained generalization of the class of real functions which is not necessarily differentiable. Chalco-Cano *et al.* [26] obtained error bounds to quadrature rules for $gH-$differentiable interval-valued functions. Further, Roy and Panda [144] introduced the concept of μ-monotonic property of interval-valued function in the higher dimension and derived some results by using generalized Hukuhara differentiability. For more deatils of interval-valued functions, we refer to [19, 20, 67, 96, 101, 144] and references therein.

An *et al.* [4] introduced (h_1, h_2)-convex interval-valued function and obtained some interval Hermite–Hadamard type inequalities. Further, Budak *et al.* [25] established the Hermite–Hadamard inequality for the convex interval-valued function and for the product of two convex interval-valued functions. Zhao *et al.* [185] introduced the notion of harmonical h-convexity for interval-valued functions and proved some new Hermite–Hadamard type inequalities for the interval Riemann integral. Recently, Zhao *et al.* [183, 186] proposed

DOI: 10.1201/9781003408284-6

the notion of interval-valued convex functions on coordinates and established Hermite–Hadamard type inequalities for these interval-valued coordinated convex functions. Further, Budak *et al.* [24] described a new concept of interval-valued fractional integrals on coordinates and investigated Hermite–Hadamard type inequalities for interval-valued coordinated convex functions using these fractional integrals. Kara *et al.* [70] proved Hermite–Hadamard-Fejér type inclusions for the product of two interval-valued convex functions on coordinates.

The organization of this chapter is as follows: In Section 6.2, we recall some basic results that are necessary for our main results. In Section 6.3, we introduce the concept of (h_1, h_2)-preinvex interval-valued function and establish the Hermite–Hadamard inequality for preinvex interval-valued functions and for the product of two preinvex interval-valued functions via interval-valued Riemann–Liouville fractional integrals. In Section 6.4, we define preinvex interval-valued functions on coordinates in a rectangle from the plane and investigate Hermite–Hadamard type inclusions for coordinated preinvex interval-valued functions. Further, we present Hermite–Hadamard type inclusions for the product of two interval-valued preinvex functions on coordinates. In Section 6.5, we define harmonically h-preinvexity of interval-valued functions and prove fractional Hermite–Hadamard type inclusions for harmonically h-preinvex interval-valued functions. In this way, these findings include several well-known results and newly obtained results of the existing literature as special cases. Some examples are given to confirm our theoretical results.

6.2   Preliminaries

In this section, we mention some definitions and related results required for this chapter. The following theorem gives a relation between interval-Riemann integrable (IR-integrable) and Riemann integrable (R-integrable) functions [110].

Theorem 6.2.1. *Let $\xi : [c, d] \to \mathbb{R}_I$ be an interval-valued function such that $\xi(\delta) = [\underline{\xi}(\delta), \overline{\xi}(\delta)]$. $\xi \in IR_{([c,d])}$ if and only if $\underline{\xi}(\delta), \overline{\xi}(\delta) \in R_{([c,d])}$ and*

$$(IR) \int_c^d \xi(\delta)d\delta = \left[(R) \int_c^d \underline{\xi}(\delta)d\delta, (R) \int_c^d \overline{\xi}(\delta)d\delta \right],$$

where $R_{([c,d])}$ denotes the collection of R-integrable functions.

The following interval-valued left-sided and right-sided Riemann–Liouville fractional integral of function are defined by Lupulescu [101] and Budak *et al.* [25], respectively.

Definition 6.2.1. *Let* $\xi : [c,d] \to \mathbb{R}_I$ *be an interval-valued function such that* $\xi(\delta) = [\underline{\xi}(\delta), \overline{\xi}(\delta)]$ *and* $\xi \in IR_{([c,d])}$. *The interval-valued left-sided Riemann–Liouville fractional integral of function* ξ *is defined by*

$$J_{c+}^{\alpha} \xi(x) = \frac{1}{\Gamma(\alpha)} (IR) \int_{c}^{x} (x - \delta)^{(\alpha-1)} \xi(\delta) d\delta, \quad x > c, \alpha > 0,$$

where $\Gamma(\alpha)$ *is the Gamma function.*

Definition 6.2.2. *Let* $\xi : [c,d] \to \mathbb{R}_I$ *be an interval-valued function such that* $\xi(\delta) = [\underline{\xi}(\delta), \overline{\xi}(\delta)]$ *and* $\xi \in IR_{([c,d])}$. *The interval-valued right-sided Riemann–Liouville fractional integral of function* ξ *is defined by*

$$J_{d-}^{\alpha} \xi(x) = \frac{1}{\Gamma(\alpha)} (IR) \int_{x}^{d} (\delta - x)^{(\alpha-1)} \xi(\delta) d\delta, \quad x < d, \alpha > 0,$$

where $\Gamma(\alpha)$ *is the Gamma function.*

The following result is given by Budak *et al.* [25].

Corollary 6.2.1. *If* $\xi : [c,d] \to \mathbb{R}_I$ *is an interval-valued function such that* $\xi(\delta) = [\underline{\xi}(\delta), \overline{\xi}(\delta)]$ *with* $\underline{\xi}(\delta), \overline{\xi}(\delta) \in R_{([c,d])}$, *then we have*

$$J_{c+}^{\alpha} \xi(x) = [J_{c+}^{\alpha} \underline{\xi}(x), J_{c+}^{\alpha} \overline{\xi}(x)]$$

and

$$J_{d-}^{\alpha} \xi(x) = [J_{d-}^{\alpha} \underline{\xi}(x) J_{d-}^{\alpha} \overline{\xi}(x)].$$

Sadowska [147] gave the following concept of convex interval-valued functions.

Definition 6.2.3. *Let* $\xi : [c,d] \to \mathbb{R}_I^+$ *be an interval-valued function such that* $\xi(x) = [\underline{\xi}(x), \overline{\xi}(x)]$. *We say that* ξ *is convex interval-valued function if*

$$\xi(\delta x + (1 - \delta)y) \supseteq \delta\xi(x) + (1 - \delta)\xi(y), \quad \forall\, \delta \in [0,1] \text{ and } \forall\, x, y \in [c,d].$$

6.3 Hermite–Hadamard Type Inclusions for Interval-Valued Preinvex Functions

In this section, first, we give the definition of interval-valued h-preinvex function and discuss some special cases of interval-valued h-preinvex functions [160].

Definition 6.3.1. *Let* $h : [a,b] \to \mathbb{R}$ *be a nonnegative function,* $(0,1) \subseteq [a,b]$ *and* $h \neq 0$. *Let* $K \subseteq \mathbb{R}$ *be a invex set with respect to* $\eta : K \times K \to \mathbb{R}$, $\xi(x) = [\underline{\xi}(x), \overline{\xi}(x)]$ *be an interval-valued function defined on* K. *We say that* ξ *is* $h-$*preinvex at x with respect to* η *if*

$$\xi(y + \delta\eta(x,y)) \supseteq h(\delta)\xi(x) + h(1 - \delta)\xi(y), \quad \forall\, \delta \in [0,1] \text{ and } \forall\, x, y \in K.$$

Now, we discuss some special cases of interval-valued $h-$preinvex functions.

(1) If $h(\delta) = 1$, then we have the definition of interval-valued $P-$preinvex functions.

(2) If $h(\delta) = \delta$, then we have the definition of interval-valued preinvex functions.

(3) If $h(\delta) = \delta^{-1}$, then we have the definition of interval-valued $Q-$preinvex functions.

(4) If $h(\delta) = \delta^s$ with $s \in (0,1)$, then we have the definition of interval-valued $s-$preinvex functions.

Example 6.3.1. $K = [1,2], \xi(x) = [x, 10 - e^x], \eta(x,y) = x - 2y$ *and* $h(\delta) = \delta$. *Then* ξ *is* $h-$*preinvex interval-valued function with respect to* η.

Theorem 6.3.1. *Let* $h : [a,b] \to \mathbb{R}$ *be a nonnegative function,* $(0,1) \subseteq [c,d]$ *and* $h \neq 0$. *Let* K *be an invex subset of* $\mathbb{R}$ *with respect to* $\eta : K \times K \to \mathbb{R}$ *and* ξ *be an interval-valued function defined on* K. *Then* ξ *is* $h-$*preinvex at* x *if and only if*
$$\underline{\xi}(y + \delta\eta(x,y)) \leq h(\delta)\underline{\xi}(x) + h(1-\delta)\underline{\xi}(y)$$
and
$$\overline{\xi}(y + \delta\eta(x,y)) \geq h(\delta)\overline{\xi}(x) + h(1-\delta)\overline{\xi}(y), \ \forall \ \delta \in [0,1] \ and \ \forall \ x \in K.$$

Now, we establish the Hermite–Hadamard inequalities for the preinvex interval-valued functions [160].

Theorem 6.3.2. *Let* $K \subseteq \mathbb{R}$ *be an open invex subset with respect to* $\eta : K \times K \to \mathbb{R}$ *and* $c, d \in K$ *with* $c < c+\eta(d,c)$. *If* $\xi : [c, c+\eta(d,c)] \to \mathbb{R}_I^+$ *is a preinvex interval-valued function such that* $\xi(\delta) = [\underline{\xi}(\delta), \overline{\xi}(\delta)]$. $\xi \in L^1[c, c+\eta(d,c)]$ *and* η *satisfies Condition* C *and* $\alpha > 0$, *then we have*
$$\xi\left(c + \frac{\eta(d,c)}{2}\right) \supseteq \frac{\Gamma(\alpha+1)}{2\eta^\alpha(d,c)}[J_{c^+}^\alpha \xi(c + \eta(d,c)) + J_{(c+\eta(d,c))^-}^\alpha \xi(c)]$$
$$\supseteq \frac{\xi(c) + \xi(c + \eta(d,c))}{2} \supseteq \frac{\xi(c) + \xi(d)}{2}. \tag{6.1}$$

Proof Since ξ is preinvex interval-valued function, we have
$$\xi\left(x + \frac{1}{2}\eta(y,x)\right) \supseteq \frac{\xi(x) + \xi(y)}{2}, \ \forall \ x,y \in [c, c+\eta(d,c)]. \tag{6.2}$$

Using $x = c + (1-\delta)\eta(d,c)$, $y = c + \delta\eta(d,c)$ and Condition C in (6.2), we get
$$\xi(c + (1-\delta)\eta(d,c) + \frac{1}{2}\eta(c + \delta\eta(d,c), c + (1-\delta)\eta(d,c)))$$
$$\supseteq \frac{\xi(c + (1-\delta)\eta(d,c)) + \xi(c + \delta\eta(d,c))}{2}.$$

This implies,

$$\xi\left(c + \frac{1}{2}\eta(d,c)\right) \supseteq \frac{\xi(c + (1-\delta)\eta(d,c)) + \xi(c + \delta\eta(d,c))}{2}. \qquad (6.3)$$

Multiplying by $\delta^{\alpha-1}$, $\alpha > 0$ on both sides in (6.3), we have

$$\delta^{\alpha-1}\xi\left(c + \frac{1}{2}\eta(d,c)\right) \supseteq \frac{\delta^{\alpha-1}}{2}[\xi(c + (1-\delta)\eta(d,c)) + \xi(c + \delta\eta(d,c))]. \quad (6.4)$$

Integrating above inequality on $[0,1]$, we get

$$(IR)\int_0^1 \delta^{\alpha-1}\xi\left(c + \frac{1}{2}\eta(d,c)\right)d\delta$$

$$\supseteq \frac{1}{2}\left[(IR)\int_0^1 \delta^{\alpha-1}\xi(c + (1-\delta)\eta(d,c))d\delta + (IR)\int_0^1 \delta^{\alpha-1}\xi(c + \delta\eta(d,c))d\delta\right].$$
$$(6.5)$$

Applying Theorem 6.2.1 in above relation, we get

$$(IR)\int_0^1 \delta^{\alpha-1}\xi\left(c + \frac{1}{2}\eta(d,c)\right)d\delta$$

$$= \left[(R)\int_0^1 \delta^{\alpha-1}\underline{\xi}\left(c + \frac{1}{2}\eta(d,c)\right)d\delta, (R)\int_0^1 \delta^{\alpha-1}\overline{\xi}\left(c + \frac{1}{2}\eta(d,c)\right)d\delta\right]$$

$$= \left[\underline{\xi}\left(c + \frac{1}{2}\eta(d,c)\right)(R)\int_0^1 \delta^{\alpha-1}d\delta, \overline{\xi}\left(c + \frac{1}{2}\eta(d,c)\right)(R)\int_0^1 \delta^{\alpha-1}d\delta\right]$$

$$= \left[\frac{1}{\alpha}\underline{\xi}\left(c + \frac{1}{2}\eta(d,c)\right), \frac{1}{\alpha}\overline{\xi}\left(c + \frac{1}{2}\eta(d,c)\right)\right]$$

$$= \frac{1}{\alpha}\xi\left(c + \frac{1}{2}\eta(d,c)\right). \qquad (6.6)$$

$$(IR)\int_0^1 \delta^{\alpha-1}\xi(c + \delta\eta(d,c))d\delta$$

$$= \left[(R)\int_0^1 \delta^{\alpha-1}\underline{\xi}(c + \delta\eta(d,c))d\delta, (R)\int_0^1 \delta^{\alpha-1}\overline{\xi}(c + \delta\eta(d,c))d\delta\right].$$

This implies,

$$(IR) \int_0^1 \delta^{\alpha-1} \xi(c + \delta\eta(d,c)) d\delta$$

$$= \left[\frac{1}{\eta^\alpha(d,c)} (R) \int_c^{c+\eta(d,c)} (v-c)^{\alpha-1} \underline{\xi}(v) dv, \right.$$

$$\left. \frac{1}{\eta^\alpha(d,c)} (R) \int_c^{c+\eta(d,c)} (v-c)^{\alpha-1} \overline{\xi}(v) dv \right]$$

$$= \frac{\Gamma(\alpha)}{\eta^\alpha(d,c)} [J^\alpha_{(c+\eta(d,c))^-} \underline{\xi}(c), J^\alpha_{(c+\eta(d,c))^-} \overline{\xi}(c)]$$

$$= \frac{\Gamma(\alpha)}{\eta^\alpha(d,c)} J^\alpha_{(c+\eta(d,c))^-} \xi(c). \tag{6.7}$$

Similarly,

$$(IR) \int_0^1 \delta^{\alpha-1} \xi(c + (1-\delta)\eta(d,c)) d\delta = \frac{\Gamma(\alpha)}{\eta^\alpha(d,c)} J^\alpha_{c^+} \xi(c + \eta(d,c)). \tag{6.8}$$

Using (6.6), (6.7) and (6.8) in (6.5), we have

$$\xi\left(c + \frac{1}{2}\eta(d,c)\right) \supseteq \frac{\Gamma(\alpha+1)}{2\eta^\alpha(d,c)} [J^\alpha_{c^+} \xi(c + \eta(d,c)) + J^\alpha_{(c+\eta(d,c))^-} \xi(c)]. \tag{6.9}$$

Now, we prove the second pair of inequality.
Since, ξ is an interval-valued preinvex function on $[c, c + \eta(d,c)]$. Therefore,

$$\xi(c + \delta\eta(d,c)) = \xi(c + \eta(d,c) + (1-\delta)\eta(c, c + \eta(d,c)))$$
$$\supseteq \delta\xi(c + \eta(d,c)) + (1-\delta)\xi(c) \tag{6.10}$$

and

$$\xi(c + (1-\delta)\eta(d,c)) = \xi(c + \eta(d,c) + \delta\eta(c, c + \eta(d,c)))$$
$$\supseteq (1-\delta)\xi(c + \eta(d,c)) + \delta\xi(c). \tag{6.11}$$

Adding (6.10) and (6.11), we have

$$\xi(c + \delta\eta(d,c)) + \xi(c + (1-\delta)\eta(d,c)) \supseteq \xi(c) + \xi(c + \eta(d,c)). \tag{6.12}$$

Multiplying by $\delta^{\alpha-1}$ and integrating on $[0,1]$, we have

$$(IR) \int_0^1 \delta^{\alpha-1} \xi(c + \delta\eta(d,c)) d\delta + (IR) \int_0^1 \delta^{\alpha-1} \xi(c + (1-\delta)\eta(d,c)) d\delta$$

$$\supseteq (IR) \int_0^1 \delta^{\alpha-1} [\xi(c) + \xi(c + \eta(d,c))]. \tag{6.13}$$

Applying Theorem 6.2.1 in above relation, we get

$$(IR) \int_0^1 \delta^{\alpha-1}[\underline{\xi}(c) + \underline{\xi}(c + \eta(d,c))]$$

$$= \left[(R) \int_0^1 \delta^{\alpha-1}[\underline{\xi}(c) + \underline{\xi}(c + \eta(d,c))]d\delta, \right.$$

$$\left. (R) \int_0^1 \delta^{\alpha-1}[\overline{\xi}(c) + \overline{\xi}(c + \eta(d,c))]d\delta \right]$$

$$= \left[[\underline{\xi}(c) + \underline{\xi}(c + \eta(d,c))](R) \int_0^1 \delta^{\alpha-1}d\delta, \right.$$

$$\left. [\overline{\xi}(c) + \overline{\xi}(c + \eta(d,c))](R) \int_0^1 \delta^{\alpha-1}d\delta \right]$$

$$= \left[\frac{1}{\alpha}[\underline{\xi}(c) + \underline{\xi}(c + \eta(d,c))], \frac{1}{\alpha}[\overline{\xi}(c) + \overline{\xi}(c + \eta(d,c))] \right]$$

$$= \frac{1}{\alpha}[\xi(c) + \xi(c + \eta(d,c))]. \tag{6.14}$$

Using (6.7), (6.8) and (6.14) in (6.13), we have

$$\frac{\Gamma(\alpha+1)}{2\eta^\alpha(d,c)}[J^\alpha_{c^+}\xi(c + \eta(d,c)) + J^\alpha_{(c+\eta(d,c))^-}\xi(c)]$$

$$\supseteq \frac{\xi(c) + \xi(c + \eta(d,c))}{2} \supseteq \frac{\xi(c) + \xi(d)}{2}. \tag{6.15}$$

From (6.9) and (6.15), we get

$$\xi\left(c + \frac{\eta(d,c)}{2}\right) \supseteq \frac{\Gamma(\alpha+1)}{2\eta^\alpha(d,c)}[J^\alpha_{c^+}\xi(c + \eta(d,c)) + J^\alpha_{(c+\eta(d,c))^-}\xi(c)]$$

$$\supseteq \frac{\xi(c) + \xi(c + \eta(d,c))}{2} \supseteq \frac{\xi(c) + \xi(d)}{2}.$$

This completes the proof.

Corollary 6.3.1. *If $\alpha = 1$, then Theorem 6.3.2 reduces to the following result:*

$$\xi\left(c + \frac{\eta(d,c)}{2}\right) \supseteq \frac{1}{\eta(d,c)} \int_c^{c+\eta(d,c)} \xi(\delta)d\delta$$

$$\supseteq \frac{\xi(c) + \xi(c + \eta(d,c))}{2} \supseteq \frac{\xi(c) + \xi(d)}{2}.$$

Remark 6.3.1. *When $\eta(d,c) = d - c$, then above theorem reduces to Theorem 3.4 of [25], i.e.*

$$\xi\left(\frac{c+d}{2}\right) \supseteq \frac{\Gamma(\alpha+1)}{2(d-c)^\alpha}[J^\alpha_{c^+}\xi(d) + J^\alpha_{d^-}\xi(c)]$$

$$\supseteq \frac{\xi(c) + \xi(d)}{2}. \tag{6.16}$$

We prove Hermite–Hadamard type inequalities for the product of two preinvex interval-valued functions [160].

Theorem 6.3.3. *Let $K \subseteq \mathbb{R}$ be an open invex subset with respect to η : $K \times K \to \mathbb{R}$ and $c, d \in K$ with $c < c + \eta(d, c)$. If $\xi, \psi : [c, c + \eta(d, c)] \to \mathbb{R}_I^+$ is a preinvex interval-valued function such that $\xi(\delta) = [\underline{\xi}(\delta), \overline{\xi}(\delta)]$ and $\psi(\delta) = [\underline{\psi}(\delta), \overline{\psi}(\delta)]$. $\xi, \psi \in L^1[c, c + \eta(d, c)]$ and η satisfies Condition C and $\alpha > 0$, then we have*

$$\frac{\Gamma(\alpha + 1)}{2\eta^\alpha(d, c)}[J_{c^+}^\alpha \xi(c + \eta(d, c))\psi(c + \eta(d, c)) + J_{(c+\eta(d,c))^-}^\alpha \xi(c)\psi(c)]$$

$$\supseteq \left(\frac{1}{2} - \frac{\alpha}{(\alpha + 1)(\alpha + 2)}\right) M(c, c + \eta(d, c))$$

$$+ \frac{\alpha}{(\alpha + 1)(\alpha + 2)} N(c, c + \eta(d, c)),$$

where $M(c, c + \eta(d, c)) = \xi(c)\psi(c) + \xi(c + \eta(d, c))\psi(c + \eta(d, c))$ and $N(c, c + \eta(d, c)) = \xi(c)\psi(c + \eta(d, c)) + \xi(c + \eta(d, c))\psi(c)$.

Proof Since ξ and ψ are two preinvex interval-valued functions for $\delta \in [0, 1]$, we have

$$\xi(c + \delta\eta(d, c)) = \xi(c + \eta(d, c) + (1 - \delta)\eta(c, c + \eta(d, c)))$$
$$\supseteq \delta\xi(c + \eta(d, c)) + (1 - \delta)\xi(c) \qquad (6.17)$$

and

$$\psi(c + \delta\eta(d, c)) = \psi(c + \eta(d, c) + (1 - \delta)\eta(c, c + \eta(d, c)))$$
$$\supseteq \delta\psi(c + \eta(d, c)) + (1 - \delta)\psi(c). \qquad (6.18)$$

Since, $\xi(x), \psi(x) \in \mathbb{R}_I^+$, $\forall x \in [c, d]$, then from (6.17) and (6.18), we have

$$\xi(c + \delta\eta(d, c))\psi(c + \delta\eta(d, c))$$
$$\supseteq \delta^2 \xi(c + \eta(d, c))\psi(c + \eta(d, c)) + (1 - \delta)^2 \xi(c)\psi(c)$$
$$+ \delta(1 - \delta)[\xi(c + \eta(d, c))\psi(c) + \xi(c)\psi(c + \eta(d, c))]. \qquad (6.19)$$

Similarly,

$$\xi(c + (1 - \delta)\eta(d, c))\psi(c + (1 - \delta)\eta(d, c))$$
$$\supseteq \delta^2 \xi(c)\psi(c) + (1 - \delta)^2 \xi(c + \eta(d, c))\psi(c + \eta(d, c))$$
$$+ \delta(1 - \delta)[\xi(c + \eta(d, c))\psi(c) + \xi(c)\psi(c + \eta(d, c))]. \qquad (6.20)$$

Adding (6.19) and (6.20), we have

$$\xi(c + \delta\eta(d,c))\psi(c + \delta\eta(d,c)) + \xi(c + (1-\delta)\eta(d,c))\psi(c + (1-\delta)\eta(d,c))$$
$$\supseteq \delta^2[\xi(c)\psi(c) + \xi(c + \eta(d,c))\psi(c + \eta(d,c))]$$
$$+ (1-\delta)^2[\xi(c)\psi(c) + \xi(c + \eta(d,c))\psi(c + \eta(d,c))]$$
$$+ 2\delta(1-\delta)[\xi(c + \eta(d,c))\psi(c) + \xi(c)\psi(c + \eta(d,c))]$$
$$= [\delta^2 + (1-\delta)^2][\xi(c)\psi(c) + \xi(c + \eta(d,c))\psi(c + \eta(d,c))]$$
$$+ 2\delta(1-\delta)[\xi(c + \eta(d,c))\psi(c) + \xi(c)\psi(c + \eta(d,c))]$$
$$= [2\delta^2 - 2\delta + 1]M(c, c + \eta(d,c)) + 2\delta(1-\delta)N(c, c + \eta(d,c)). \qquad (6.21)$$

Multiplying by $\delta^{\alpha-1}$ on both sides and integrating on $[0,1]$, we have

$$(IR)\int_0^1 \delta^{\alpha-1}\xi(c + \delta\eta(d,c))\psi(c + \delta\eta(d,c))d\delta$$

$$+ (IR)\int_0^1 \delta^{\alpha-1}\xi(c + (1-\delta)\eta(d,c))\psi(c + (1-\delta)\eta(d,c))d\delta$$

$$\supseteq (IR)\int_0^1 [2\delta^{\alpha+1} - 2\delta^\alpha + \delta^{\alpha-1}]M(c, c + \eta(d,c))d\delta$$

$$+ (IR)\int_0^1 2[\delta^\alpha - \delta^{\alpha+1}]N(c, c + \eta(d,c))d\delta. \qquad (6.22)$$

Since,

$$(IR)\int_0^1 \delta^{\alpha-1}\xi(c + \delta\eta(d,c))\psi(c + \delta\eta(d,c))d\delta$$

$$= \frac{\Gamma(\alpha)}{\eta^\alpha(d,c)}J^\alpha_{(c+\eta(d,c))^-}\xi(c)\psi(c), \qquad (6.23)$$

$$(IR)\int_0^1 \delta^{\alpha-1}\xi(c + (1-\delta)\eta(d,c))\psi(c + (1-\delta)\eta(d,c))d\delta$$

$$= \frac{\Gamma(\alpha)}{\eta^\alpha(d,c)}J^\alpha_{c^+}\xi(c + \eta(d,c))\psi(c + \eta(d,c)), \qquad (6.24)$$

$$(IR)\int_0^1 [2\delta^{\alpha+1} - 2\delta^\alpha + \delta^{\alpha-1}]M(c, c + \eta(d,c))d\delta$$

$$= \frac{2}{\alpha}\left(\frac{1}{2} - \frac{\alpha}{(\alpha+1)(\alpha+2)}\right)M(c, c + \eta(d,c)) \qquad (6.25)$$

and

$$(IR)\int_0^1 2[\delta^\alpha - \delta^{\alpha+1}]N(c, c + \eta(d,c))d\delta$$

$$= \frac{2}{(\alpha+1)(\alpha+2)}N(c, c + \eta(d,c)). \qquad (6.26)$$

Using (6.23), (6.24), (6.25) and (6.26) in (6.22), we have

$$\frac{\Gamma(\alpha)}{\eta^\alpha(d,c)} [J^\alpha_{(c+\eta(d,c))^-}\xi(c)\psi(c) + J^\alpha_{c^+}\xi(c+\eta(d,c))\psi(c+\eta(d,c))]$$

$$\supseteq \frac{2}{\alpha}\left(\frac{1}{2} - \frac{\alpha}{(\alpha+1)(\alpha+2)}\right) M(c,c+\eta(d,c))$$

$$+ \frac{2}{(\alpha+1)(\alpha+2)} N(c,c+\eta(d,c)). \tag{6.27}$$

This implies,

$$\frac{\Gamma(\alpha+1)}{2\eta^\alpha(d,c)} [J^\alpha_{c^+}\xi(c+\eta(d,c))\psi(c+\eta(d,c)) + J^\alpha_{(c+\eta(d,c))^-}\xi(c)\psi(c)]$$

$$\supseteq \left(\frac{1}{2} - \frac{\alpha}{(\alpha+1)(\alpha+2)}\right) M(c,c+\eta(d,c))$$

$$+ \frac{\alpha}{(\alpha+1)(\alpha+2)} N(c,c+\eta(d,c)).$$

Corollary 6.3.2. *If* $\alpha = 1$, *then above theorem reduces to the following result:*

$$\frac{1}{\eta(d,c)} \int_c^{c+\eta(d,c)} \xi(\delta)\psi(\delta)d\delta \supseteq \frac{1}{3}M(c,c+\eta(d,c)) + \frac{1}{6}N(c,c+\eta(d,c)).$$

Remark 6.3.2. *When* $\eta(d,c) = d-c$, *then above theorem reduces to Theorem 3.5 of [25], i.e.*

$$\frac{\Gamma(\alpha+1)}{2(d-c)^\alpha} [J^\alpha_{c^+}\xi(d)\psi(d) + J^\alpha_{d^-}\xi(c)\psi(c)]$$

$$\supseteq \left(\frac{1}{2} - \frac{\alpha}{(\alpha+1)(\alpha+2)}\right) M(c,d) + \frac{\alpha}{(\alpha+1)(\alpha+2)} N(c,d),$$

where $M(c,d) = \xi(c)\psi(c) + \xi(d)\psi(d)$ *and*
$N(c,d) = \xi(c)\psi(d) + \xi(d)\psi(c)$.

Theorem 6.3.4. *Let* $K \subseteq \mathbb{R}$ *be an open invex subset with respect to* $\eta :$ $K \times K \to \mathbb{R}$ *and* $c,d \in K$ *with* $c < c + \eta(d,c)$. *If* $\xi, \psi : [c, c+\eta(d,c)] \to \mathbb{R}_I^+$ *is a preinvex interval-valued function such that* $\xi(\delta) = [\underline{\xi}(\delta), \overline{\xi}(\delta)]$ *and* $\psi(\delta) = [\underline{\psi}(\delta), \overline{\psi}(\delta)]$. $\xi, \psi \in L^1[c, c+\eta(d,c)]$ *and* η *satisfies Condition C and* $\alpha > 0$, *then we have*

$$2\xi\left(c + \frac{1}{2}\eta(d,c)\right)\psi\left(c + \frac{1}{2}\eta(d,c)\right)$$

$$\supseteq \frac{\Gamma(\alpha+1)}{2\eta^\alpha(d,c)} [J^\alpha_{c^+}\xi(c+\eta(d,c))\psi(c+\eta(d,c)) + J^\alpha_{(c+\eta(d,c))^-}\xi(c)\psi(c)]$$

$$+ \left(\frac{1}{2} - \frac{\alpha}{(\alpha+1)(\alpha+2)}\right) N(c,c+\eta(d,c)) + \frac{\alpha}{(\alpha+1)(\alpha+2)} M(c,c+\eta(d,c)),$$

where $M(c,c+\eta(d,c))$ *and* $N(c,c+\eta(d,c))$ *are defined as previous.*

Proof Since ξ is a preinvex interval-valued function, we have

$$\xi\left(x + \frac{1}{2}\eta(y,x)\right) \supseteq \frac{\xi(x) + \xi(y)}{2}, \ \forall \ x, y \in [c, c + \eta(d,c)].$$

Using $x = c + (1 - \delta)\eta(d,c)$, $y = c + \delta\eta(d,c)$ and Condition C in above, we get

$$\xi(c + (1 - \delta)\eta(d,c) + \frac{1}{2}\eta(c + \delta\eta(d,c), c + (1 - \delta)\eta(d,c)))$$
$$\supseteq \frac{\xi(c + (1 - \delta)\eta(d,c)) + \xi(c + \delta\eta(d,c))}{2}.$$

This implies,

$$\xi\left(c + \frac{1}{2}\eta(d,c)\right) \supseteq \frac{\xi(c + (1 - \delta)\eta(d,c)) + \xi(c + \delta\eta(d,c))}{2}. \tag{6.28}$$

Similarly,

$$\psi\left(c + \frac{1}{2}\eta(d,c)\right) \supseteq \frac{\psi(c + (1 - \delta)\eta(d,c)) + \psi(c + \delta\eta(d,c))}{2}. \tag{6.29}$$

From (6.28) and (6.29), we get

$$\xi\left(c + \frac{1}{2}\eta(d,c)\right)\psi\left(c + \frac{1}{2}\eta(d,c)\right)$$
$$\supseteq \frac{1}{4}[\xi(c + (1 - \delta)\eta(d,c)) + \xi(c + \delta\eta(d,c))]$$
$$\times [\psi(c + (1 - \delta)\eta(d,c)) + \psi(c + \delta\eta(d,c))]$$
$$= \frac{1}{4}[\xi(c + (1 - \delta)\eta(d,c))\psi(c + (1 - \delta)\eta(d,c))$$
$$+ \xi(c + \delta\eta(d,c))\psi(c + \delta\eta(d,c))$$
$$+ \xi(c + (1 - \delta)\eta(d,c))\psi(c + \delta\eta(d,c))$$
$$+ \xi(c + \delta\eta(d,c))\psi(c + (1 - \delta)\eta(d,c))]. \tag{6.30}$$

Since ξ and $\psi \in \mathbb{R}_{\mathbb{I}}^+$, $\forall x \in [c, c + \eta(d,c)]$ are two preinvex interval-valued functions for $\delta \in [0,1]$, we have

$$\xi(c + (1 - \delta)\eta(d,c))\psi(c + \delta\eta(d,c))$$
$$\supseteq \delta^2\xi(c)\psi(c + \eta(d,c)) + (1 - \delta)^2\xi(c + \eta(d,c))\psi(c)$$
$$+ \delta(1 - \delta)[\xi(c + \eta(d,c))\psi(c + \eta(d,c)) + \xi(c)\psi(c)]. \tag{6.31}$$

Similarly,

$$\xi(c + \delta\eta(d,c))\psi(c + (1 - \delta)\eta(d,c))$$
$$\supseteq \delta^2\xi(c + \eta(d,c))\psi(c) + (1 - \delta)^2\xi(c)\psi(c + \eta(d,c))$$
$$+ \delta(1 - \delta)[\xi(c + \eta(d,c))\psi(c + \eta(d,c)) + \xi(c)\psi(c)]. \tag{6.32}$$

Adding (6.31) and (6.32), we obtain

$$\xi(c + (1 - \delta)\eta(d, c))\psi(c + \delta\eta(d, c)) + \xi(c + \delta\eta(d, c))\psi(c + (1 - \delta)\eta(d, c))$$
$$\supseteq [2\delta^2 - 2\delta + 1]N(c, c + \eta(d, c)) + 2\delta(1 - \delta)M(c, c + \eta(d, c)). \qquad (6.33)$$

From (6.30) and (6.33), we have

$$\xi\left(c + \frac{1}{2}\eta(d, c)\right)\psi\left(c + \frac{1}{2}\eta(d, c)\right) \supseteq \frac{1}{4}[(2\delta^2 - 2\delta + 1)N(c, c + \eta(d, c))$$
$$+ 2\delta(1 - \delta)M(c, c + \eta(d, c))] + \frac{1}{4}[\xi(c + (1 - \delta)\eta(d, c))\psi(c + (1 - \delta)\eta(d, c))$$
$$+ \xi(c + \delta\eta(d, c))\psi(c + \delta\eta(d, c))].$$

Multiplying by $\delta^{\alpha-1}$ on both sides in above, then integrating on $[0, 1]$, we obtain

$$(IR) \int_0^1 \xi\left(c + \frac{1}{2}\eta(d, c)\right)\psi\left(c + \frac{1}{2}\eta(d, c)\right)\delta^{\alpha-1}d\delta$$

$$\supseteq \frac{1}{4}(IR) \int_0^1 \delta^{\alpha-1}(2\delta^2 - 2\delta + 1)N(c, c + \eta(d, c))d\delta$$

$$+ \frac{1}{2}(IR) \int_0^1 \delta^\alpha(1 - \delta)M(c, c + \eta(d, c))d\delta$$

$$+ \frac{1}{4}(IR) \int_0^1 \xi(c + (1 - \delta)\eta(d, c))\psi(c + (1 - \delta)\eta(d, c))\delta^{\alpha-1}d\delta$$

$$+ \frac{1}{4}(IR) \int_0^1 \xi(c + \delta\eta(d, c))\psi(c + \delta\eta(d, c))\delta^{\alpha-1}d\delta.$$

This implies,

$$2\xi\left(c + \frac{1}{2}\eta(d, c)\right)\psi\left(c + \frac{1}{2}\eta(d, c)\right)$$

$$\supseteq \frac{\Gamma(\alpha + 1)}{2\eta^\alpha(d, c)}[J^\alpha_{c+}\xi(c + \eta(d, c))\psi(c + \eta(d, c)) + J^\alpha_{(c+\eta(d,c))-}\xi(c)\psi(c)]$$

$$+ \left(\frac{1}{2} - \frac{\alpha}{(\alpha + 1)(\alpha + 2)}\right)N(c, c + \eta(d, c)) + \frac{\alpha}{(\alpha + 1)(\alpha + 2)}M(c, c + \eta(d, c)).$$

This completes the proof.

Corollary 6.3.3. *If $\alpha = 1$, then Theorem (6.3.4) reduces to the following result:*

$$2\xi\left(c + \frac{1}{2}\eta(d, c)\right)\psi\left(c + \frac{1}{2}\eta(d, c)\right)$$

$$\supseteq \frac{1}{\eta(d, c)} \int_c^{c+\eta(d,c)} \xi(\delta)\psi(\delta)d\delta + \frac{1}{3}N(c, c + \eta(d, c)) + \frac{1}{6}M(c, c + \eta(d, c)).$$

Remark 6.3.3. *When $\eta(d, c) = d - c$, then above theorem reduces to Theorem 3.6 of [25], i.e.*

$$2\xi\left(\frac{c+d}{2}\right)\psi\left(\frac{c+d}{2}\right)$$
$$\supseteq \frac{\Gamma(\alpha+1)}{2(d-c)^\alpha}[J_{c+}^\alpha \xi(d)\psi(d) + J_{d-}^\alpha \xi(c)\psi(c)]$$
$$+ \left(\frac{1}{2} - \frac{\alpha}{(\alpha+1)(\alpha+2)}\right)N(c,d) + \frac{\alpha}{(\alpha+1)(\alpha+2)}M(c,d),$$

where $M(c,d)$ and $N(c,d)$ are defined as previous.

6.4 Hermite–Hadamard Type Inclusions for Interval-Valued Coordinated Preinvex Functions

In this section, first, we give the definition of interval-valued coordinated preinvex function [93].

Definition 6.4.1. *Let $K_1 \times K_2$ be an invex set with respect to η_1 and η_2, $\xi = [\underline{\xi}, \overline{\xi}]$ be an interval valued function defined on $K_1 \times K_2$. The function ξ is said to be interval-valued coordinated preinvex function with respect to η_1 and η_2 if the partial mappings $\xi_v : K_1 \to \mathbb{R}_I^+$, $\xi_v(w) = \xi(w, v)$ and $\xi_u : K_2 \to \mathbb{R}_I^+$, $\xi_u(z) = \xi(u, z)$ are interval-valued preinvex functions with respect to η_1 and η_2, respectively, for all $u \in K_1$ and $v \in K_2$.*

Remark 6.4.1. *From the definition of interval-valued coordinated preinvex functions, it follows that if ξ is an interval-valued coordinated preinvex function, then*

$$\xi(u + \delta_1\eta_1(w, u), v + \delta_2\eta_2(z, v)) \supseteq (1 - \delta_1)(1 - \delta_2)\xi(u, v) + (1 - \delta_1)\delta_2\xi(u, z)$$
$$+ \delta_1(1 - \delta_2)\xi(w, v) + \delta_1\delta_2\xi(w, z),$$

for all $(u, v), (u, z), (w, v), (w, z) \in K_1 \times K_2$ and $\delta_1, \delta_2 \in [0, 1]$.

If $\eta_1(w, u) = w - u$ and $\eta_2(z, v) = z - v$, then the definition of interval-valued coordinated preinvex function reduces to the definition of interval-valued coordinated convex function proposed by Zhao *et al.* [183].

Example 6.4.1. *An interval-valued function $\xi : [0, 1] \times [\frac{1}{2}, 1] \to \mathbb{R}_I^+$ defined as $\xi(u, v) = [u + v, (2 - u)(2 - v)]$ is an interval-valued coordinated preinvex function with respect to $\eta_1(w, u) = w - u - 1$ and $\eta_2(z, v) = z - 2v$ for all $u, w \in [0, 1]$ and $v, z \in [\frac{1}{2}, 1]$.*

Now, we establish Hermite–Hadamard type inclusions for interval-valued preinvex functions on coordinates [93]. Throughout this section, we will not include the symbols (R), (IR) and (ID) before the integral sign.

Theorem 6.4.1. *Let $K_1 \times K_2$ be an invex set with respect to η_1 and η_2. If $\xi : K_1 \times K_2 \to \mathbb{R}_I^+$ is an interval-valued coordinated preinvex function with respect to η_1 and η_2 such that $\xi = [\underline{\xi}, \overline{\xi}]$ and $a < a + \eta_1(b, a)$, $c < c + \eta_2(d, c)$, where $a, b \in K_1$ and $c, d \in K_2$. If η_1, η_2 satisfy Condition C, then we have*

$$\xi\left(a + \frac{1}{2}\eta_1(b, a), c + \frac{1}{2}\eta_2(d, c)\right)$$

$$\supseteq \frac{1}{\eta_1(b, a)\eta_2(d, c)} \int_a^{a+\eta_1(b,a)} \int_c^{c+\eta_2(d,c)} \xi(u, v) dv du$$

$$\supseteq \frac{1}{4}[\xi(a, c) + \xi(b, c) + \xi(a, d) + \xi(b, d)].$$

Proof Since ξ is an interval-valued preinvex function on coordinates with respect to η_1 and η_2, we have

$$\xi(a + \delta_1\eta_1(b, a), c + \delta_2\eta_2(d, c)) \supseteq (1 - \delta_1)(1 - \delta_2)\xi(a, c) + (1 - \delta_1)\delta_2\xi(a, d)$$

$$+ \delta_1(1 - \delta_2)\xi(b, c) + \delta_1\delta_2\xi(b, d). \qquad (6.34)$$

Integrating (6.34) with respect to (δ_1, δ_2) over $[0, 1] \times [0, 1]$, we get

$$\int_0^1 \int_0^1 \xi(a + \delta_1\eta_1(b, a), c + \delta_2\eta_2(d, c)) d\delta_2 d\delta_1$$

$$\supseteq \int_0^1 \int_0^1 (1 - \delta_1)(1 - \delta_2)\xi(a, c) d\delta_2 d\delta_1 + \int_0^1 \int_0^1 (1 - \delta_1)\delta_2\xi(a, d) d\delta_2 d\delta_1$$

$$+ \int_0^1 \int_0^1 \delta_1(1 - \delta_2)\xi(b, c) d\delta_2 d\delta_1 + \int_0^1 \int_0^1 \delta_1\delta_2\xi(b, d) d\delta_2 d\delta_1.$$

This implies that

$$\frac{1}{\eta_1(b, a)\eta_2(d, c)} \int_a^{a+\eta_1(b,a)} \int_c^{c+\eta_2(d,c)} \xi(u, v) dv du$$

$$\supseteq \frac{1}{4}[\xi(a, c) + \xi(a, d) + \xi(b, c) + \xi(b, d)]. \qquad (6.35)$$

Using the definition of an interval-valued coordinated preinvex function and Condition C for η_1, η_2, we get

$$\xi\left(a + \frac{1}{2}\eta_1(b,a), c + \frac{1}{2}\eta_2(d,c)\right)$$

$$= \xi(a + \delta_1\eta_1(b,a) + \frac{1}{2}\eta_1(a + (1-\delta_1)\eta_1(b,a), a + \delta_1\eta_1(b,a)), \ c + \delta_2\eta_2(d,c)$$

$$+ \frac{1}{2}\eta_2(c + (1-\delta_2)\eta_2(d,c), c + \delta_2\eta_2(d,c)))$$

$$\supseteq \frac{1}{4}[\xi(a + \delta_1\eta_1(b,a), c + \delta_2\eta_2(d,c))$$

$$+ \xi(a + \delta_1\eta_1(b,a), c + (1-\delta_2)\eta_2(d,c))$$

$$+ \xi(a + (1-\delta_1)\eta_1(b,a), c + \delta_2\eta_2(d,c))$$

$$+ \xi(a + (1-\delta_1)\eta_1(b,a), c + (1-\delta_2)\eta_2(d,c))] \tag{6.36}$$

Thus, integrating (6.36) with respect to (δ_1, δ_2) over $[0,1] \times [0,1]$, we get

$$\int_0^1 \int_0^1 \xi\left(a + \frac{1}{2}\eta_1(b,a), c + \frac{1}{2}\eta_2(d,c)\right) d\delta_2 d\delta_1$$

$$\supseteq \frac{1}{4}\int_0^1 \int_0^1 [\xi(a + \delta_1\eta_1(b,a), c + \delta_2\eta_2(d,c))$$

$$+ \xi(a + \delta_1\eta_1(b,a), c + (1-\delta_2)\eta_2(d,c))$$

$$+ \xi(a + (1-\delta_1)\eta_1(b,a), c + \delta_2\eta_2(d,c))$$

$$+ \xi(a + (1-\delta_1)\eta_1(b,a), c + (1-\delta_2)\eta_2(d,c))] d\delta_2 d\delta_1.$$

This implies

$$\xi\left(a + \frac{1}{2}\eta_1(b,a), c + \frac{1}{2}\eta_2(d,c)\right)$$

$$\supseteq \frac{1}{\eta_1(b,a)\eta_2(d,c)} \int_a^{a+\eta_1(b,a)} \int_c^{c+\eta_2(d,c)} \xi(u,v)dv du. \tag{6.37}$$

From (6.35) and (6.37), we get the desired result.

Theorem 6.4.2. *Let $K_1 \times K_2$ be an invex set with respect to η_1 and η_2. If $\xi : [a, a + \eta_1(b,a)] \times [c, c + \eta_2(d,c)] \to \mathbb{R}_I^+$ is an interval-valued coordinated preinvex function with respect to η_1 and η_2 such that $\xi = [\underline{\xi}, \overline{\xi}]$ and $a < a + \eta_1(b,a)$, $c < c + \eta_2(d,c)$, where $a, b \in K_1$ and $c, d \in K_2$. If η_1, η_2 satisfy*

Condition C, then we have

$$\frac{1}{\eta_1(b,a)} \int_a^{a+\eta_1(b,a)} \xi\left(u, c + \frac{1}{2}\eta_2(d,c)\right) du$$

$$+ \frac{1}{\eta_2(d,c)} \int_c^{c+\eta_2(d,c)} \xi\left(a + \frac{1}{2}\eta_1(b,a), v\right) dv$$

$$\supseteq \frac{2}{\eta_1(b,a)\eta_2(d,c)} \int_a^{a+\eta_1(b,a)} \int_c^{c+\eta_2(d,c)} \xi(u,v) dv du$$

$$\supseteq \frac{1}{2}\left[\frac{1}{\eta_1(b,a)} \int_a^{a+\eta_1(b,a)} (\xi(u,c) + \xi(u, c + \eta_2(d,c))) du \right.$$

$$\left. + \frac{1}{\eta_2(d,c)} \int_c^{c+\eta_2(d,c)} (\xi(a,v) + \xi(a + \eta_1(b,a), v)) dv \right]. \qquad (6.38)$$

Proof Since ξ is an interval-valued preinvex function on coordinates $[a, a + \eta_1(b,a)] \times [c, c + \eta_2(d,c)]$, then $\xi_u : [c, c + \eta_2(d,c)] \to \mathbb{R}_I^+$, $\xi_u(v) = \xi(u,v)$ is an interval-valued preinvex function on $[c, c + \eta_2(d,c)]$ for all $u \in [a, a + \eta_1(b,a)]$. From Corollary 6.3.1, we have

$$\xi_u\left(c + \frac{1}{2}\eta_2(d,c)\right) \supseteq \frac{1}{\eta_2(d,c)} \int_c^{c+\eta_2(d,c)} \xi_u(v) dv \supseteq \frac{\xi_u(r) + \xi_u(c + \eta_2(d,c))}{2}.$$

This implies

$$\xi\left(u, c + \frac{1}{2}\eta_2(d,c)\right) \supseteq \frac{1}{\eta_2(d,c)} \int_c^{c+\eta_2(d,c)} \xi(u,v) dv$$

$$\supseteq \frac{\xi(u,c) + \xi(u, c + \eta_2(d,c))}{2}. \qquad (6.39)$$

Integrating (6.39) over $[a, a + \eta_1(b,a)]$ with respect to u, then dividing by $\eta_1(b,a)$, we get

$$\frac{1}{\eta_1(b,a)} \int_a^{a+\eta_1(b,a)} \xi\left(u, c + \frac{1}{2}\eta_2(d,c)\right) du$$

$$\supseteq \frac{1}{\eta_1(b,a)\eta_2(d,c)} \int_a^{a+\eta_1(b,a)} \int_c^{c+\eta_2(d,c)} \xi(u,v) dv du$$

$$\supseteq \frac{1}{2\eta_1(b,a)} \int_a^{a+\eta_1(b,a)} (\xi(u,c) + \xi(u, c + \eta_2(d,c))) du. \qquad (6.40)$$

Similarly, $\xi_v : [a, a + \eta_1(a, b)] \to \mathbb{R}_I^+, \xi_v(u) = \xi(u, v)$ is interval-valued preinvex function on $[a, a + \eta_1(a, b)]$ for all $v \in [c, c + \eta_2(d, c)]$. Then, we have

$$\frac{1}{\eta_2(d, c)} \int_c^{c+\eta_2(d,c)} \xi\left(a + \frac{1}{2}\eta_1(b, a), v\right) dv$$

$$\supseteq \frac{1}{\eta_1(b, a)\eta_2(d, c)} \int_a^{a+\eta_1(b,a)} \int_c^{c+\eta_2(d,c)} \xi(u, v) dv du$$

$$\supseteq \frac{1}{2\eta_2(d, c)} \int_c^{c+\eta_2(d,c)} (\xi(a, v) + \xi(a + \eta_1(b, a), v)) dv. \qquad (6.41)$$

By adding (6.40) and (6.41), we have

$$\frac{1}{\eta_1(b, a)} \int_a^{a+\eta_1(b,a)} \xi\left(u, c + \frac{1}{2}\eta_2(d, c))\right) du$$

$$+ \frac{1}{\eta_2(d, c)} \int_c^{c+\eta_2(d,c)} \xi\left(a + \frac{1}{2}\eta_1(b, a), v\right) dv$$

$$\supseteq \frac{2}{\eta_1(b, a)\eta_2(d, c)} \int_a^{a+\eta_1(b,a)} \int_c^{c+\eta_2(d,c)} \xi(u, v) dv du$$

$$\supseteq \frac{1}{2} \left[\frac{1}{\eta_1(b, a)} \int_a^{a+\eta_1(b,a)} (\xi(u, c) + \xi(u, c + \eta_2(d, c))) du \right.$$

$$\left. + \frac{1}{\eta_2(d, c)} \int_c^{c+\eta_2(d,c)} (\xi(a, v) + \xi(a + \eta_1(b, a), v)) dv \right].$$

This completes the proof.

Example 6.4.2. Let $[a, a + \eta_1(b, a)] = [\frac{1}{4}, \frac{1}{2}]$, $[c, c + \eta_2(d, c)] = [\frac{1}{4}, \frac{1}{2}]$ and $\eta_1(b, a) = b - 2a$, $\eta_2(d, c) = d - 2c$. Let $\xi : [\frac{1}{4}, \frac{1}{2}] \times [\frac{1}{4}, \frac{1}{2}] \to \mathbb{R}_I^+$ be defined by $\xi(u, v) = [uv, (1 - u)(1 - v)]$ $\forall u \in [\frac{1}{4}, \frac{1}{2}]$ and $v \in [\frac{1}{4}, \frac{1}{2}]$. Then all assumptions of Theorem 6.4.2 are satisfied.

Theorem 6.4.3. Let $K_1 \times K_2$ be an invex set with respect to η_1 and η_2. If $\xi : [a, a + \eta_1(b, a)] \times [c, c + \eta_2(d, c)] \to \mathbb{R}_I^+$ is an interval-valued coordinated preinvex function with respect to η_1 and η_2 such that $\xi = [\underline{\xi}, \overline{\xi}]$ and $a < a + \eta_1(b, a)$, $c < c + \eta_2(d, c)$, where $a, b \in K_1$ and $c, d \in K_2$. If η_1, η_2 satisfy

Condition C, then we have

$$\xi\left(a + \frac{1}{2}\eta_1(b,a), c + \frac{1}{2}\eta_2(d,c)\right)$$

$$\supseteq \frac{1}{2}\left[\frac{1}{\eta_1(b,a)}\int_a^{a+\eta_1(b,a)} \xi\left(u, c + \frac{1}{2}\eta_2(d,c)\right) du\right.$$

$$\left. + \frac{1}{\eta_2(d,c)}\int_c^{c+\eta_2(d,c)} \xi\left(a + \frac{1}{2}\eta_1(b,a), v\right) dv\right]$$

$$\supseteq \frac{1}{\eta_1(b,a)\eta_2(d,c)}\int_a^{a+\eta_1(b,a)}\int_c^{c+\eta_2(d,c)} \xi(u,v)\,dv\,du$$

$$\supseteq \frac{1}{4}\left[\frac{1}{\eta_1(b,a)}\int_a^{a+\eta_1(b,a)} (\xi(u,c) + \xi(u, c+\eta_2(d,c)))du\right.$$

$$\left. + \frac{1}{\eta_2(d,c)}\int_c^{c+\eta_2(d,c)} (\xi(a,v) + \xi(a+\eta_1(b,a), v))dv\right]$$

$$\supseteq \frac{1}{4}[\xi(a,c) + \xi(a+\eta_1(b,a), c) + \xi(a, c+\eta_2(d,c))$$

$$+ \xi(a+\eta_1(b,a), c+\eta_2(d,c))]$$

$$\supseteq \frac{1}{4}[\xi(a,c) + \xi(b,c) + \xi(a,d) + \xi(b,d)].$$

Proof Since ξ is an interval-valued preinvex function on coordinates $[a, a + \eta_1(b,a)] \times [c, c+\eta_2(d,c)]$, then from Corollary 6.3.1 we get

$$\xi\left(a + \frac{1}{2}\eta_1(b,a), c + \frac{1}{2}\eta_2(d,c)\right)$$

$$\supseteq \frac{1}{\eta_1(b,a)}\int_a^{a+\eta_1(b,a)} \xi\left(u, c + \frac{1}{2}\eta_2(d,c)\right) du, \qquad (6.42)$$

$$\xi\left(a + \frac{1}{2}\eta_1(b,a), c + \frac{1}{2}\eta_2(d,c)\right)$$

$$\supseteq \frac{1}{\eta_2(d,c)}\int_c^{c+\eta_2(d,c)} \xi\left(a + \frac{1}{2}\eta_1(b,a), v\right) dv. \qquad (6.43)$$

Adding (6.42) and (6.43), we have

$$\xi\left(a + \frac{1}{2}\eta_1(b,a),\ c + \frac{1}{2}\eta_2(d,c))\right)$$

$$\supseteq \frac{1}{2}\left[\frac{1}{\eta_1(b,a)}\int_a^{a+\eta_1(b,a)}\xi\left(u, c + \frac{1}{2}\eta_2(d,c))\right) du\right.$$

$$\left. + \frac{1}{\eta_2(d,c)}\int_c^{c+\eta_2(d,c)}\xi\left(a + \frac{1}{2}\eta_1(b,a), v\right) dv\right]. \qquad (6.44)$$

Again from Corollary 6.3.1, we get

$$\frac{1}{\eta_1(b,a)}\int_a^{a+\eta_1(b,a)}\xi(u,c)du \supseteq \frac{\xi(a,c) + \xi(a + \eta_1(b,a),c)}{2}, \qquad (6.45)$$

$$\frac{1}{\eta_1(b,a)}\int_a^{a+\eta_1(b,a)}\xi(u, c + \eta_2(d,c))du$$

$$\supseteq \frac{\xi(a, c + \eta_2(d,c)) + \xi(a + \eta_1(b,a), c + \eta_2(d,c))}{2}, \qquad (6.46)$$

$$\frac{1}{\eta_2(d,c)}\int_c^{c+\eta_2(d,c)}\xi(a,v)dv \supseteq \frac{\xi(a,c) + \xi(a, c + \eta_2(d,c))}{2}, \qquad (6.47)$$

$$\frac{1}{\eta_2(d,c)}\int_c^{c+\eta_2(d,c)}\xi(a + \eta_1(b,a), v)dv$$

$$\supseteq \frac{\xi(a + \eta_1(b,a),c) + \xi(a + \eta_1(b,a), c + \eta_2(d,c))}{2}. \qquad (6.48)$$

Adding (6.45)–(6.48), we get

$$\frac{1}{\eta_1(b,a)}\int_a^{a+\eta_1(b,a)}(\xi(u,c) + \xi(u, c + \eta_2(d,c)))du$$

$$+ \frac{1}{\eta_2(d,c)}\int_c^{c+\eta_2(d,c)}(\xi(a,v) + \xi(a + \eta_1(b,a), v))dv$$

$$\supseteq \xi(a,c) + \xi(a + \eta_1(b,a),c) + \xi(a, c + \eta_2(d,c))$$

$$+ \xi(a + \eta_1(b,a), c + \eta_2(d,c)). \qquad (6.49)$$

By Corollary 6.3.1, we also have

$$\xi(a,c) + \xi(a + \eta_1(b,a),c) + \xi(a, c + \eta_2(d,c))$$

$$+ \xi(a + \eta_1(b,a), c + \eta_2(d,c))$$

$$\supseteq \xi(a,c) + \xi(b,c) + \xi(a,d) + \xi(b,d). \qquad (6.50)$$

From (6.38), (6.44), (6.49), and (6.50), we get the desired result.

Remark 6.4.2. *If we put $\eta_1(b,a) = b - a$ and $\eta_2(d,c) = d - c$ in Theorem 6.4.3, we obtain Theorem 7 of [183], i.e.*

$$\xi\left(\frac{a+b}{2}, \frac{c+d}{2}\right)$$

$$\supseteq \frac{1}{2}\left[\frac{1}{(b-a)}\int_a^b \xi\left(u, \frac{c+d}{2}\right)du + \frac{1}{(d-c)}\int_c^d \xi\left(\frac{a+b}{2}, v\right)dv\right]$$

$$\supseteq \frac{1}{(b-a)(d-c)}\int_a^b \int_c^d \xi(u,v)dvdu$$

$$\supseteq \frac{1}{4}\left[\frac{1}{(b-a)}\int_a^b (\xi(u,c) + \xi(u,d))du\right.$$

$$\left.+\frac{1}{(d-c)}\int_c^d (\xi(a,v) + \xi(b,v))dv\right]$$

$$\supseteq \frac{1}{4}[\xi(a,c) + \xi(b,c) + \xi(a,d) + \xi(b,d)].$$

Next, we prove Hermite–Hadamard type inclusions for the product of two interval-valued coordinated preinvex functions [93].

Theorem 6.4.4. *Let $K_1 \times K_2$ be an invex set with respect to η_1 and η_2. If $\xi, \psi : [a, a + \eta_1(b,a)] \times [c, c + \eta_2(d,c)] \to \mathbb{R}_I^+$ are interval-valued coordinated preinvex functions with respect to η_1 and η_2 such that $\xi = [\underline{\xi}, \overline{\xi}]$, $\psi = [\underline{\psi}, \overline{\psi}]$ and $a < a + \eta_1(b,a)$, $c < c + \eta_2(d,c)$, where $a, b \in K_1$ and $c, d \in K_2$. If η_1, η_2 satisfy Condition C, then*

$$\frac{1}{\eta_1(b,a)\eta_2(d,c)}\int_a^{a+\eta_1(b,a)} \int_c^{c+\eta_2(d,c)} \xi(u,v)\psi(u,v)dvdu$$

$$\supseteq \frac{1}{9}N_1(a,b,c,d) + \frac{1}{18}N_2(a,b,c,d) + \frac{1}{18}N_3(a,b,c,d) + \frac{1}{36}N_4(a,b,c,d),$$

where

$N_1(a,b,c,d) = \xi(a,c)\psi(a,c) + \xi(a + \eta_1(b,a),c)\psi(a + \eta_1(b,a),c) + \xi(a,c + \eta_2(d,c))\psi(a,c+\eta_2(d,c))+\xi(a+\eta_1(b,a),c+\eta_2(d,c))\psi(a+\eta_1(b,a),c+\eta_2(d,c)),$

$N_2(a,b,c,d) = \xi(a,c)\psi(a + \eta_1(b,a),c) + \xi(a + \eta_1(b,a),c)\psi(a,c) + \xi(a,c+\eta_2(d,c))\psi(a+\eta_1(b,a),c+\eta_2(d,c))+\xi(a+\eta_1(b,a),c+\eta_2(d,c))\psi(a,c+\eta_2(d,c)),$

$N_3(a,b,c,d) = \xi(a,c)\psi(a,c + \eta_2(d,c)) + \xi(a + \eta_1(b,a),c)\psi(a + \eta_1(b,a),c+\eta_2(d,c))+\xi(a,c+\eta_2(d,c))\psi(a,c)+\xi(a+\eta_1(b,a),c+\eta_2(d,c))\psi(a+\eta_1(b,a),c),$

$N_4(a,b,c,d) = \xi(a,c)\psi(a + \eta_1(b,a),c + \eta_2(d,c)) + \xi(a + \eta_1(b,a),c)\psi(a,c+\eta_2(d,c))+\xi(a,c+\eta_2(d,c))\psi(a+\eta_1(b,a),c)+\xi(a+\eta_1(b,a),c+\eta_2(d,c))\psi(a,c).$

Proof Since ξ and ψ are interval-valued coordinated preinvex functions on $[a, a + \eta_1(b, a)] \times [c, c + \eta_2(d, c)]$, we have

$$\xi_u(v) : [c, c + \eta_2(d, c)] \to \mathbb{R}_I^+, \ \xi_u(v) = \xi(u, v)$$

and

$$\psi_u(v) : [c, c + \eta_2(d, c)] \to \mathbb{R}_I^+, \ \psi_u(v) = \psi(u, v)$$

are interval-valued preinvex functions on $[c, c + \eta_2(d, c)]$ for all $u \in [a, a + \eta_1(b, a)]$.

Similarly,

$$\xi_v(u) : [a, a + \eta_1(b, a)] \to \mathbb{R}_I^+, \ \xi_v(u) = \xi(u, v)$$

and

$$\psi_v(u) : [a, a + \eta_1(b, a)] \to \mathbb{R}_I^+, \ \psi_v(u) = \psi(u, v)$$

are interval-valued preinvex functions on $[a, a + \eta_1(b, a)]$ for all $v \in [c, c + \eta_2(d, c)]$.

From Corollary (6.3.2), we get

$$\frac{1}{\eta_2(d, c)} \int_c^{c + \eta_2(d, c)} \xi_u(v) \psi_u(v) dv$$

$$\supseteq \frac{1}{3} [\xi_u(c) \psi_u(c) + \xi_u(c + \eta_2(d, c)) \psi_u(c + \eta_2(d, c))]$$

$$+ \frac{1}{6} [\xi_u(c) \psi_u(c + \eta_2(d, c)) + \xi_u(c + \eta_2(d, c)) \psi_u(c)].$$

This implies

$$\frac{1}{\eta_2(d, c)} \int_c^{c + \eta_2(d, c)} \xi(u, v) \psi(u, v) dv$$

$$\supseteq \frac{1}{3} [\xi(u, c) \psi(u, c) + \xi(u, c + \eta_2(d, c)) \psi(u, c + \eta_2(d, c))]$$

$$+ \frac{1}{6} [\xi(u, c) \psi(u, c + \eta_2(d, c)) + \xi(u, c + \eta_2(d, c)) \psi(u, c)]. \tag{6.51}$$

Integrating (6.51) with respect to u over $[a, a + \eta_1(b, a)]$ and after then dividing by $\eta_1(b, a)$, we find

$$\frac{1}{\eta_1(b, a) \eta_2(d, c)} \int_a^{a + \eta_1(b, a)} \int_c^{c + \eta_2(d, c)} \xi(u, v) \psi(u, v) dv du$$

$$\supseteq \frac{1}{3 \eta_1(b, a)} \int_a^{a + \eta_1(b, a)} [\xi(u, c) \psi(u, c) + \xi(u, c + \eta_2(d, c)) \psi(u, c + \eta_2(d, c))] du$$

$$+ \frac{1}{6 \eta_1(b, a)} \int_a^{a + \eta_1(b, a)} [\xi(u, c) \psi(u, c + \eta_2(d, c)) + \xi(u, c + \eta_2(d, c)) \psi(u, c)] du.$$

$$\tag{6.52}$$

Again from Corollary 6.3.2, we have

$$\frac{1}{\eta_1(b,a)} \int_a^{a+\eta_1(b,a)} \xi(u,c)\psi(u,c)du$$

$$\supseteq \frac{1}{3}[\xi(a,c)\psi(a,c) + \xi(a+\eta_1(b,a),c)\psi(a+\eta_1(b,a),c]$$

$$+ \frac{1}{6}[\xi(a,c)\psi(a+\eta_1(b,a),c) + \xi(a+\eta_1(b,a),c)\psi(a,c)], \qquad (6.53)$$

$$\frac{1}{\eta_1(b,a)} \int_a^{a+\eta_1(b,a)} \xi(u,c+\eta_2(d,c))\psi(u,c+\eta_2(d,c))du$$

$$\supseteq \frac{1}{3}[\xi(a,c+\eta_2(d,c))\psi(a,c+\eta_2(d,c))$$

$$+ \xi(a+\eta_1(b,a),c+\eta_2(d,c))\psi(a+\eta_1(b,a),c+\eta_2(d,c))]$$

$$+ \frac{1}{6}[\xi(a,c+\eta_2(d,c))\psi(a+\eta_1(b,a),c+\eta_2(d,c))$$

$$+ \xi(a+\eta_1(b,a),c+\eta_2(d,c))\psi(a,c+\eta_2(d,c))], \qquad (6.54)$$

$$\frac{1}{\eta_1(b,a)} \int_a^{a+\eta_1(b,a)} \xi(u,c)\psi(u,c+\eta_2(d,c))du$$

$$\supseteq \frac{1}{3}[\xi(a,c))\psi(a,c+\eta_2(d,c)) + \xi(a+\eta_1(b,a),c)\psi(a+\eta_1(b,a),c+\eta_2(d,c))]$$

$$+ \frac{1}{6}[\xi(a,c)\psi(a+\eta_1(b,a),c+\eta_2(d,c)) + \xi(a+\eta_1(b,a),r)\psi(a,c+\eta_2(d,c))], \qquad (6.55)$$

$$\frac{1}{\eta_1(b,a)} \int_a^{a+\eta_1(b,a)} \xi(u,c+\eta_2(d,c))\psi(u,c)du$$

$$\supseteq \frac{1}{3}[\xi(a,c+\eta_2(d,c))\psi(a,c) + \xi(a+\eta_1(b,a),c+\eta_2(d,c))\psi(a+\eta_1(b,a),c]$$

$$+ \frac{1}{6}[\xi(a,c+\eta_2(d,c))\psi(a+\eta_1(b,a),c) + \xi(p+\eta_1(b,a),c+\eta_2(d,c))\psi(a,c)]. \qquad (6.56)$$

Substituting (6.53)–(6.56) into (6.52), we obtain the desired result. Similarly, we can obtain the same result by using Corollary 6.3.2 for the product $\xi_v(u)\psi_v(u)$ on $[a, a+\eta_1(b,a)]$.

Remark 6.4.3. *If we put* $\eta_1(b, a) = b - a$ *and* $\eta_2(d, c) = d - c$ *in Theorem 6.4.4, we obtain Theorem 8 of [183], i.e.,*

$$\frac{1}{(b-a)(d-c)} \int_a^b \int_c^d \xi(u,v)\psi(u,v)dvdu$$

$$\supseteq \frac{1}{9}N_1^*(a,b,c,d) + \frac{1}{18}N_2^*(a,b,c,d) + \frac{1}{18}N_3^*(a,b,c,d)$$

$$+ \frac{1}{36}N_4^*(a,b,c,d),$$

where

$$N_1^*(a,b,c,d) = \xi(a,c)\psi(a,c) + \xi(b,c)\psi(b,c) + \xi(a,d)\ \psi(a,d) + \xi(b,d)\psi(b,d),$$

$$N_2^*(a,b,c,d) = \xi(a,c)\psi(b,c) + \xi(b,c)\psi(a,c) + \xi(a,d)\ \psi(b,d)) + \xi(b,d)\psi(a,d),$$

$$N_3^*(a,b,c,d) = \xi(a,c)\psi(a,d) + \xi(b,c)\psi(b,d) + \xi(a,d)\psi(a,c) + \xi(b,d)\psi(b,c),$$

$$N_4^*(a,b,c,d) = \xi(a,c)\psi(b,d) + \xi(b,c)\psi(a,d) + \xi(a,d)\psi(b,c) + \xi(b,d)\psi(a,c).$$

Theorem 6.4.5. *Let* $K_1 \times K_2$ *be an invex set with respect to* η_1 *and* η_2. *If* $\xi, \psi : [a, a + \eta_1(b,a)] \times [c, c + \eta_2(d,c)] \to \mathbb{R}_I^+$ *are interval-valued coordinated preinvex functions with respect to* η_1 *and* η_2 *such that* $\xi = [\underline{\xi}, \overline{\xi}]$, $\psi = [\underline{\psi}, \overline{\psi}]$ *and* $a < a + \eta_1(b,a)$, $c < c + \eta_2(d,c)$, *where* $a, b \in K_1$ *and* $c, d \in K_2$. *If* η_1, η_2 *satisfy Condition C, then we have*

$$4\xi\left(a + \frac{1}{2}\eta_1(b,a), c + \frac{1}{2}\eta_2(d,c)\right)\psi\left(a + \frac{1}{2}\eta_1(b,a), c + \frac{1}{2}\eta_2(d,c)\right)$$

$$\supseteq \frac{1}{\eta_1(b,a)\eta_2(d,c)} \int_a^{a+\eta_1(b,a)} \int_c^{c+\eta_2(d,c)} \xi(u,v)\psi(u,v)dvdu$$

$$+ \frac{5}{36}N_1(a,b,c,d) + \frac{7}{36}N_2(a,b,c,d) + \frac{7}{36}N_3(a,b,c,d) + \frac{2}{9}N_4(a,b,c,d),$$

where $N_1(a,b,c,d)$, $N_2(a,b,c,d)$, $N_3(a,b,c,d)$, *and* $N_4(a,b,c,d)$ *are defined as previous.*

Proof Since ξ and ψ are interval-valued coordinated preinvex functions, therefore from Corollary 6.3.3, we have

$$2\xi\left(a + \frac{1}{2}\eta_1(b,a), c + \frac{1}{2}\eta_2(d,c)\right)\psi\left(a + \frac{1}{2}\eta_1(b,a), c + \frac{1}{2}\eta_2(d,c)\right)$$

$$\supseteq \frac{1}{\eta_1(b,a)} \int_a^{a+\eta_1(b,a)} \xi(u, c + \frac{1}{2}\eta_2(d,c))\psi(u, c + \frac{1}{2}\eta_2(d,c))du$$

$$+ \frac{1}{6}\left[\xi(a, c + \frac{1}{2}\eta_2(d,c))\psi(a, c + \frac{1}{2}\eta_2(d,c))\right.$$

$$\left. +\xi(a + \eta_1(b,a), c + \frac{1}{2}\eta_2(d,c))\psi(a + \eta_1(b,a), c + \frac{1}{2}\eta_2(d,c))\right]$$

$$+\frac{1}{3}\left[\xi(a,c+\frac{1}{2}\eta_2(d,c))\psi(a+\eta_1(b,a),c+\frac{1}{2}\eta_2(d,c))\right.$$

$$\left.+\xi(a+\eta_1(b,a),c+\frac{1}{2}\eta_2(d,c))\psi(a,c+\frac{1}{2}\eta_2(d,c))\right] \qquad (6.57)$$

and

$$2\xi\left(a+\frac{1}{2}\eta_1(b,a),c+\frac{1}{2}\eta_2(d,c)\right)\psi\left(a+\frac{1}{2}\eta_1(b,a),c+\frac{1}{2}\eta_2(d,c)\right)$$

$$\geqq\frac{1}{\eta_2(d,c)}\int_c^{c+\eta_2(d,c)}\xi(c+\frac{1}{2}\eta_2(d,c),v)\psi(a+\frac{1}{2}\eta_1(b,a),v)dv$$

$$+\frac{1}{6}\left[\xi(a+\frac{1}{2}\eta_1(b,a),c)\psi(a+\frac{1}{2}\eta_1(b,a),c)\right.$$

$$\left.+\xi(a+\frac{1}{2}\eta_1(b,a),c+\eta_2(d,c))\psi(a+\frac{1}{2}\eta_1(b,a),c+\eta_2(d,c))\right]$$

$$+\frac{1}{3}\left[\xi(a+\frac{1}{2}\eta_1(b,a),c)\psi(a+\frac{1}{2}\eta_1(b,a),c+\eta_2(d,c))\right.$$

$$\left.+\xi(a+\frac{1}{2}\eta_1(b,a),c+\eta_2(d,c))\psi(a+\frac{1}{2}\eta_1(b,a),c)\right]. \qquad (6.58)$$

Adding (6.57) and (6.58), then multiplying both sides of the resultant one by two, we find

$$8\xi\left(a+\frac{1}{2}\eta_1(b,a),c+\frac{1}{2}\eta_2(d,c)\right)\psi\left(a+\frac{1}{2}\eta_1(b,a),c+\frac{1}{2}\eta_2(d,c)\right)$$

$$\geqq\frac{2}{\eta_1(b,a)}\int_a^{a+\eta_1(b,a)}\xi(u,c+\frac{1}{2}\eta_2(d,c))\psi(u,c+\frac{1}{2}\eta_2(d,c))du$$

$$+\frac{2}{\eta_2(d,c)}\int_c^{c+\eta_2(d,c)}\xi(c+\frac{1}{2}\eta_2(d,c),v)\psi(a+\frac{1}{2}\eta_1(b,a),v)dv$$

$$+\frac{1}{6}\left[2\xi(a,c+\frac{1}{2}\eta_2(d,c))\psi(a,c+\frac{1}{2}\eta_2(d,c))\right.$$

$$+2\xi(a+\eta_1(b,a),c+\frac{1}{2}\eta_2(d,c))\psi(a+\eta_1(b,a),c+\frac{1}{2}\eta_2(d,c))$$

$$+2\xi(a+\frac{1}{2}\eta_1(b,a),c)\psi(a+\frac{1}{2}\eta_1(b,a),c)$$

$$\left.+2\xi(a+\frac{1}{2}\eta_1(b,a),c+\eta_2(d,c))\psi(a+\frac{1}{2}\eta_1(b,a),c+\eta_2(d,c))\right]$$

$$+\frac{1}{3}\left[2\xi(a,c+\frac{1}{2}\eta_2(d,c))\psi(a+\eta_1(b,a),c+\frac{1}{2}\eta_2(d,c))\right.$$

$$+2\xi(a+\eta_1(b,a),c+\frac{1}{2}\eta_2(d,c))\psi(a,c+\frac{1}{2}\eta_2(d,c))$$

$$+ 2\xi(a + \frac{1}{2}\eta_1(b,a),c)\psi(a + \frac{1}{2}\eta_1(b,a),c + \eta_2(d,c))$$

$$+ 2\xi(a + \frac{1}{2}\eta_1(b,a),c + \eta_2(d,c))\psi(a + \frac{1}{2}\eta_1(b,a),c)\Big].$$ (6.59)

Now, from Corollary 6.3.3, we have

$$2\xi(a,c + \frac{1}{2}\eta_2(d,c))\psi(a,c + \frac{1}{2}\eta_2(d,c))$$

$$\supseteq \frac{1}{\eta_2(d,c)} \int_c^{c+\eta_2(d,c)} \xi(a,v)\psi(a,v)dv$$

$$+ \frac{1}{6}[\xi(a,c)\psi(a,c) + \xi(a,c + \eta_2(d,c))\psi(a,c + \eta_2(d,c))]$$

$$+ \frac{1}{3}[\xi(a,c)\psi(a,c + \eta_2(d,c)) + \xi(a,c + \eta_2(d,c))\psi(a,c)],$$ (6.60)

$$2\xi(a + \eta_1(b,a),c + \frac{1}{2}\eta_2(d,c))\psi(a + \eta_1(b,a),c + \frac{1}{2}\eta_2(d,c))$$

$$\supseteq \frac{1}{\eta_2(d,c)} \int_c^{c+\eta_2(d,c)} \xi(a + \eta_1(b,a),v)\psi(a + \eta_1(b,a),v)dv$$

$$+ \frac{1}{6}[\xi(a + \eta_1(b,a),c)\psi(a + \eta_1(b,a),c)$$

$$+ \xi(a + \eta_1(b,a),c + \eta_2(d,c))\psi(a + \eta_1(b,a),c + \eta_2(d,c))]$$

$$+ \frac{1}{3}[\xi(a + \eta_1(b,a),c)\psi(a + \eta_1(b,a),c + \eta_2(d,c))$$

$$+ \xi(a + \eta_1(b,a),c + \eta_2(d,c))\psi(a + \eta_1(b,a),c)],$$ (6.61)

$$2\xi\left(a + \frac{1}{2}\eta_1(b,a),c\right)\psi\left(a + \frac{1}{2}\eta_1(b,a),c\right)$$

$$\supseteq \frac{1}{\eta_1(b,a)} \int_a^{a+\eta_1(b,a)} \xi(u,c)\psi(u,c)du$$

$$+ \frac{1}{6}[\xi(a,c)\psi(a,c) + \xi(a + \eta_1(b,a),c)\psi(a + \eta_1(b,a),c)]$$

$$+ \frac{1}{3}[\xi(a,c)\psi(a + \eta_1(b,a),c) + \xi(a + \eta_1(b,a),c)\psi(a,c)],$$ (6.62)

$$2\xi\left(a + \frac{1}{2}\eta_1(b,a), c + \eta_2(d,c)\right)\psi\left(a + \frac{1}{2}\eta_1(b,a), c + \eta_2(d,c)\right)$$

$$\supseteq \frac{1}{\eta_1(b,a)}\int_a^{a+\eta_1(b,a)} \xi(u, c + \eta_2(d,c))\psi(u, c + \eta_2(d,c))du$$

$$+ \frac{1}{6}[\xi(a, c + \eta_2(d,c))\psi(a, c + \eta_2(d,c))$$
$$+ \xi(a + \eta_1(b,a), c + \eta_2(d,c))\psi(a + \eta_1(b,a), c + \eta_2(d,c))]$$
$$+ \frac{1}{3}[\xi(a, c + \eta_2(d,c))\psi(a + \eta_1(b,a), c + \eta_2(d,c))$$
$$+ \xi(a + \eta_1(b,a), c + \eta_2(d,c))\psi(a, c + \eta_2(d,c))], \tag{6.63}$$

$$2\xi(a, c + \frac{1}{2}\eta_2(d,c))\psi(a + \eta_1(b,a), c + \frac{1}{2}\eta_2(d,c))$$

$$\supseteq \frac{1}{\eta_2(d,c)}\int_c^{c+\eta_2(d,c)} \xi(a, v)\psi(a + \eta_1(b,a), v)dv$$

$$+ \frac{1}{6}[\xi(a, c)\psi(a + \eta_1(b,a), c) + \xi(a, c + \eta_2(d,c))\psi(a + \eta_1(b,a), c + \eta_2(d,c))]$$
$$+ \frac{1}{3}[\xi(a, c)\psi(a + \eta_1(b,a), c + \eta_2(d,c)) + \xi(a, c + \eta_2(d,c))\psi(a + \eta_1(b,a), c)], \tag{6.64}$$

$$2\xi(a + \eta_1(b,a), c + \frac{1}{2}\eta_2(d,c))\psi(a, c + \frac{1}{2}\eta_2(d,c))$$

$$\supseteq \frac{1}{\eta_2(d,c)}\int_c^{c+\eta_2(d,c)} \xi(a + \eta_1(b,a), v)\psi(a, v)dv$$

$$+ \frac{1}{6}[\xi(a + \eta_1(b,a), c)\psi(a, c) + \xi(a + \eta_1(b,a), c + \eta_2(d,c))\psi(a, c + \eta_2(d,c))]$$
$$+ \frac{1}{3}[\xi(a + \eta_1(b,a), c)\psi(a, c + \eta_2(d,c)) + \xi(a + \eta_1(b,a), c + \eta_2(d,c))\psi(a, c)], \tag{6.65}$$

$$2\xi\left(a + \frac{1}{2}\eta_1(b,a), c\right)\psi\left(a + \frac{1}{2}\eta_1(b,a), c + \eta_2(d,c)\right)$$

$$\supseteq \frac{1}{\eta_1(b,a)}\int_a^{a+\eta_1(b,a)} \xi(u, c)\psi(u, c + \eta_2(d,c))du$$

$$+ \frac{1}{6}[\xi(a, c)\psi(a, c + \eta_2(d,c)) + \xi(a + \eta_1(b,a), c)\psi(a + \eta_1(b,a), c + \eta_2(d,c))]$$
$$+ \frac{1}{3}[\xi(a, c)\psi(a + \eta_1(b,a), c + \eta_2(d,c)) + \xi(a + \eta_1(b,a), c)\psi(a, c + \eta_2(d,c))], \tag{6.66}$$

$$2\xi\left(a+\frac{1}{2}\eta_1(b,a),c+\eta_2(d,c)\right)\psi\left(a+\frac{1}{2}\eta_1(b,a),c\right)$$

$$\supseteq\frac{1}{\eta_1(b,a)}\int_a^{a+\eta_1(b,a)}\xi(u,c+\eta_2(d,c))\psi(u,c)du$$

$$+\frac{1}{6}[\xi(a,c+\eta_2(d,c))\psi(a,c)+\xi(a+\eta_1(b,a),c+\eta_2(d,c))\psi(a+\eta_1(b,a),c)]$$

$$+\frac{1}{3}[\xi(a,c+\eta_2(d,c))\psi(a+\eta_1(b,a),c)+\xi(a+\eta_1(b,a),c+\eta_2(d,c))\psi(a,c)].$$

$$(6.67)$$

Using (6.60)–(6.67) in (6.59), we get

$$8\xi\left(a+\frac{1}{2}\eta_1(b,a),c+\frac{1}{2}\eta_2(d,c)\right)\psi\left(a+\frac{1}{2}\eta_1(b,a),c+\frac{1}{2}\eta_2(d,c)\right)$$

$$\supseteq\frac{2}{\eta_1(b,a)}\int_a^{a+\eta_1(b,a)}\xi\left(u,c+\frac{1}{2}\eta_2(d,c)\right)\psi\left(u,c+\frac{1}{2}\eta_2(d,c)\right)du$$

$$+\frac{2}{\eta_2(d,c)}\int_c^{c+\eta_2(d,c)}\xi\left(c+\frac{1}{2}\eta_2(d,c),v\right)\psi\left(a+\frac{1}{2}\eta_1(b,a),v\right)dv$$

$$+\frac{1}{6\eta_2(d,c)}\int_c^{c+\eta_2(d,c)}(\xi(a,v)\psi(a,v)+\xi(a+\eta_1(b,a),v)\psi(a+\eta_1(b,a),v))dv$$

$$+\frac{1}{3\eta_2(d,c)}\int_c^{c+\eta_2(d,c)}(\xi(a,v)\psi(a+\eta_1(b,a),v)+\xi(a+\eta_1(b,a),v)\psi(a,v))dv$$

$$+\frac{1}{6\eta_1(b,a)}\int_a^{a+\eta_1(b,a)}(\xi(u,c)\psi(u,c)+\xi(u,c+\eta_2(d,c))\psi(u,c+\eta_2(d,c)))du$$

$$+\frac{1}{3\eta_1(b,a)}\int_a^{a+\eta_1(b,a)}(\xi(u,c)\psi(u,c+\eta_2(d,c))+\xi(u,c+\eta_2(d,c))\psi(u,c))du$$

$$+\frac{1}{18}N_1(a,b,c,d)+\frac{1}{9}N_2(a,b,c,d)+\frac{1}{9}N_3(a,b,c,d)+\frac{2}{9}N_4(a,b,c,d).\quad(6.68)$$

Again from Corollary 6.3.3, we have

$$\frac{2}{\eta_2(d,c)}\int_c^{c+\eta_2(d,c)}\xi\left(a+\frac{1}{2}\eta_1(b,a),v\right)\psi\left(a+\frac{1}{2}\eta_1(b,a),v\right)dv$$

$$\supseteq\frac{1}{\eta_1(b,a)\eta_2(d,c)}\int_a^{a+\eta_1(b,a)}\int_c^{c+\eta_2(d,c)}\xi(u,v)\psi(u,v)dvdu$$

$$+\frac{1}{6\eta_2(d,c)}\int_c^{c+\eta_2(d,c)}(\xi(a,v)\psi(a,v)+\xi(a+\eta_1(b,a),v)\psi(a+\eta_1(b,a),v))dv$$

$$+\frac{1}{3\eta_2(d,c)}\int_c^{c+\eta_2(d,c)}(\xi(a,v)\psi(a+\eta_1(b,a),v)+\xi(a+\eta_1(b,a),v)\psi(a,v))dv,$$

$$(6.69)$$

$$\frac{2}{\eta_1(b,a)} \int_a^{a+\eta_1(b,a)} \xi\left(u, c + \frac{1}{2}\eta_2(d,c)\right)\psi\left(u, c + \frac{1}{2}\eta_2(d,c)\right)du$$

$$\supseteq \frac{1}{\eta_1(b,a)\eta_2(d,c)} \int_a^{a+\eta_1(b,a)} \int_c^{c+\eta_2(d,c)} \xi(u,v)\psi(u,v)dvdu$$

$$+ \frac{1}{6\eta_1(b,a)} \int_a^{a+\eta_1(b,a)} (\xi(u,c)\psi(u,c) + \xi(u, c + \eta_2(d,c))\psi(u, c + \eta_2(d,c)))du$$

$$+ \frac{1}{3\eta_1(b,a)} \int_a^{a+\eta_1(b,a)} (\xi(u,c)\psi(u, c + \eta_2(d,c)) + \xi(u, c + \eta_2(d,c))\psi(u,c))du. \tag{6.70}$$

Using (6.69) and (6.70) in (6.68), we get

$$8\xi\left(a + \frac{1}{2}\eta_1(b,a), c + \frac{1}{2}\eta_2(d,c)\right)\psi\left(a + \frac{1}{2}\eta_1(b,a), c + \frac{1}{2}\eta_2(d,c)\right)$$

$$\supseteq \frac{2}{\eta_1(b,a)\eta_2(d,c)} \int_a^{a+\eta_1(b,a)} \int_c^{c+\eta_2(d,c)} \xi(u,v)\psi(u,v)dvdu$$

$$+ \frac{1}{3\eta_2(d,c)} \int_c^{c+\eta_2(d,c)} (\xi(a,v)\psi(a,v) + \xi(a + \eta_1(b,a), v)\psi(a + \eta_1(b,a), v)$$

$$+ 2\xi(a,v)\psi(a + \eta_1(b,a), v) + 2\xi(a + \eta_1(b,a), v)\psi(a,v))dv$$

$$+ \frac{1}{3\eta_1(b,a)} \int_a^{a+\eta_1(b,a)} (\xi(u,c)\psi(u,c) + \xi(u, c + \eta_2(d,c))\psi(u, c + \eta_2(d,c))$$

$$+ 2\xi(u,c)\psi(u, c + \eta_2(d,c)) + 2\xi(u, c + \eta_2(d,c))\psi(u,c))du$$

$$+ \frac{1}{18}N_1(a,b,c,d) + \frac{1}{9}N_2(a,b,c,d) + \frac{1}{9}N_3(a,b,c,d) + \frac{2}{9}N_4(a,b,c,d). \tag{6.71}$$

Applying Corollary 6.3.3 for each integral in right side of (6.71), we obtain our desired result.

Remark 6.4.4. *If we put $\eta_1(b,a) = b - a$ and $\eta_2(d,c) = d - c$ in Theorem 6.4.5, we obtain Theorem 9 of [183], i.e.,*

$$4\xi\left(\frac{a+b}{2}, \frac{c+d}{2}\right)\psi\left(\frac{a+b}{2}, \frac{c+d}{2}\right)$$

$$\supseteq \frac{1}{(b-a)(d-c)} \int_a^b \int_c^d \xi(u,v)\psi(u,v)dvdu$$

$$+ \frac{5}{36}N_1^*(a,b,c,d) + \frac{7}{36}N_2^*(a,b,c,d) + \frac{7}{36}N_3^*(a,b,c,d) + \frac{2}{9}N_4^*(a,b,c,d),$$

where $N_1^*(a,b,c,d)$, $N_2^*(a,b,c,d)$, $N_3^*(a,b,c,d)$, and $N_4^*(a,b,c,d)$ are defined as previous.

6.5 Hermite–Hadamard Type Fractional Inclusions for Harmonically h-Preinvex Interval-Valued Functions

In this section, first, we define harmonically h-preinvex interval-valued function and discuss some special cases of harmonically h-preinvex interval-valued function [91].

Definition 6.5.1. *Let* $h : [0,1] \subseteq J \to \mathbb{R}$ *be a nonnegative function such that* $h \not\equiv 0$, *and* $K \subseteq \mathbb{R} \backslash \{0\}$ *be a harmonic invex set with respect to* $\eta(.,.)$. *Let* $\xi : K \subseteq \mathbb{R} \backslash \{0\} \to \mathbb{R}_I^+$ *be an interval-valued function on set* K, *then* ξ *is called harmonically h-preinvex interval-valued function with respect to* $\eta(.,.)$ *if*

$$\xi \left(\frac{x(x + \eta(y,x))}{x + (1-\delta)\eta(y,x)} \right) \supseteq h(1-\delta)\xi(x) + h(\delta)\xi(y), \quad \forall \, \delta \in [0,1] \text{ and } \forall \, x, y \in K.$$

Now, we consider some special cases of harmonically h-preinvex interval-valued functions.

(1) For $h(\delta) = 1$, function ξ is called a harmonically P−preinvex interval-valued function.

(2) For $h(\delta) = \delta$, function ξ is called a harmonically preinvex interval-valued function.

(3) If $h(\delta) = \delta^s$, $s \in (0,1)$, then we get the definition of Breckner type of s−harmonically preinvex interval-valued functions.

(4) If $h(\delta) = \delta^{-s}$, $s \in (0,1)$, then we get the definition of Godunova-Levin type of s−harmonically preinvex interval-valued functions.

Example 6.5.1. *Let* $K = [1,2] \subset \mathbb{R} \backslash \{0\}$, $\xi(x) = \left[1 - \frac{1}{2x^2}, 1 + \frac{1}{2x} \right]$, $\eta(y,x) = y - 2x$, $h(\delta) = \delta$ *then* ξ *is harmonically h-preinvex interval-valued function on* K.

Now, we establish fractional inclusion of Hermite–Hadamard for harmonically h-preinvex interval-valued functions [91].

Theorem 6.5.1. *Let* $h : [0,1] \to \mathbb{R}$ *be a nonnegative function such that* $h(\frac{1}{2}) \neq 0$. *Let* $\xi : K = [c, c + \eta(d,c)] \subseteq \mathbb{R} \backslash \{0\} \to \mathbb{R}_I^+$ *be a harmonically h-preinvex interval-valued function such that* $\xi = [\underline{\xi}, \overline{\xi}]$ *and* $c, d \in K$ *with* $c < c + \eta(d,c)$. *If* $\xi \in L[c, c + \eta(d,c)]$, $\alpha > 0$ *and* η *holds Condition C, then*

$$\frac{1}{\alpha h(\frac{1}{2})} \xi \left(\frac{2c(c + \eta(d,c))}{2c + \eta(d,c)} \right) \supseteq \Gamma(\alpha) \left(\frac{c(c + \eta(d,c))}{\eta(d,c)} \right)^\alpha \left[J^\alpha_{\left(\frac{1}{c + \eta(d,c)} \right)^+} (\xi o \Omega) \left(\frac{1}{c} \right) \right.$$

$$+ J^\alpha_{\left(\frac{1}{c} \right)^-} (\xi o \Omega) \left. \left(\frac{1}{c + \eta(d,c)} \right) \right]$$

$$\supseteq [\xi(c) + \xi(c + \eta(d,c))] \int_0^1 \delta^{\alpha-1}[h(\delta) + h(1-\delta)]d\delta,$$

where $\Omega(x) = \frac{1}{x}$ and $\xi o \Omega$ is defined by $\xi o \Omega(x) = \xi(\Omega(x)), \ \forall\, x \in \left[\frac{1}{c+\eta(d,c)}, \frac{1}{c}\right].$

Proof Since ξ is harmonically h-preinvex interval-valued function on $[c, c + \eta(d, c)]$, we have

$$\frac{1}{h(\frac{1}{2})}\xi\left(\frac{2x(x+\eta(y,x))}{2x+\eta(y,x)}\right) \supseteq \xi(x) + \xi(y), \ \forall\, x,y \in [c, c+\eta(d,c))]. \quad (6.72)$$

Let $x = \frac{c(c+\eta(d,c))}{c+(1-\delta)\eta(d,c)}$ and $y = \frac{c(c+\eta(d,c))}{c+\delta\eta(d,c)}$. Then, using Condition C in (6.72), we get

$$\frac{1}{h(\frac{1}{2})}\xi\left(\frac{2c(c+\eta(d,c))}{2c+\eta(d,c)}\right) \supseteq \xi\left(\frac{c(c+\eta(d,c))}{c+(1-\delta)\eta(d,c)}\right) + \xi\left(\frac{c(c+\eta(d,c))}{c+\delta\eta(d,c)}\right). \quad (6.73)$$

Multiplying (6.73) by $\delta^{\alpha-1}$, $\alpha > 0$ and integrating over $[0,1]$ with respect to δ, we have

$$\frac{1}{h(\frac{1}{2})}(IR)\int_0^1 \delta^{\alpha-1}\xi\left(\frac{2c(c+\eta(d,c))}{2c+\eta(d,c)}\right)d\delta$$
$$\supseteq (IR)\int_0^1 \delta^{\alpha-1}\xi\left(\frac{c(c+\eta(d,c))}{c+(1-\delta)\eta(d,c)}\right)d\delta$$
$$+ (IR)\int_0^1 \delta^{\alpha-1}\xi\left(\frac{c(c+\eta(d,c))}{c+\delta\eta(d,c)}\right)d\delta. \quad (6.74)$$

Applying Theorem 6.2.1 in above relation, we get

$$(IR)\int_0^1 \delta^{\alpha-1}\xi\left(\frac{2c(c+\eta(d,c))}{2c+\eta(d,c)}\right)d\delta$$
$$= \left[(R)\int_0^1 \delta^{\alpha-1}\underline{\xi}\left(\frac{2c(c+\eta(d,c))}{2c+\eta(d,c)}\right)d\delta,\right.$$
$$(R)\int_0^1 \delta^{\alpha-1}\overline{\xi}\left(\frac{2c(c+\eta(d,c))}{2c+\eta(d,c)}\right)d\delta\right]$$
$$= \left[\frac{1}{\alpha}\underline{\xi}\left(\frac{2c(c+\eta(d,c))}{2c+\eta(d,c)}\right), \frac{1}{\alpha}\overline{\xi}\left(\frac{2c(c+\eta(d,c))}{2c+\eta(d,c)}\right)\right]$$
$$= \frac{1}{\alpha}\xi\left(\frac{2c(c+\eta(d,c))}{2c+\eta(d,c)}\right), \quad (6.75)$$

$$(IR)\int_0^1 \delta^{\alpha-1}\xi\left(\frac{c(c+\eta(d,c))}{c+(1-\delta)\eta(d,c)}\right)d\delta$$
$$= \left[(R)\int_0^1 \delta^{\alpha-1}\underline{\xi}\left(\frac{c(c+\eta(d,c))}{c+(1-\delta)\eta(d,c)}\right)d\delta,\right.$$
$$(R)\int_0^1 \delta^{\alpha-1}\overline{\xi}\left(\frac{c(c+\eta(d,c))}{c+(1-\delta)\eta(d,c)}\right)d\delta\right] \quad (6.76)$$

$$= \Gamma(\alpha) \left(\frac{c(c + \eta(d, c))}{\eta(d, c)} \right)^\alpha \left[J^\alpha_{\left(\frac{1}{c + \eta(d,c)} \right)^+ \underline{\xi} o\Omega} \left(\frac{1}{c} \right), J^\alpha_{\left(\frac{1}{c + \eta(d,c)} \right)^+ \overline{\xi} o\Omega} \left(\frac{1}{c} \right) \right]$$

$$= \Gamma(\alpha) \left(\frac{c(c + \eta(d, c))}{\eta(d, c)} \right)^\alpha J^\alpha_{\left(\frac{1}{c + \eta(d,c)} \right)^+ (\xi o\Omega)} \left(\frac{1}{c} \right). \tag{6.77}$$

Similarly,

$$(IR) \int_0^1 t^{\alpha-1} \xi \left(\frac{c(c + \eta(d, c))}{c + \delta\eta(d, c)} \right) d\delta$$

$$= \Gamma(\alpha) \left(\frac{c(c + \eta(d, c))}{\eta(d, c)} \right)^\alpha J^\alpha_{\left(\frac{1}{c} \right)^- (\xi o\Omega)} \left(\frac{1}{c + \eta(d, c)} \right). \tag{6.78}$$

Using (6.75), (6.77) and (6.78) in (6.74), we have

$$\frac{1}{\alpha h(\frac{1}{2})} \xi \left(\frac{2c(c + \eta(d, c))}{2c + \eta(d, c)} \right)$$

$$\supseteq \Gamma(\alpha) \left(\frac{c(c + \eta(d, c))}{\eta(d, c)} \right)^\alpha \left[J^\alpha_{\left(\frac{1}{c + \eta(d,c)} \right)^+ (\xi o\Omega)} \left(\frac{1}{c} \right) \right.$$

$$\left. + J^\alpha_{\left(\frac{1}{c} \right)^- (\xi o\Omega)} \left(\frac{1}{c + \eta(d, c)} \right) \right]. \tag{6.79}$$

Since, ξ is an harmonically h-preinvex interval-valued function on $[c, c + \eta(d, c)]$, we have

$$\xi \left(\frac{c(c + \eta(d, c))}{c + (1 - \delta)\eta(d, c)} \right) = \xi \left(\frac{(c + \eta(d, c))(c + \eta(d, c) + \eta(c, c + \eta(d, c)))}{c + \eta(d, c) + \delta\eta(c, c + \eta(d, c))} \right)$$

$$\supseteq h(\delta)\xi(c + \eta(d, c)) + h(1 - \delta)\xi(c) \tag{6.80}$$

and

$$\xi \left(\frac{c(c + \eta(d, c))}{c + \delta\eta(d, c)} \right) = \xi \left(\frac{(c + \eta(d, c))(c + \eta(d, c)) + \eta(c, c + \eta(d, c)))}{c + \eta(d, c)) + (1 - \delta)\eta(c, c + \eta(d, c))} \right)$$

$$\supseteq h(1 - \delta)\xi(c + \eta(d, c)) + h(\delta)\xi(c). \tag{6.81}$$

Adding (6.80) and (6.81), we have

$$\xi \left(\frac{c(c + \eta(d, c))}{c + (1 - \delta)\eta(d, c)} \right) + \xi \left(\frac{c(c + \eta(d, c))}{c + \delta\eta(d, c)} \right)$$

$$\supseteq [h(\delta) + h(1 - \delta)][\xi(c) + \xi(c + \eta(d, c))]. \tag{6.82}$$

Multiplying (6.82) by $\delta^{\alpha-1}$ and integrating over $[0,1]$ with respect to δ, we have

$$(IR)\int_0^1 \delta^{\alpha-1}\xi\left(\frac{c(c+\eta(d,c))}{c+(1-\delta)\eta(d,c)}\right)d\delta$$

$$+(IR)\int_0^1 \delta^{\alpha-1}\xi\left(\frac{c(c+\eta(d,c))}{c+\delta\eta(d,c)}\right)d\delta$$

$$\supseteq (IR)\int_0^1 \delta^{\alpha-1}[h(\delta)+h(1-\delta)][\xi(c)+\xi(c+\eta(d,c))]d\delta.$$

This implies

$$\Gamma(\alpha)\left(\frac{c(c+\eta(d,c))}{\eta(d,c)}\right)^\alpha\left[J^\alpha_{(\frac{1}{c+\eta(d,c)})^+}(\xi o\Omega)\left(\frac{1}{c}\right)+J^\alpha_{(\frac{1}{c})^-}(\xi o\Omega)\left(\frac{1}{c+\eta(d,c)}\right)\right]$$

$$\supseteq [\xi(c)+\xi(c+\eta(d,c))]\int_0^1 \delta^{\alpha-1}[h(\delta)+h(1-\delta)]d\delta. \qquad (6.83)$$

From (6.79) and (6.83), we get

$$\frac{1}{\alpha h(\frac{1}{2})}\xi\left(\frac{2c(c+\eta(d,c))}{2c+\eta(d,c)}\right)$$

$$\supseteq \Gamma(\alpha)\left(\frac{c(c+\eta(d,c))}{\eta(d,c)}\right)^\alpha\left[J^\alpha_{(\frac{1}{c+\eta(d,c)})^+}(\xi o\Omega)\left(\frac{1}{c}\right)\right.$$

$$\left.+J^\alpha_{(\frac{1}{c})^-}(\xi o\Omega)\left(\frac{1}{c+\eta(d,c)}\right)\right]$$

$$\supseteq [\xi(c)+\xi(c+\eta(d,c))]\int_0^1 \delta^{\alpha-1}[h(\delta)+h(1-\delta)]d\delta.$$

Example 6.5.2. Let $K = [c, c+\eta(d,c))] = [1,2]$, $\eta(d,c) = d - 2c$. Let $\alpha = 1$ and $h(\delta) = \delta \; \forall \; \delta \in [0,1]$, $\xi : K \to \mathbb{R}_I^+$ be defined by

$$\xi(x) = \left[-\frac{1}{x}+2, \frac{1}{x}+2\right], \quad \forall \; x \in K.$$

Then all assumptions of Theorem 6.5.1 are satisfied.

Now we present some particular cases of Theorem 6.5.1.

Corollary 6.5.1. *If $\alpha = 1$, then Theorem 6.5.1 gives the following result:*

$$\frac{1}{h(\frac{1}{2})}\xi\left(\frac{2c(c+\eta(d,c))}{2c+\eta(d,c)}\right) \supseteq \frac{2c(c+\eta(d,c))}{\eta(d,c)}\int_c^{c+\eta(d,c)}\frac{\xi(x)}{x^2}dx$$

$$\supseteq [\xi(c)+\xi(c+\eta(d,c))]\int_0^1 [h(\delta)+h(1-\delta)]d\delta.$$

Corollary 6.5.2. *If $h(\delta) = \delta$, then Theorem 6.5.1 gives the following result:*

$$\xi\left(\frac{2c(c + \eta(d, c))}{2c + \eta(d, c))}\right)$$

$$\supseteq \frac{\Gamma(\alpha + 1)}{2}\left(\frac{c(c + \eta(d, c))}{\eta(d, c)}\right)^\alpha\left[J^\alpha_{\left(\frac{1}{c + \eta(d,c)}\right)^+}\xi\left(\frac{1}{c}\right) + J^\alpha_{\left(\frac{1}{c}\right)^-}\xi\left(\frac{1}{c + \eta(d, c)}\right)\right]$$

$$\supseteq \frac{\xi(c) + \xi(c + \eta(d, c))}{2}.$$

Remark 6.5.1. *If we put $\eta(d, c) = d - c$ in above theorem, we obtain Theorem 5 of [162], i.e.*

$$\frac{1}{\alpha h(\frac{1}{2})}\xi\left(\frac{2cd}{c + d}\right)$$

$$\supseteq \Gamma(\alpha)\left(\frac{cd}{d - c}\right)^\alpha\left[J^\alpha_{\left(\frac{1}{d}\right)^+}(\xi o\Omega)\left(\frac{1}{c}\right) + J^\alpha_{\left(\frac{1}{c}\right)^-}(\xi o\Omega)\left(\frac{1}{d}\right)\right]$$

$$\supseteq [\xi(c) + \xi(d)]\int_0^1 \delta^{\alpha - 1}[h(\delta) + h(1 - \delta)]d\delta.$$

Remark 6.5.2. *If we put $\eta(d, c) = d - c$ and $\alpha = 1$ in above theorem, we obtain Theorem 1 of [185], i.e.,*

$$\frac{1}{h(\frac{1}{2})}\xi\left(\frac{2cd}{c + d}\right)$$

$$\supseteq \left(\frac{cd}{d - c}\right)^\alpha\left[J^\alpha_{\left(\frac{1}{d}\right)^+}(\xi o\Omega)\left(\frac{1}{c}\right) + J^\alpha_{\left(\frac{1}{c}\right)^-}(\xi o\Omega)\left(\frac{1}{d}\right)\right]$$

$$\supseteq [\xi(c) + \xi(d)]\int_0^1 [h(\delta) + h(1 - \delta)]d\delta.$$

Remark 6.5.3. *If we put $\eta(d, c) = d - c$ and $h(\delta) = \delta$ in above theorem, we obtain Theorem 3.6 of [99], i.e.,*

$$\frac{2}{\alpha}\xi\left(\frac{2cd}{c + d}\right)$$

$$\supseteq \Gamma(\alpha)\left(\frac{cd}{d - c}\right)^\alpha\left[J^\alpha_{\left(\frac{1}{d}\right)^+}(\xi o\Omega)\left(\frac{1}{c}\right) + J^\alpha_{\left(\frac{1}{c}\right)^-}(\xi o\Omega)\left(\frac{1}{d}\right)\right]$$

$$\supseteq \frac{1}{\alpha}[\xi(c) + \xi(d)].$$

Next, we prove fractional inclusions of Hermite–Hadamard-type for the product of two harmonically h-preinvex interval-valued functions [91].

Theorem 6.5.2. *Let $h_1, h_2 : [0, 1] \to \mathbb{R}$ be nonnegative functions and $h_1, h_2 \not\equiv 0$ Let $\xi, \psi : K = [c, c + \eta(d, c)] \subseteq \mathbb{R}\backslash\{0\} \to \mathbb{R}_I^+$ be two harmonically h_1- and h_2-preinvex interval-valued functions, respectively, such that $\xi = [\underline{\xi}, \overline{\xi}]$, $\psi = [\underline{\psi}, \overline{\psi}]$*

and $c, d \in K$ with $c < c + \eta(d,c))$. If $\xi\psi \in L^1[c, c + \eta(d,c)]$, $\alpha > 0$ and η holds Condition C, then

$$
\Gamma(\alpha) \left(\frac{c(c + \eta(d,c))}{\eta(d,c)} \right)^\alpha \left[J^\alpha_{\left(\frac{1}{c+\eta(d,c)}\right)^+} (\xi o \Omega) \left(\frac{1}{c} \right) (\psi o \Omega) \left(\frac{1}{c} \right) \right.
$$

$$
\left. + J^\alpha_{\left(\frac{1}{c}\right)^-} (\xi o \Omega) \left(\frac{1}{c + \eta(d,c)} \right) (\psi o \Omega) \left(\frac{1}{c + \eta(d,c)} \right) \right]
$$

$$
\supseteq F(c, c + \eta(d,c)) \int_0^1 [\delta^{\alpha-1} + (1-\delta)^{\alpha-1}] h_1(\delta) h_2(\delta) d\delta
$$

$$
+ G(c, c + \eta(d,c)) \int_0^1 [\delta^{\alpha-1} + (1-\delta)^{\alpha-1}] h_1(1-\delta) h_2(\delta)] d\delta, \qquad (6.84)
$$

where $F(c, c + \eta(d,c)) = \xi(c)\psi(c) + \xi(c + \eta(d,c))\psi(c + \eta(d,c))$,
$G(c, c + \eta(d,c)) = \xi(c)\psi(c + \eta(d,c)) + \xi(c + \eta(d,c))\psi(c)$
and $\Omega(x) = \frac{1}{x}$.

Proof Since ξ and ψ are two harmonically h_1- and h_2-preinvex interval-valued functions on $[c, c + \eta(d,c))]$, respectively. Therefore,

$$
\xi\left(\frac{c(c+\eta(d,c))}{c + (1-\delta)\eta(d,c)} \right) = \xi\left(\frac{(c+\eta(d,c))(c+\eta(d,c) + \eta(c, c+\eta(d,c)))}{c + \eta(d,c) + \delta\eta(c, c+\eta(d,c))} \right)
$$

$$
\supseteq h_1(\delta)\xi(c + \eta(d,c)) + h_1(1-\delta)\xi(c) \qquad (6.85)
$$

and

$$
\psi\left(\frac{c(c+\eta(d,c))}{c + (1-\delta)\eta(d,c)} \right) = \psi\left(\frac{(c+\eta(d,c))(c+\eta(d,c) + \eta(c, c+\eta(d,c)))}{c + \eta(d,c) + \delta\eta(c, c+\eta(d,c))} \right)
$$

$$
\supseteq h_2(\delta)\psi(c + \eta(d,c)) + h_2(1-\delta)\psi(c). \qquad (6.86)
$$

Since, $\xi(x), \psi(x) \in \mathbb{R}^+_I$, $\forall\, x \in [c, c + \eta(d,c)]$, then from (6.85) and (6.86), we obtain

$$
\xi\left(\frac{c(c+\eta(d,c))}{c + (1-\delta)\eta(d,c)} \right) \psi\left(\frac{c(c+\eta(d,c))}{c + (1-\delta)\eta(d,c)} \right)
$$
$$
\supseteq h_1(\delta)h_2(\delta)\xi(c+\eta(d,c))\psi(c+\eta(d,c)) + h_1(1-\delta)h_2(1-\delta)\xi(c)\psi(c)
$$
$$
+ h_1(\delta)h_2(1-\delta)\xi(c+\eta(d,c))\psi(c) + h_1(1-\delta)h_2(\delta)\xi(c)\psi(c+\eta(d,c)).
$$
$$
(6.87)
$$

Similarly,

$$
\xi\left(\frac{c(c+\eta(d,c))}{c + \delta\eta(d,c)} \right) \psi\left(\frac{c(c+\eta(d,c))}{c + \delta\eta(d,c)} \right)
$$
$$
\supseteq h_1(1-\delta)h_2(1-\delta)\xi(c+\eta(d,c))\psi(c+\eta(d,c)) + h_1(\delta)h_2(\delta)\xi(c)\psi(c)
$$
$$
+ h_1(1-\delta)h_2(\delta)\xi(c+\eta(d,c))\psi(c) + h_1(\delta)h_2(1-\delta)\xi(c)\psi(c+\eta(d,c)).
$$
$$
(6.88)
$$

Adding (6.87) and (6.88), we have

$$\xi\left(\frac{c(c+\eta(d,c))}{c+(1-\delta)\eta(d,c)}\right)\psi\left(\frac{c(c+\eta(d,c))}{c+(1-\delta)\eta(d,c)}\right)$$
$$+\xi\left(\frac{c(c+\eta(d,c))}{c+\delta\eta(d,c)}\right)\psi\left(\frac{c(c+\eta(d,c))}{c+\delta\eta(d,c)}\right)$$

$$\supseteq [h_1(\delta)h_2(\delta)+h_1(1-\delta)h_2(1-\delta)][\xi(c)\psi(c)+\xi(c+\eta(d,c))\psi(c+\eta(d,c))]$$
$$+[h_1(\delta)h_2(1-\delta)+h_1(1-\delta)h_2(\delta)][\xi(c+\eta(d,c))\psi(c)+\xi(c)\psi(c+(d,r))]$$
$$=F(c,c+\eta(d,c))[h_1(\delta)h_2(\delta)+h_1(1-\delta)h_2(1-\delta)]$$
$$+G(c,c+\eta(d,c))[h_1(1-\delta)h_2(\delta)+h_1(\delta)h_2(1-\delta)]. \tag{6.89}$$

Multiplying (6.89) by $\delta^{\alpha-1}$ and integrating over $[0,1]$ with respect to δ, we have

$$(IR)\int_0^1 \delta^{\alpha-1}\xi\left(\frac{c(c+\eta(d,c))}{c+(1-\delta)\eta(d,c)}\right)\psi\left(\frac{c(c+\eta(d,c))}{c+(1-\delta)\eta(d,c)}\right)d\delta$$
$$+(IR)\int_0^1 \delta^{\alpha-1}\xi\left(\frac{c(c+\eta(d,c))}{c+\delta\eta(d,c)}\right)\psi\left(\frac{c(c+\eta(d,c))}{c+\delta\eta(d,c)}\right)d\delta$$
$$\supseteq (IR)\int_0^1 \delta^{\alpha-1}F(c,c+\eta(d,c))[h_1(\delta)h_2(\delta)+h_1(1-\delta)h_2(1-\delta)]d\delta$$
$$+(IR)\int_0^1 \delta^{\alpha-1}G(c,c+\eta(d,c))[h_1(1-\delta)h_2(\delta)+h_1(\delta)h_2(1-\delta)]d\delta. \tag{6.90}$$

Since,

$$(IR)\int_0^1 \delta^{\alpha-1}\xi\left(\frac{c(c+\eta(d,c))}{c+(1-\delta)\eta(d,c)}\right)\psi\left(\frac{c(c+\eta(d,c))}{c+(1-\delta)\eta(d,c)}\right)d\delta$$
$$=\Gamma(\alpha)\left(\frac{c(c+\eta(d,c))}{\eta(d,c)}\right)^{\alpha}J^{\alpha}_{\left(\frac{1}{c+\eta(d,c)}\right)^+}(\xi o\Omega)\left(\frac{1}{c}\right)(\psi o\Omega)\left(\frac{1}{c}\right), \tag{6.91}$$

$$(IR)\int_0^1 \delta^{\alpha-1}\xi\left(\frac{c(c+\eta(d,c))}{c+\delta\eta(d,c)}\right)\psi\left(\frac{c(c+\eta(d,c))}{c+\delta\eta(d,c)}\right)d\delta$$
$$=\Gamma(\alpha)\left(\frac{c(c+\eta(d,c))}{\eta(d,c)}\right)^{\alpha}\left[J^{\alpha}_{\left(\frac{1}{c}\right)^-}(\xi o\Omega)\left(\frac{1}{c+\eta(d,c)}\right)\right.$$
$$\left.\times(\psi o\Omega)\left(\frac{1}{c+\eta(d,c)}\right)\right]. \tag{6.92}$$

Using (6.91), (6.92) in (6.90), we have

$$\Gamma(\alpha)\left(\frac{c(c+\eta(d,c))}{\eta(d,c)}\right)^{\alpha}\left[J^{\alpha}_{\left(\frac{1}{c+\eta(d,c)}\right)^+}(\xi o\Omega)\left(\frac{1}{c}\right)(\psi o\Omega)\left(\frac{1}{c}\right)\right.$$

$$\left.+J^{\alpha}_{\left(\frac{1}{c}\right)^-}(\xi o\Omega)\left(\frac{1}{c+\eta(d,c)}\right)(\psi o\Omega)\left(\frac{1}{c+\eta(d,c)}\right)\right]$$

$$\supseteq F(c,c+\eta(d,c))\int_0^1[\delta^{\alpha-1}+(1-\delta)^{\alpha-1}]h_1(\delta)h_2(\delta)d\delta$$

$$+\,G(c,c+\eta(d,c))\int_0^1[\delta^{\alpha-1}+(1-\delta)^{\alpha-1}]h_1(1-\delta)h_2(\delta)]d\delta.$$

Corollary 6.5.3. *If $\alpha = 1$, then Theorem 6.5.2 gives the following result:*

$$\frac{c(c+\eta(d,c))}{\eta(d,c)}\int_c^{c+\eta(d,c))}\frac{\xi(x)\psi(x)}{x^2}dx$$

$$\supseteq F(c,c+\eta(d,c))\int_0^1 h_1(\delta)h_2(\delta)d\delta + G(c,c+\eta(d,c))\int_0^1 h_1(1-\delta)h_2(\delta)]d\delta.$$

Corollary 6.5.4. *If $h_1(\delta) = h_2(\delta) = \delta$, then Theorem 6.5.2 gives the following result:*

$$\Gamma(\alpha)\left(\frac{c(c+\eta(d,c))}{\eta(d,c)}\right)^{\alpha}\left[J^{\alpha}_{\left(\frac{1}{c+\eta(d,c)}\right)^+}(\xi o\Omega)\left(\frac{1}{c}\right)(\psi o\Omega)\left(\frac{1}{c}\right)\right.$$

$$\left.+J^{\alpha}_{\left(\frac{1}{c}\right)^-}(\xi o\Omega)\left(\frac{1}{c+\eta(d,c)}\right)(\psi o\Omega)\left(\frac{1}{c+\eta(d,c)}\right)\right]$$

$$\supseteq F(c,c+\eta(d,c))\int_0^1\delta^2[\delta^{\alpha-1}+(1-\delta)^{\alpha-1}]d\delta$$

$$+\,G(c,c+\eta(d,c))\int_0^1\delta(1-\delta)[\delta^{\alpha-1}+(1-\delta)^{\alpha-1}]d\delta$$

$$=\frac{\alpha^2+\alpha+2}{\alpha(\alpha+1)(\alpha+2)}F(c,c+\eta(d,c))+\frac{2}{(\alpha+1)(\alpha+2)}G(c,c+\eta(d,c)).$$

Remark 6.5.4. *If we put $\eta(d,c) = d-c$ in Theorem 6.5.2, we obtain Theorem 6 of [162], i.e.,*

$$\Gamma(\alpha)\left(\frac{cd}{d-c}\right)^{\alpha}\left[J^{\alpha}_{\left(\frac{1}{d}\right)^+}(\xi o\Omega)\left(\frac{1}{c}\right)(\psi o\Omega)\left(\frac{1}{c}\right)\right.$$

$$\left.+J^{\alpha}_{\left(\frac{1}{c}\right)^-}(\xi o\Omega)\left(\frac{1}{d}\right)(\psi o\Omega)\left(\frac{1}{d}\right)\right]$$

$$\supseteq F(c,d)\int_0^1[\delta^{\alpha-1}+(1-\delta)^{\alpha-1}]h_1(\delta)h_2(\delta)d\delta$$

$$+\,G(c,d)\int_0^1[\delta^{\alpha-1}+(1-\delta)^{\alpha-1}]h_1(1-\delta)h_2(\delta)]d\delta.$$

Remark 6.5.5. *If we put* $\eta(d,c) = d - c$ *and* $\alpha = 1$ *in Theorem 6.5.2, we obtain Theorem 3 of [185], i.e.,*

$$\left(\frac{cd}{d-c)}\right)^{\alpha}\left[J^{\alpha}_{\left(\frac{1}{d}\right)^+}(\xi o\Omega)\left(\frac{1}{c}\right)(\psi o\Omega)\left(\frac{1}{c}\right)\right.$$

$$\left.+J^{\alpha}_{\left(\frac{1}{c}\right)^-}(\xi o\Omega)\left(\frac{1}{d}\right)(\psi o\Omega)\left(\frac{1}{d}\right)\right]$$

$$\supseteq 2F(c,d)\int_0^1 h_1(\delta)h_2(\delta)d\delta + 2G(c,d)\int_0^1 h_1(1-\delta)h_2(\delta)]d\delta.$$

Theorem 6.5.3. *Let* $h_1, h_2 : [0,1] \to \mathbb{R}$ *be nonnegative functions and* $h_1(\frac{1}{2})h_2(\frac{1}{2}) \neq 0$. *Let* $\xi, \psi : K = [c, c+\eta(d,c)] \subseteq \mathbb{R}\backslash\{0\} \to \mathbb{R}^+_I$ *be two harmonically* h_1- *and* h_2-*preinvex interval-valued functions, respectively, such that* $\xi = [\underline{\xi}, \overline{\xi}], \psi = [\underline{\psi}, \overline{\psi}]$ *and* $c, d \in K$ *with* $c < c+\eta(d,c))$. *If* $\xi\psi \in L[c, c+\eta(d,c)]$, $\alpha > 0$ *and* η *holds Condition C, then*

$$\frac{1}{\alpha h_1(\frac{1}{2})h_2(\frac{1}{2})}\xi\left(\frac{2c(c+\eta(d,c))}{2c+\eta(d,c)}\right)\psi\left(\frac{2c(c+\eta(d,c))}{2c+\eta(d,c)}\right)$$

$$\supseteq \Gamma(\alpha)\left(\frac{c(c+\eta(d,c))}{\eta(d,c)}\right)^{\alpha}\left[J^{\alpha}_{\left(\frac{1}{c+\eta(d,c)}\right)^+}(\xi o\Omega)\left(\frac{1}{c}\right)(\psi o\Omega)\left(\frac{1}{c}\right)\right.$$

$$\left.+J^{\alpha}_{\left(\frac{1}{c}\right)^-}(\xi o\Omega)\left(\frac{1}{c+\eta(d,c)}\right)(\psi o\Omega)\left(\frac{1}{c+\eta(d,c)}\right)\right]$$

$$+ F(c, c+\eta(d,c))\int_0^1 (\delta^{\alpha-1} + (1-\delta)^{\alpha-1})h_1(\delta)h_2(1-\delta)d\delta$$

$$+ G(c, c+\eta(d,c))\int_0^1 (\delta^{\alpha-1} + (1-\delta)^{\alpha-1})h_1(\delta)h_2(\delta)d\delta,$$

where $F(c, c+\eta(d,c))$ *and* $G(c, c+\eta(d,c))$ *are defined as previous.*

Proof Since ξ is harmonically h_1-preinvex interval-valued function on $[c, c+\eta(d,c))]$, we have

$$\frac{1}{h_1(\frac{1}{2})}\xi\left(\frac{2x(x+\eta(y,x))}{2x+\eta(y,x)}\right) \supseteq \xi(x) + \xi(y), \ \forall \ x, y \in [c, c+\eta(d,c)]. \quad (6.93)$$

Let $u = \frac{c(c+\eta(d,c))}{c+(1-\delta)\eta(d,c)}$ and $v = \frac{c(c+\eta(d,c))}{c+\delta\eta(d,c)}$. Then, using Condition C in (6.93), we get

$$\frac{1}{h_1(\frac{1}{2})}\xi\left(\frac{2c(c+\eta(d,c))}{2c+\eta(d,c)}\right) \supseteq \xi\left(\frac{c(c+\eta(d,c))}{c+(1-\delta)\eta(d,c)}\right) + \xi\left(\frac{c(c+\eta(d,c))}{c+\delta\eta(d,c)}\right). \quad (6.94)$$

Similarly,

$$\frac{1}{h_2(\frac{1}{2})}\psi\left(\frac{2c(c+\eta(d,c))}{2c+\eta(d,c)}\right) \supseteq \psi\left(\frac{c(c+\eta(d,c))}{c+(1-\delta)\eta(d,c)}\right) + \psi\left(\frac{c(c+\eta(d,c))}{c+\delta\eta(d,c)}\right).$$

$$(6.95)$$

From (6.94) and (6.95), we get

$$\frac{1}{h_1(\frac{1}{2})h_2(\frac{1}{2})}\xi\left(\frac{2c(c+\eta(d,c))}{2c+\eta(d,c)}\right)\psi\left(\frac{2c(c+\eta(d,c))}{2c+\eta(d,c)}\right)$$

$$\supseteq \left[\xi\left(\frac{c(c+\eta(d,c))}{c+(1-\delta)\eta(d,c)}\right) + \xi\left(\frac{c(c+\eta(d,c))}{c+\delta\eta(d,c)}\right)\right]$$

$$\times \left[\psi\left(\frac{c(c+\eta(d,c))}{c+(1-\delta)\eta(d,c)}\right) + \psi\left(\frac{c(c+\eta(d,c))}{c+\delta\eta(d,c)}\right)\right]$$

$$= \xi\left(\frac{c(c+\eta(d,c))}{c+(1-\delta)\eta(d,c)}\right)\psi\left(\frac{c(c+\eta(d,c))}{c+(1-\delta)\eta(d,c)}\right)$$

$$+ \xi\left(\frac{c(c+\eta(d,c))}{c+\delta\eta(d,c)}\right)\psi\left(\frac{c(c+\eta(d,c))}{c+\delta\eta(d,c)}\right)$$

$$+ \left[\xi\left(\frac{c(c+\eta(d,c))}{c+(1-\delta)\eta(d,c)}\right)\psi\left(\frac{c(c+\eta(d,c))}{c+\delta\eta(d,c)}\right)\right.$$

$$\left. +\xi\left(\frac{c(c+\eta(d,c))}{c+\delta\eta(d,c)}\right)\psi\left(\frac{c(c+\eta(d,c))}{c+(1-\delta)\eta(d,c)}\right)\right].$$

$$(6.96)$$

Since $\xi(x)$ and $\psi(x) \in X_I^+$, $\forall x \in [c, c+\eta(d,c)]$ are two harmonically h_1- and h_2- preinvex interval-valued functions, respectively. Therefore,

$$\xi\left(\frac{c(c+\eta(d,c))}{c+(1-\delta)\eta(d,c)}\right)\psi\left(\frac{c(c+\eta(d,c))}{c+\delta\eta(d,c)}\right)$$

$$\supseteq h_1(\delta)h_2(\delta)\xi(c+\eta(d,c))\psi(c) + h_1(1-\delta)h_2(1-\delta)\xi(c)\psi(c+\eta(d,c))$$

$$+ h_1(\delta)h_2(1-\delta)\xi(c+\eta(d,c))\psi(c+\eta(d,c)) + h_1(1-\delta)h_2(\delta)\xi(c)\psi(c).$$

$$(6.97)$$

Similarly,

$$\xi\left(\frac{c(c+\eta(d,c))}{c+\delta\eta(d,c)}\right)\psi\left(\frac{c(c+\eta(d,c))}{c+(1-\delta)\eta(d,c)}\right)$$

$$\supseteq h_1(\delta)h_2(\delta)\xi(c))\psi(c+\eta(d,c)) + h_1(1-\delta)h_2(1-\delta)\xi(c+\eta(d,c))\psi(c)$$

$$+ h_1(\delta)h_2(1-\delta)\xi(c)\psi(c) + h_1(1-\delta)h_2(\delta)\xi(c+\eta(d,c))\psi(c+\eta(d,c)).$$

$$(6.98)$$

Adding (6.97) and (6.98), we obtain

$$\xi\left(\frac{c(c+\eta(d,c))}{c+(1-\delta)\eta(d,c)}\right)\psi\left(\frac{c(c+\eta(d,c))}{c+\delta\eta(d,c)}\right)$$

$$+\xi\left(\frac{c(c+\eta(d,c))}{c+\delta\eta(d,c)}\right)\psi\left(\frac{c(c+\eta(d,c))}{c+(1-\delta)\eta(d,c)}\right)$$

$$\supseteq G(c,c+\eta(d,c))[h_1(\delta)h_2(\delta)+h_1(1-\delta)h_2(1-\delta)]$$

$$+F(c,c+\eta(d,c))[h_1(1-\delta)h_2(\delta)+h_1(\delta)h_2(1-\delta)]. \tag{6.99}$$

From (6.96) and (6.99), we have

$$\frac{1}{h_1(\frac{1}{2})h_2(\frac{1}{2})}\xi\left(\frac{2c(c+\eta(d,c))}{2c+\eta(d,c)}\right)\psi\left(\frac{2c(c+\eta(d,c))}{2c+\eta(d,c)}\right)$$

$$\supseteq \xi\left(\frac{c(c+\eta(d,c))}{c+(1-\delta)\eta(d,c)}\right)\psi\left(\frac{c(c+\eta(d,c))}{c+(1-\delta)\eta(d,c)}\right)$$

$$+\xi\left(\frac{c(c+\eta(d,c))}{c+\delta\eta(d,c)}\right)\psi\left(\frac{c(c+\eta(d,c))}{c+\delta\eta(d,c)}\right)$$

$$+G(c,c+\eta(d,c))[h_1(\delta)h_2(\delta)+h_1(1-\delta)h_2(1-\delta)]d\delta$$

$$+F(c,c+\eta(d,c))[h_1(1-\delta)h_2(\delta)+h_1(\delta)h_2(1-\delta)]d\delta. \tag{6.100}$$

Multiplying (6.100) by $\delta^{\alpha-1}$, then integrating over $[0,1]$ with respect to δ, we get

$$\frac{1}{h_1(\frac{1}{2})h_2(\frac{1}{2})}(IR)\int_0^1\delta^{\alpha-1}\xi\left(\frac{2c(c+\eta(d,c))}{2c+\eta(d,c)}\right)\psi\left(\frac{2c(c+\eta(d,c))}{2c+\eta(d,c)}\right)d\delta$$

$$\supseteq (IR)\int_0^1\delta^{\alpha-1}\xi\left(\frac{c(c+\eta(d,c))}{c+(1-\delta)\eta(d,c)}\right)\psi\left(\frac{c(c+\eta(d,c))}{c+(1-\delta)\eta(d,c)}\right)d\delta$$

$$+(IR)\int_0^1\delta^{\alpha-1}\xi\left(\frac{c(c+\eta(d,c))}{c+\delta\eta(d,c)}\right)\psi\left(\frac{c(c+\eta(d,c))}{c+\delta\eta(d,c)}\right)d\delta$$

$$+G(c,c+\eta(d,c))(IR)\int_0^1\delta^{\alpha-1}[h_1(\delta)h_2(\delta)+h_1(1-\delta)h_2(1-\delta)]d\delta$$

$$+F(c,c+\eta(d,c))(IR)\int_0^1\delta^{\alpha-1}[h_1(1-\delta)h_2(\delta)+h_1(\delta)h_2(1-\delta)]d\delta.$$

This implies,

$$\frac{1}{\alpha h_1(\frac{1}{2})h_2(\frac{1}{2})}\xi\left(\frac{2c(c+\eta(d,c))}{2c+\eta(d,c)}\right)\psi\left(\frac{2c(c+\eta(d,c))}{2c+\eta(d,c)}\right)$$

$$\supseteq \Gamma(\alpha)\left(\frac{c(c+\eta(d,c))}{\eta(d,c)}\right)^\alpha\left[J^\alpha_{(\frac{1}{c+\eta(d,c)})^+}(\xi o \Omega)\left(\frac{1}{c}\right)\psi o \Omega\left(\frac{1}{c}\right)\right.$$

$$\left.+J^\alpha_{(\frac{1}{c})^-}(\xi o \Omega)\left(\frac{1}{c+\eta(d,c)}\right)(\psi o \Omega)\left(\frac{1}{c+\eta(d,c)}\right)\right]$$

$$+F(c,c+\eta(d,c))\int_0^1(\delta^{\alpha-1}+(1-\delta)^{\alpha-1})h_1(\delta)h_2(1-\delta)]d\delta$$

$$+G(c,c+\eta(d,c))\int_0^1(\delta^{\alpha-1}+(1-\delta)^{\alpha-1})h_1(\delta)h_2(\delta)d\delta.$$

Corollary 6.5.5. *If $\alpha = 1$, then Theorem 6.5.3 gives the following result:*

$$\frac{1}{2h_1(\frac{1}{2})h_2(\frac{1}{2})}\xi\left(\frac{2c(c+\eta(d,c))}{2c+\eta(d,c)}\right)\psi\left(\frac{2c(c+\eta(d,c))}{2c+\eta(d,c)}\right)$$

$$\supseteq \frac{c(c+\eta(d,c))}{\eta(d,c)}\int_c^{c+\eta(d,c)}\frac{\xi(x)\psi(x)}{x^2}dx$$

$$+F(c,c+\eta(d,c))\int_0^1 h_1(\delta)h_2(1-\delta)d\delta$$

$$+G(c,c+\eta(d,c))\int_0^1 h_1(\delta)h_2(\delta)d\delta.$$

Corollary 6.5.6. *If $h_1(\delta) = h_2(\delta) = \delta$, then Theorem 6.5.3 gives the following result:*

$$\frac{4}{\alpha}\xi\left(\frac{2c(c+\eta(d,c))}{2c+\eta(d,c)}\right)\psi\left(\frac{2c(c+\eta(d,c))}{2c+\eta(d,c)}\right)$$

$$\supseteq \Gamma(\alpha)\xi\left(\frac{c(c+\eta(d,c))}{\eta(d,c)}\right)^\alpha\left[J^\alpha_{(\frac{1}{c+\eta(d,c)})^+}(\xi o \Omega)\left(\frac{1}{c}\right)\psi o \Omega\left(\frac{1}{c}\right)\right.$$

$$\left.+J^\alpha_{(\frac{1}{c})^-}(\xi o \Omega)\left(\frac{1}{c+\eta(d,c)}\right)(\psi o \Omega)\left(\frac{1}{c+\eta(d,c)}\right)\right]$$

$$+F(c,c+\eta(d,c))\int_0^1\delta(1-\delta)(\delta^{\alpha-1}+(1-\delta)^{\alpha-1})d\delta$$

$$+G(c,c+\eta(d,c))\int_0^1\delta^2(\delta^{\alpha-1}+(1-\delta)^{\alpha-1})d\delta$$

$$= \Gamma(\alpha)\xi \left(\frac{c(c + \eta(d,c))}{\eta(d,c)} \right)^{\alpha} \left[J^{\alpha}_{\left(\frac{1}{c+\eta(d,c)} \right)^+} (\xi o\Omega) \left(\frac{1}{c} \right) \psi o\Omega \left(\frac{1}{c} \right) \right.$$

$$\left. + J^{\alpha}_{\left(\frac{1}{c} \right)^-} (\xi o\Omega) \left(\frac{1}{c + \eta(d,c)} \right) (\psi o\Omega) \left(\frac{1}{c + \eta(d,c)} \right) \right]$$

$$+ \frac{2}{(\alpha+1)(\alpha+2)} F(c, c + \eta(d,c)) + \frac{\alpha^2 + \alpha + 2}{\alpha(\alpha+1)(\alpha+2)} G(c, c + \eta(d,c)).$$

Remark 6.5.6. *If we put $\eta(d,c) = d - c$ in above theorem, we obtain Theorem 7 of [162], i.e.,*

$$\frac{1}{\alpha h_1(\frac{1}{2}) h_2(\frac{1}{2})} \xi \left(\frac{2cd}{c+d} \right) \psi \left(\frac{2cd}{c+d} \right)$$

$$\supseteq \Gamma(\alpha) \left(\frac{cd}{d-c} \right)^{\alpha} \left[J^{\alpha}_{\left(\frac{1}{d} \right)^+} (\xi o\Omega) \left(\frac{1}{c} \right) (\psi o\Omega) \left(\frac{1}{c} \right) \right.$$

$$\left. + J^{\alpha}_{\left(\frac{1}{c} \right)^-} (\xi o\Omega) \left(\frac{1}{d} \right) (\psi o\Omega) \left(\frac{1}{d} \right) \right]$$

$$+ F(c,d) \int_0^1 (\delta^{\alpha-1} + (1-\delta)^{\alpha-1}) h_1(\delta) h_2(1-\delta) d\delta$$

$$+ G(c,d) \int_0^1 (\delta^{\alpha-1} + (1-\delta)^{\alpha-1}) h_1(\delta) h_2(\delta) d\delta.$$

Remark 6.5.7. *If we put $\eta(d,c) = d - c$ and $\alpha = 1$ in above theorem, we obtain Theorem 4 of [185], i.e.,*

$$\frac{1}{h_1(\frac{1}{2}) h_2(\frac{1}{2})} \xi \left(\frac{2cd}{c+d} \right) \psi \left(\frac{2cd}{c+d} \right)$$

$$\supseteq \left(\frac{cd}{d-c} \right) \left[J^{\alpha}_{\left(\frac{1}{d} \right)^+} (\xi o\Omega) \left(\frac{1}{c} \right) (\psi o\Omega) \left(\frac{1}{c} \right) \right.$$

$$\left. + J^{\alpha}_{\left(\frac{1}{c} \right)^-} (\xi o\Omega) \left(\frac{1}{d} \right) (\psi o\Omega) \left(\frac{1}{d} \right) \right]$$

$$+ 2F(c,d) \int_0^1 h_1(\delta) h_2(1-\delta) d\delta + 2G(c,d) \int_0^1 h_1(\delta) h_2(\delta) d\delta.$$

Chapter 7

Some Inequalities for Multidimensional General h-Harmonic Preinvex and Strongly Generalized Convex Stochastic Processes

7.1 Introduction

Dragomir [45] established Hermite–Hadamard inequality for convex functions on the coordinates on a rectangle from the plane $\mathbb{R}^2$. It is well known that every convex mapping $\xi : [a, b] \times [c, d] \to \mathbb{R}$ is convex on the coordinates, but the converse is not true in general. Noor *et al.* [126] introduced harmonic preinvex function and obtained several new Hermite–Hadamard type inequalities for harmonic preinvex functions. Further, Noor *et al.* [121] defined the relative harmonic preinvex functions and derived some integral inequalities Hermite–Hadamard, Simpson's and trapezoidal for the relative harmonic preinvex function.

If T is some index set and S is the common sample space of the random variable, then the collection of random variables $\{\xi_t(s), t \in T, s \in S\}$ is called stochastic process. Allen [3] introduced basic theory of stochastic processes and applied these methods to biological problems, such as enzyme kinetics, population extinction, the spread of epidemics and the genetics of inbreeding.

Nikodem [116] introduced some powerful properties of convex stochastic processes. He proved that a convex stochastic processes $\xi : K \times \Omega \to \overline{\mathbb{R}}$ is continuous if and only if for all $x, y \in K$ and $\delta \in [0, 1]$

$$\xi(\delta x + (1 - \delta)y, .) \leq \delta \xi(x, .) + (1 - \delta)\xi(y, .) \quad \text{almost everywhere,}$$

where $\overline{\mathbb{R}}$ denotes the extended real line. Kuhn [89] studied the convex stochastic programs with a generalized non convex dependence on the random parameters. Kuhn [89] proved that, under certain conditions, the saddle structure can be restored by adding specific random variables to the profit functions. These random variables are referred to as 'correction terms'.

DOI: 10.1201/9781003408284-7

Kotrys [88] defined the strongly convex stochastic processes and derived the Hermite–Hadamard inequality, Jensen inequality and Kuhn and Bernstein theorem. A stochastic process $\xi : K \times \Omega \to \mathbb{R}$ is strongly δ−convex with modulus $\mu(.)$ if and only if the stochastic process $\phi : K \times \Omega \to \mathbb{R}$ defined by $\phi(x, .) = \xi(x, .) - \mu(.)x^2$ is δ−convex [88]. Further, Karahan and Okur [71] obtained some generalized Hermite–Hadamard inequalities for convex stochastic process on the coordinates. Further, Okur and Aliyev [129] introduced general preinvex and multidimensional general preinvex stochastic processes, and derived Hermite–Hadamard inequality for these processes. Further, Ibrahim [63] introduced the concept of strongly h−convex stochastic process. Jung *et al.* [69] introduced the notion of η-convex stochastic processes and established Jensen, Hermite–Hadamard and Ostrowski type inequalities. Further, Fu *et al.* [51] derived Hölder-İşcan and Improved power mean integral inequalities, and proved Hermite–Hadamard type integral inequalities for n−polynomial stochastic processes.

The organization of the chapter is as follows: In Section 7.2, we recall some basic results that are necessary for our main results. In Section 7.3, we introduce the concept of general h−harmonic preinvexity for real-valued stochastic processes $(Gh-HP_{\eta\varphi}SP)$ and discuss some special cases of our definition. We prove Hermite–Hadamard type inequalities for $Gh-HP_{\eta\varphi}SP$. Further, in Section 7.4, we define multidimensional general h−harmonic preinvex stochastic process and special cases in favor of the definition. Also, we obtain Hermite–Hadamard type inequalities for multidimensional general h−harmonic preinvex stochastic processes. In Section 7.5, we introduce the concept of strongly η-convex stochastic processes. Further, we prove the Hermite–Hadamard inequality, Ostrowski inequality and some other interesting inequalities.

7.2   Preliminaries

In this section, we mention some definitions and related results required for this chapter. We suppose that K be a nonempty closed set. Also suppose that $\eta(.,.) : K \times K \to \mathbb{R}$ and $\varphi : K \to K$ be arbitrary functions.

Noor *et al.* [121] derived the following Hermite–Hadamard inequality for relative harmonic preinvex functions.

Theorem 7.2.1. *Let* $\xi : K = [c, c+\eta(d,c)] \subseteq \mathbb{R}\backslash\{0\} \to \mathbb{R}$ *be relative harmonic preinvex function with* $c < c + \eta(d,c)$. *If* $\xi \in L[c, c + \eta(d,c)]$ *and Condition C holds, then*

$$\frac{1}{2h(\frac{1}{2})}\xi\left(\frac{2c(c+\eta(d,c))}{2c+\eta(d,c)}\right) \leq \frac{c(c+\eta(d,c))}{\eta(d,c)}\int_c^{c+\eta(d,c)}\frac{\xi(x)}{x^2}dx$$

$$\leq [\xi(c) + \xi(d)]\int_0^1 h(\delta)d\delta. \qquad (7.1)$$

Okur and Aliyev [129] introduced the concepts of general invex set and preinvex functions.

Definition 7.2.1. *A set K is said to be general invex set with respect to $\eta(.,.)$ and φ, if $\varphi(x) + \delta\eta(\varphi(y),\varphi(x)) \in K, \forall\, x,y \in \mathbb{R} : \varphi(x),\varphi(y) \in K, \delta \in [0,1]$.*

Definition 7.2.2. *A function ξ on K is said to be general preinvex with respect to arbitrary function η and ϕ, if*

$$\xi(\varphi(x) + \delta\eta(\varphi(y),\varphi(x))) \le (1-\delta)\xi(\varphi(x)) + \delta\xi(\varphi(y)), \forall\, x,y \in \mathbb{R} :$$
$$\varphi(x),\varphi(y) \in K, \delta \in [0,1]. \tag{7.2}$$

Okur and Aliyev [129] gave the following Condition C for $GP_{\eta\varphi}SP$.

Definition 7.2.3. *Let $\eta : K \times K \to \mathbb{R}$ hold the following criterions:*

$$\eta(\varphi(x),\varphi(x) + \delta\eta(\varphi(y)\varphi(x))) = -\delta\eta(\varphi(y),\varphi(x));$$

$$\eta(\varphi(y),\varphi(x) + \delta\eta(\varphi(y)\varphi(x))) = (1-\delta)\eta(\varphi(y),\varphi(x));$$

$\forall\, x,y \in \mathbb{R} : \varphi(x),\varphi(y) \in K, \delta \in [0,1]$.

The following Hermite–Hadamard inequality for general preinvex functions is obtained by Awan *et al.* [10].

Theorem 7.2.2. *Let $\xi : K = [\varphi(c),\varphi(c) + \eta(\varphi(d),\varphi(c))] \to \mathbb{R}$ be a general preinvex function with $\eta(\varphi(d),\varphi(c)) > 0$. If $\eta(.,.)$ satisfies the Condition C, then we have*

$$\xi\left(\frac{2\varphi(c) + \eta(\varphi(d),\varphi(c))}{2}\right) \le \frac{1}{\eta(\varphi(d),\varphi(c))} \int_{\varphi(c)}^{\varphi(c)+\eta(\varphi(d),\varphi(c))} \xi(\varphi(x))d\varphi(x)$$
$$\le \frac{\xi(\varphi(c)) + \xi(\varphi(d))}{2}. \tag{7.3}$$

Okur and Aliyev [129] defined multidimensional general preinvex stochastic processes.

Definition 7.2.4. *A stochastic process $\xi : K_{\eta\varphi} \times \Omega \to \mathbb{R}$ is said to be multidimensional general preinvex ($MGP_{\eta\varphi}SP$) with respect to η and φ on $K_{\eta\varphi}$ if the following inequality holds almost everywhere*

$$\xi((\varphi(y) + \delta\eta(\varphi(x),\varphi(y))),.) \le \delta\xi(\varphi(x),.) + (1-\delta)\xi(\varphi(y),.),$$

for all $\varphi(x),\varphi(y) \in K_{\eta\varphi}$ and $\delta \in [0,1]$, where $K_{\eta\varphi} \subseteq \mathbb{R}^n$ be a non empty closed set, and $\eta : K_{\eta\varphi} \times K_{\eta\varphi} \to \mathbb{R}^n$ and $\varphi : K_{\eta\varphi} \to K_{\eta\varphi}$ be arbitrary functions.

Kotrys [87] proved the Hermite–Hadamard inequality for convex stochastic processes.

Theorem 7.2.3. *Let $\xi : I \times \Omega \to \mathbb{R}$ is convex and mean square continuous in the interval I. Then for any $c, d \in I$, we have*

$$\xi\left(\frac{c+d}{2}, .\right) \leq \frac{1}{d-c}\int_c^d \xi(u,.)du \leq \frac{\xi(c,.)+\xi(d,.)}{2} \quad (a.e.).$$

The following definition of strongly convex stochastic processes is given by Kotrys [88].

Definition 7.2.5. *Let $\mu : \Omega \to \mathbb{R}$ denote a positive random variable. The stochastic process $\xi : I \times \Omega \to \mathbb{R}$ is called strongly convex with modulus $\mu(.) > 0$, if for all $\delta \in [0,1]$ and $x, y \in I$ the inequality*
$$\xi(\delta x + (1-\delta)y, .) \leq \delta \xi(x,.) + (1-\delta)\xi(y,.) - \mu(.)\delta(1-\delta)(x-y)^2 \quad (a.e.) \text{ is}$$
satisfied.

Kotrys [88] proved the following Hermite–Hadamard inequality for strongly Jensen convex stochastic processes.

Theorem 7.2.4. *Let $\xi : I \times \Omega \to \mathbb{R}$ be a stochastic process, which is strongly Jensen convex with modulus $\mu(.)$ and mean square continuous in the interval I. Then for any $c, d \in I$, we have*

$$\xi\left(\frac{c+d}{2}, .\right) + \frac{\mu(.)}{12}(d-c)^2 \leq \frac{1}{d-c}\int_c^d \xi(u,.)du \leq \frac{\xi(c,.)+\xi(d,.)}{2}$$
$$- \frac{\mu(.)}{6}(d-c)^2 \quad (a.e.).$$

Jung *et al.* [69] introduced the concept of η-convex stochastic processes.

Definition 7.2.6. *Let (Ω, A, P) be a probability space and $I \subseteq \mathbb{R}$ be an interval, then $\xi : I \times \Omega \to \mathbb{R}$ is an η-convex stochastic process if*

$$\xi(\delta x + (1-\delta)y,.) \leq \xi(y,.) + \delta\eta(\xi(x,.),\xi(y,.)) \quad (a.e.), \ \forall x, y \in I \text{ and } \delta \in [0,1].$$

Fu *et al.* [51] derived the following Hölder-İşcan and Improved power mean integral inequalities.

Theorem 7.2.5. *(Hölder-İşcan integral inequality) Let $\xi, \phi : [c,d] \times \Omega \to \mathbb{R}$ be real stochastic process and $|\xi'|^p$, $|\phi'|^p$ be mean square integrable on $[c,d]$. If $p > 1$ and $\frac{1}{p} + \frac{1}{q} = 1$, then the following inequality holds almost everywhere:*

$$\int_c^d |\xi(x,.)\phi(x,.)|dx$$
$$\leq \frac{1}{d-c}\left[\left(\int_c^d (d-x)|\xi(x,.)|^p dx\right)^{1/p}\left(\int_c^d (d-x)|\phi(x,.)|^q dx\right)^{1/q}\right.$$
$$\left. + \left(\int_c^d (x-c)|\xi(x,.)|^p dx\right)^{1/p}\left(\int_c^d (x-c)|\phi(x,.)|^q dx\right)^{1/q}\right].$$

Theorem 7.2.6. *(Improved power mean integral inequality) Let $\xi, \phi : [c,d] \times \Omega \to \mathbb{R}$ be real stochastic process and $|\xi|$, $|\xi||\phi|^q$ be mean square integrable on $[c,d]$. If $q \geq 1$, then the following inequality holds almost everywhere:*

$$\int_c^d |\xi(x,.)\phi(x,.)| dx \leq \frac{1}{d-c} \left[\left(\int_c^d (d-x)|\xi(x,.)| dx \right)^{1-\frac{1}{q}} \right.$$

$$\times \left(\int_c^d (d-x)|\xi(x,.)||\phi(x,.)|^q dx \right)^{1/q}$$

$$\left. + \left(\int_c^d (x-c)|\xi(x,.)| dx \right)^{1-\frac{1}{q}} \left(\int_c^d (x-c)|\xi(x,.)||\phi(x,.)|^q dx \right)^{1/q} \right].$$

The following lemmas are obtained by Gonzales *et al.* [53] and Fu *et al.* [51], respectively.

Lemma 7.2.1. *Let $\xi : I \times \Omega \to \mathbb{R}$ be a stochastic process which is mean square differentiable on I^0. If ξ' is mean square integrable on $[c,d]$, where $c,d \in I$ with $c < d$, then the following equality holds*

$$\xi(t,.) - \frac{1}{d-c} \int_c^d \xi(u,.) du = \frac{(x-c)^2}{d-c} \int_0^1 t\xi'(tx + (1-t)c,.) dt$$

$$- \frac{(d-x)^2}{d-c} \int_0^1 t\xi'(tx + (1-t)d,.) dt, \ (a.e), \ for \ each \ x \in [c,d].$$

Lemma 7.2.2. *Let $\xi : I \times \Omega \to \mathbb{R}$ be a mean square differentiable stochastic process on I^0 and ξ' is mean square integrable on $[c,d]$, where $c,d \in I, c < d$. Then we have almost everywhere*

$$\frac{\xi(c,.) + \xi(d,.)}{2} - \frac{1}{d-c} \int_c^d \xi(u,.) du = \frac{d-c}{2} \int_0^1 (1-2\delta)\xi'(\delta c + (1-\delta)d,.) d\delta.$$

Jung *et al.* [69] proved the Hermite–Hadamard inequality for η-convex stochastic processes.

Theorem 7.2.7. *Suppose that $\xi : [c,d] \times \Omega \to \mathbb{R}$ is an η-convex stochastic process such that η is bounded above on $\xi[c,d] \times \xi[c,d]$, then the following inequalities hold almost everywhere:*

$$\xi\left(\frac{c+d}{2},.\right) - \frac{M_\eta}{2} \leq \frac{1}{d-c} \int_c^d \xi(u,.) du$$

$$\leq \frac{\xi(c,.) + \xi(d,.)}{2} + \frac{1}{4}(\eta(\xi(c,.), \xi(d,.)) + \eta(\xi(d,.), \xi(c,.)))$$

$$\leq \frac{\xi(c,.) + \xi(d,.)}{2} + M_\eta.$$

Now we are ready to discuss our main results. First, we discuss general h-harmonically preinvex stochastic process and second multidimensional general h-harmonically preinvex stochastic process.

7.3   General h–Harmonically Preinvex Stochastic Process $(Gh - HP_{\eta\varphi}SP)$

We give the definition of general h–harmonically preinvex stochastic process and discuss some special cases of our new definition. Further, we present Hermite–Hadamard type inequalities for these stochastic processes [155].

Definition 7.3.1. *Let $h : (0,1) \to \mathbb{R}$ be a nonnegative function, $h \not\equiv 0$ and $\xi : K \times \Omega \subseteq \mathbb{R}\backslash\{0\} \times \Omega \to \mathbb{R}$ be a stochastic process on the general harmonic invex set K with respect to η and φ. Then the stochastic process ξ is called general h–harmonic preinvex stochastic process $(Gh - HP_{\eta\varphi}SP)$ with respect to η and φ if and only if*

$$\xi\left(\frac{\varphi(x)(\varphi(x) + \eta(\varphi(y), \varphi(x)))}{\varphi(x) + (1-\delta)\eta(\varphi(y), \varphi(x))}, . \right) \leq h(1-\delta)\xi(\varphi(x), .) + h(\delta)\xi(\varphi(y), .),$$

$$\forall \; x, y \in \mathbb{R} : \varphi(x), \varphi(y) \in K, \delta \in [0,1].$$
$$(7.4)$$

Note that for $\delta = \frac{1}{2}$, we have Jensen type $Gh - HP_{\eta\varphi}SP$ with respect to η and φ

$$\xi\left(\frac{2\varphi(x)(\varphi(x) + \eta(\varphi(y)\varphi(x)))}{2\varphi(x) + \eta(\varphi(y), \varphi(x))}, . \right) \leq h\left(\frac{1}{2}\right)[\xi(\varphi(x), .) + \xi(\varphi(y), .)],$$

$$\forall \; x, y \in \mathbb{R} : \varphi(x), \varphi(y) \in K. \quad (7.5)$$

We now discuss some special cases of Definition 7.3.1:

1 If $h(\delta) = \delta$, then Definition 7.3.1 reduces to the definition of general harmonic preinvex stochastic process $(GHP_{\eta\varphi}SP)$.

2 If $h(\delta) = \delta^s$, then Definition 7.3.1 reduces to the definition of Breckner type of general s-harmonic preinvex stochastic process.

3 If $h(\delta) = \delta^{-s}$, then Definition 7.3.1 reduces to the definition of Godunova-Levin type of general s-harmonic preinvex stochastic process.

4 If $h(\delta) = 1$, then Definition 7.3.1 reduces to the definition of general harmonic P-preinvex stochastic process.

Lemma 7.3.1. *Let (Ω, ζ, P) be an arbitrary probability space and $\xi : K \times \Omega \subseteq (\mathbb{R}\backslash\{0\} \times \Omega) \to \mathbb{R}$ be a stochastic process on the general harmonic invex set K with respect to η and φ. If φ is a continuous function on K, then ξ can be known as*

1 mean square continuous on K,

2 monotonic if it is increasing and decreasing,

3 mean-square differentiable at a point $\varphi(x) \in K$,

4 mean-square integrable on $[\varphi(x), \varphi(x) + \eta(\varphi(y), \varphi(x))] \subseteq K$.

Theorem 7.3.1. *Let $h : (0,1) \to \mathbb{R}$. Under the assumptions of Lemma 7.3.1, let $\xi : [\varphi(c), \varphi(c) + \eta(\varphi(d), \varphi(c))] \times \Omega \to \mathbb{R}$ be $Gh - HP_{\eta\varphi}SP$ with respect to η and φ. If η fulfills the criterions of Condition C, then the following inequalities hold almost everywhere*

$$\frac{1}{2h(\frac{1}{2})}\xi\left(\frac{2\varphi(c)(\varphi(c) + \eta(\varphi(d), \varphi(c)))}{2\varphi(c) + \eta(\varphi(d), \varphi(c))}, .\right)$$

$$\leq \frac{\varphi(c)(\varphi(c) + \eta(\varphi(d), \varphi(c)))}{\eta(\varphi(d), \varphi(c))} \int_{\varphi(c)}^{\varphi(c) + \eta(\varphi(d), \varphi(c))} \frac{\xi(z, .)}{z^2} dz$$

$$\leq [\xi(\varphi(c), .) + \xi(\varphi(d), .)] \int_0^1 h(\delta)d\delta.$$

Proof From definition of $Gh - HP_{\eta\varphi}SP$ for $\delta = \frac{1}{2}$, we have

$$\xi\left(\frac{2\varphi(x)(\varphi(x) + \eta(\varphi(y), \varphi(x)))}{2\varphi(x) + \eta(\varphi(y), \varphi(x))}, .\right) \leq h\left(\frac{1}{2}\right)[\xi(\varphi(x), .) + \xi(\varphi(y), .)],$$

$$\forall\ x, y \in \mathbb{R} : \varphi(x), \varphi(y) \in K.$$

Applying $\varphi(x) = \left(\frac{\varphi(c)(\varphi(c)+\eta(\varphi(d),\varphi(c)))}{\varphi(c)+\delta\eta(\varphi(d),\varphi(c))}\right), \varphi(y) = \left(\frac{\varphi(c)(\varphi(c)+\eta(\varphi(d),\varphi(c)))}{\varphi(c)+(1-\delta)\eta(\varphi(d),\varphi(c))}\right)$ and Condition C in above inequality, we obtain

$$\xi\left(\frac{2\varphi(c)(\varphi(c) + \eta(\varphi(d), \varphi(c)))}{2\varphi(c) + \eta(\varphi(d), \varphi(c))}, .\right)$$

$$\leq h\left(\frac{1}{2}\right)\left[\xi\left(\frac{\varphi(c)(\varphi(c) + \eta(\varphi(d), \varphi(c)))}{\varphi(c) + \delta\eta(\varphi(d), \varphi(c))}, .\right)\right.$$

$$\left.+ \xi\left(\frac{\varphi(c)(\varphi(c) + \eta(\varphi(d), \varphi(c)))}{\varphi(c) + (1 - \delta)\eta(\varphi(d), \varphi(c))}, .\right)\right].$$

Integrating both sides of the above inequality with respect to δ over $[0, 1]$, we get

$$\xi\left(\frac{2\varphi(c)(\varphi(c) + \eta(\varphi(d), \varphi(c)))}{2\varphi(c) + \eta(\varphi(d), \varphi(c))}, .\right)$$

$$\leq h\left(\frac{1}{2}\right)\left[\int_0^1 \xi\left(\frac{\varphi(c)(\varphi(c) + \eta(\varphi(d), \varphi(c)))}{\varphi(c) + \delta\eta(\varphi(d), \varphi(c))}, .\right) d\delta\right.$$

$$\left.+ \int_0^1 \xi\left(\frac{\varphi(c)(\varphi(c) + \eta(\varphi(d), \varphi(c)))}{\varphi(c) + (1 - \delta)\eta(\varphi(d), \varphi(c))}, .\right) d\delta\right].$$

This implies,

$$\frac{1}{2h(\frac{1}{2})}\xi\left(\frac{2\varphi(c)(\varphi(c)+\eta(\varphi(d),\varphi(c)))}{2\varphi(c)+\eta(\varphi(d),\varphi(c))},.\right)$$

$$\leq \frac{\varphi(c)(\varphi(c)+\eta(\varphi(d),\varphi(c)))}{\eta(\varphi(d),\varphi(c))}\int_{\varphi(c)}^{\varphi(c)+\eta(\varphi(d),\varphi(c))}\frac{\xi(z,.)}{z^2}dz. \qquad (7.6)$$

From the definition of $Gh - HP_{\eta\varphi}SP$, we have

$$\xi\left(\frac{\varphi(c)(\varphi(c)+\eta(\varphi(d),\varphi(c)))}{\varphi(c)+(1-\delta)\eta(\varphi(d),\varphi(c))},.\right) \leq h(1-\delta)\xi(\varphi(c),.)+h(\delta)\xi(\varphi(d),.).$$

Integrating with respect to δ over $[0,1]$, we have

$$\frac{\varphi(c)(\varphi(c)+\eta(\varphi(d),\varphi(c)))}{\eta(\varphi(d),\varphi(c))}\int_{\varphi(c)}^{\varphi(c)+\eta(\varphi(d),\varphi(c))}\frac{\xi(z,.)}{z^2}dz$$

$$\leq [\xi(\varphi(c),.)+\xi(\varphi(d),.)]\int_0^1 h(\delta)d\delta. \qquad (7.7)$$

From (7.6) and (7.7), we get

$$\frac{1}{2h(\frac{1}{2})}\xi\left(\frac{2\varphi(c)(\varphi(c)+\eta(\varphi(d),\varphi(c)))}{2\varphi(c)+\eta(\varphi(d),\varphi(c))},.\right)$$

$$\leq \frac{\varphi(c)(\varphi(c)+\eta(\varphi(d),\varphi(c)))}{\eta(\varphi(d),\varphi(c))}\int_{\varphi(c)}^{\varphi(c)+\eta(\varphi(d),\varphi(c))}\frac{\xi(z,.)}{z^2}dz$$

$$\leq [\xi(\varphi(c),.)+\xi(\varphi(d),.)]\int_0^1 h(\delta)d\delta.$$

This completes the proof.

Remark 7.3.1. *If $h(\delta) = \delta$ and ξ is a $GP_{\eta\varphi}SP$, then above theorem reduces to Theorem 2.1 of [129], i.e.*

$$\xi\left(\frac{2\varphi(c)+\eta(\varphi(d),\varphi(c))}{2},.\right)$$

$$\leq \frac{1}{\eta(\varphi(d),\varphi(c))}\int_{\varphi(c)}^{\varphi(c)+\eta(\varphi(d),\varphi(c))}\xi(\varphi(x),.)d\varphi(x)$$

$$\leq \frac{\xi(\varphi(c),.)+\xi(\varphi(d),.)}{2}.$$

Theorem 7.3.2. *Let* $h : (0,1) \to \mathbb{R}$. *Under the assumptions of Lemma 7.3.1, let* $\xi : [\varphi(c), \varphi(c) + \eta(\varphi(d), \varphi(c))] \times \Omega \to \mathbb{R}$ *be* $Gh - HP_{\eta\varphi}SP$ *with respect to* η *and* φ, *then the following inequality holds almost everywhere*

$$\frac{\varphi(c)(\varphi(c) + \eta(\varphi(d), \varphi(c)))}{\eta(\varphi(d), \varphi(c))}$$

$$\times \int_{\varphi(c)}^{\varphi(c)+\eta(\varphi(d),\varphi(c))} \frac{\xi(z,.)\xi\left(\frac{z\varphi(c)(\varphi(c)+\eta(\varphi(d),\varphi(c)))}{(z-\varphi(c))(\varphi(c)+\eta(\varphi(d),\varphi(c)))+z\varphi(c)},.\right)}{z^2} dz$$

$$\leq 2\xi(\varphi(c),.)\xi(\varphi(d),.) \int_0^1 h^2(\delta)d\delta$$

$$+ [\xi^2(\varphi(c),.) + \xi^2(\varphi(d),.)] \int_0^1 h(\delta)h(1-\delta)d\delta.$$

Proof Since ξ is $Gh - HP_{\eta\varphi}SP$, therefore

$$\xi\left(\frac{\varphi(c)(\varphi(c) + \eta(\varphi(d), \varphi(c)))}{\varphi(c) + (1-\delta)\eta(\varphi(d), \varphi(c))},.\right) \leq h(1-\delta)\xi(\varphi(c),.) + h(\delta)\xi(\varphi(d),.),$$
(7.8)

and

$$\xi\left(\frac{\varphi(c)(\varphi(c) + \eta(\varphi(d), \varphi(c)))}{\varphi(c) + \delta\eta(\varphi(d), \varphi(c))},.\right) \leq h(\delta)\xi(\varphi(c),.) + h(1-\delta)\xi(\varphi(d),.),$$
$$\forall \delta \in [0,1]. \quad (7.9)$$

From (7.8) and (7.9), we obtain

$$\xi\left(\frac{\varphi(c)(\varphi(c) + \eta(\varphi(d), \varphi(c)))}{\varphi(c) + (1-\delta)\eta(\varphi(d), \varphi(c))},.\right)\xi\left(\frac{\varphi(c)(\varphi(c) + \eta(\varphi(d), \varphi(c)))}{\varphi(c) + \delta\eta(\varphi(d), \varphi(c))},.\right)$$
$$\leq h^2(1-\delta)\xi(\varphi(c),.)\xi(\varphi(d),.) + h^2(\delta)\xi(\varphi(c),.)\xi(\varphi(d),.)$$
$$+ h(\delta)h(1-\delta)[\xi^2(\varphi(c),.) + \xi^2(\varphi(d),.)].$$

Integrating the above inequality with respect to δ over $[0,1]$, we have

$$\int_0^1 \xi\left(\frac{\varphi(c)(\varphi(c) + \eta(\varphi(d), \varphi(c)))}{\varphi(c) + (1-\delta)\eta(\varphi(d), \varphi(c))},.\right)\xi\left(\frac{\varphi(c)(\varphi(c) + \eta(\varphi(d), \varphi(c)))}{\varphi(c) + \delta\eta(\varphi(d), \varphi(c))},.\right)d\delta$$

$$\leq \xi(\varphi(c),.)\xi(\varphi(d),.) \int_0^1 [h^2(1-\delta) + h^2(\delta)]d\delta$$

$$+ [\xi^2(\varphi(c),.) + \xi^2(\varphi(d),.)] \int_0^1 h(\delta)h(1-\delta)d\delta.$$

Set, $z = \left(\frac{\varphi(c)(\varphi(c)+\eta(\varphi(d),\varphi(c)))}{\varphi(c)+(1-\delta)\eta(\varphi(d),\varphi(c))}, . . \right)$ in the left hand side of the above inequality, we obtain

$$\frac{\varphi(c)(\varphi(c) + \eta(\varphi(d), \varphi(c)))}{\eta(\varphi(d), \varphi(c))}$$
$$\times \int_{\varphi(c)}^{\varphi(c)+\eta(\varphi(d),\varphi(c))} \xi(z,.)\xi \left(\frac{\frac{z\varphi(c)(\varphi(c)+\eta(\varphi(d),\varphi(c)))}{(z-\varphi(c))(\varphi(c)+\eta(\varphi(d),\varphi(c)))+z\varphi(c)},.}{z^2} \right) dz$$
$$\leq 2\xi(\varphi(c),.)\xi(\varphi(d),.) \int_0^1 h^2(\delta)d\delta$$
$$+ [\xi^2(\varphi(c),.) + \xi^2(\varphi(d),.)] \int_0^1 h(\delta)h(1-\delta)d\delta.$$

This completes the proof.

Theorem 7.3.3. *Let* $h : (0,1) \to \mathbb{R}$. *Under the assumptions of Lemma 7.3.1, let* $\xi, \psi : [\varphi(c), \varphi(c) + \eta(\varphi(d), \varphi(c)) \times \Omega \to \mathbb{R}$ *be* $Gh - HP_{\eta\varphi}SP$ *with respect to* η *and* φ, *then the following inequality holds almost everywhere*

$$\frac{\varphi(c)(\varphi(c) + \eta(\varphi(d), \varphi(c)))}{\eta(\varphi(d), \varphi(c))} \int_{\varphi(c)}^{\varphi(c)+\eta(\varphi(d),\varphi(c))} \frac{\xi(z,.)\psi(z,.)}{z^2} dz$$
$$\leq M(c,d) \int_0^1 h^2(\delta)d\delta + N(c,d) \int_0^1 h(\delta)h(1-\delta)d\delta,$$

where,

$$M(c,d) = \xi(\varphi(c),.)\psi(\varphi(c),.) + \xi(\varphi(d),.)\psi(\varphi(d),.)$$

and

$$N(c,d) = \xi(\varphi(c),.)\psi(\varphi(d),.) + \xi(\varphi(d),.)\psi(\varphi(c),.).$$

Proof Since ξ and ψ are $Gh - HP_{\eta\varphi}SP$, we have

$$\xi \left(\frac{\varphi(c)(\varphi(c) + \eta(\varphi(d), \varphi(c)))}{\varphi(c) + (1 - \delta)\eta(\varphi(d), \varphi(c))}, . \right) \leq h(1-\delta)\xi(\varphi(c),.) + h(\delta)\xi(\varphi(d),.),$$
$$(7.10)$$

and

$$\psi \left(\frac{\varphi(c)(\varphi(c) + \eta(\varphi(d), \varphi(c)))}{\varphi(c) + (1 - \delta)\eta(\varphi(d), \varphi(c))}, . \right) \leq h(1-\delta)\psi(\varphi(c),.) + h(\delta)\psi(\varphi(d),.),$$
$$\forall \delta \in [0,1].$$
$$(7.11)$$

Multiplying (7.10) and (7.11), it follows that

$$\xi\left(\frac{\varphi(c)(\varphi(c)+\eta(\varphi(d),\varphi(c)))}{\varphi(c)+(1-\delta)\eta(\varphi(d),\varphi(c))},.\right)\psi\left(\frac{\varphi(c)(\varphi(c)+\eta(\varphi(d),\varphi(c)))}{\varphi(c)+(1-\delta)\eta(\varphi(d),\varphi(c))},.\right)$$
$$\leq h^2(1-\delta)\xi(\varphi(c),.)\psi(\varphi(c),.)+h^2(\delta)\xi(\varphi(d),.)\psi(\varphi(d),.)$$
$$+h(\delta)h(1-\delta)[\xi(\varphi(c),.)\psi(\varphi(d),.)+\xi(\varphi(d),.)\psi(\varphi(c),.)].$$

Integrating the above inequality over $[0,1]$ with respect to δ, we get

$$\int_0^1 \xi\left(\frac{\varphi(c)(\varphi(c)+\eta(\varphi(d),\varphi(c)))}{\varphi(c)+(1-\delta)\eta(\varphi(d),\varphi(c))},.\right)\psi\left(\frac{\varphi(c)(\varphi(c)+\eta(\varphi(d),\varphi(c)))}{\varphi(c)+(1-\delta)\eta(\varphi(d),\varphi(c))},.\right)d\delta$$

$$\leq \xi(\varphi(c),.)\psi(\varphi(c),.)\int_0^1 h^2(1-\delta)d\delta+\xi(\varphi(d),.)\psi(\varphi(d),.)\int_0^1 h^2(\delta)d\delta$$

$$+[\xi(\varphi(c),.)\psi(\varphi(d),.)+\xi(\varphi(d),.)\psi(\varphi(c),.)]\int_0^1 h(\delta)h(1-\delta)d\delta.$$

This implies,

$$\frac{\varphi(c)(\varphi(c)+\eta(\varphi(d),\varphi(c)))}{\eta(\varphi(d),\varphi(c))}\int_{\varphi(c)}^{\varphi(c)+\eta(\varphi(d),\varphi(c))}\frac{\xi(z,.)\psi(z,.)}{z^2}dz$$

$$\leq [\xi(\varphi(c),.)\psi(\varphi(c),.)+\xi(\varphi(d),.)\psi(\varphi(d),.)]\int_0^1 h^2(\delta)d\delta$$

$$+[\xi(\varphi(c),.)\psi(\varphi(d),.)+\xi(\varphi(d),.)\psi(\varphi(c),.)]\int_0^1 h(\delta)h(1-\delta)d\delta$$

$$= M(c,d)\int_0^1 h^2(\delta)d\delta + N(c,d)\int_0^1 h(\delta)h(1-\delta)d\delta.$$

This completes the proof.

Lemma 7.3.2. *Let $\xi : K \subset \mathbb{R}\backslash\{0\} \times \Omega \to \mathbb{R}$ be a mean-square differentiable general stochastic process on K^0 and $\varphi(c),\varphi(d) \in K^0$ with $\varphi(c) < \varphi(c) + \eta(\varphi(d),\varphi(c))$. Then, we have almost everywhere*

$$\frac{\varphi(c)(\varphi(c)+\eta(\varphi(d),\varphi(c)))\eta(\varphi(d),\varphi(c))}{2}$$
$$\times \int_0^1 \frac{(1-2\delta)}{(\varphi(c)+\delta\eta(\varphi(d),\varphi(c)))^2}\xi'\left(\frac{\varphi(c)(\varphi(c)+\eta(\varphi(d),\varphi(c)))}{\varphi(c)+\delta\eta(\varphi(d),\varphi(c))},.\right)d\delta$$

$$= \frac{1}{2}[\xi(\varphi(c),.)+\xi(\varphi(c)+\eta(\varphi(d),\varphi(c)),.)]$$
$$- \frac{\varphi(c)(\varphi(c)+\eta(\varphi(d),\varphi(c)))}{\eta(\varphi(d),\varphi(c))}\int_{\varphi(c)}^{\varphi(c)+\eta(\varphi(d),\varphi(c))}\frac{\xi(z,.)}{z^2}dz. \qquad (7.12)$$

Proof

$$
L.H.S. = \frac{\varphi(c)(\varphi(c) + \eta(\varphi(d), \varphi(c)))\eta(\varphi(d), \varphi(c))}{2}
$$

$$
\times \int_0^1 \frac{(1 - 2\delta)}{(\varphi(c) + \delta\eta(\varphi(d), \varphi(c)))^2} \xi'\left(\frac{\varphi(c)(\varphi(c) + \eta(\varphi(d), \varphi(c)))}{\varphi(c) + \delta\eta(\varphi(d), \varphi(c))}, . . \right) d\delta
$$

$$
= \int_0^1 -\frac{(1 - 2\delta)}{2}\left[-\frac{\varphi(c)(\varphi(c) + \eta(\varphi(d), \varphi(c)))\eta(\varphi(d), \varphi(c))}{(\varphi(c) + \delta\eta(\varphi(d), \varphi(c)))^2} \right.
$$

$$
\left. \times \xi'\left(\frac{\varphi(c)(\varphi(c) + \eta(\varphi(d), \varphi(c)))}{\varphi(c) + \delta\eta(\varphi(d), \varphi(c))}, . \right)\right] d\delta
$$

$$
= \frac{2\delta - 1}{2}\xi\left(\frac{\varphi(c)(\varphi(c) + \eta(\varphi(d), \varphi(c)))}{\varphi(c) + \delta\eta(\varphi(d), \varphi(c))}, . . \right)\Big|_0^1
$$

$$
- \int_0^1 \xi\left(\frac{\varphi(c)(\varphi(c) + \eta(\varphi(d), \varphi(c)))}{\varphi(c) + \delta\eta(\varphi(d), \varphi(c))}, . . \right) d\delta.
$$

Set $z = \frac{\varphi(c)(\varphi(c) + \eta(\varphi(d), \varphi(c)))}{\varphi(c) + \delta\eta(\varphi(d), \varphi(c))}$, we obtain

$$
L.H.S. = \frac{1}{2}[\xi(\varphi(c), .) + \xi(\varphi(c) + \eta(\varphi(d), \varphi(c)), .)]
$$

$$
- \frac{\varphi(c)(\varphi(c) + \eta(\varphi(d), \varphi(c)))}{\eta(\varphi(d), \varphi(c))} \int_{\varphi(c)}^{\varphi(c)+\eta(\varphi(d),\varphi(c))} \frac{\xi(z, .)}{z^2} dz.
$$

This completes the proof.

Now we prove some integral inequalities with the help of Lemma 7.3.2.

Theorem 7.3.4. *Let $h : (0, 1) \to \mathbb{R}$ and $\xi : K \times \Omega \subset \mathbb{R}\backslash\{0\} \times \Omega \to \mathbb{R}$ be a mean-square differentiable general stochastic process on K^0 with respect to η and φ with $\varphi(c) < \varphi(c) + \eta(\varphi(d), \varphi(c))$. If $|\xi'|^q$ is general h−harmonically preinvex on $[\varphi(c), \varphi(c) + \eta(\varphi(d), \varphi(c))]$ for $q \geq 1$, then*

$$
\left| \frac{\xi(\varphi(c), .) + \xi(\varphi(c) + \eta(\varphi(d), \varphi(c)), .)}{2} \right.
$$

$$
\left. - \frac{\varphi(c)(\varphi(c) + \eta(\varphi(d), \varphi(c)))}{\eta(\varphi(d), \varphi(c))} \int_{\varphi(c)}^{\varphi(c)+\eta(\varphi(d),\varphi(c))} \frac{\xi(z, .)}{z^2} dz \right|
$$

$$
\leq \frac{\varphi(c)(\varphi(c) + \eta(\varphi(d), \varphi(c)))\eta(\varphi(d), \varphi(c))}{2}
$$

$$
\times \delta_1^{\frac{q-1}{q}}[\delta_2|\xi'(\varphi(c), .)|^q + \delta_3|\xi'(\varphi(d), .)|^q]^{\frac{1}{q}},
$$

where

$$\delta_1 = \int_0^1 \frac{|1-2\delta|}{(\varphi(c) + \delta\eta(\varphi(d),\varphi(c)))^2} d\delta$$

$$= \frac{1}{\varphi(c)(\varphi(c) + \eta(\varphi(d),\varphi(c)))}$$

$$- \frac{2}{\eta^2(\varphi(d),\varphi(c))} log\left(\frac{\left(\varphi(c) + \frac{1}{2}\eta(\varphi(d),\varphi(c))\right)^2}{\varphi(c)(\varphi(c) + \eta(\varphi(d),\varphi(c)))}\right),$$

$$\delta_2 = \int_0^1 \frac{|1-2\delta|h(\delta)}{(\varphi(c) + \delta\eta(\varphi(d),\varphi(c)))^2} d\delta \quad and$$

$$\delta_3 = \int_0^1 \frac{|1-2\delta|h(1-\delta)}{(\varphi(c) + \delta\eta(\varphi(d),\varphi(c)))^2} d\delta.$$

Proof Recall Lemma 7.3.2:

$$\left| \frac{\xi(\varphi(c),.) + \xi(\varphi(c) + \eta(\varphi(d),\varphi(c)),.)}{2} \right.$$

$$\left. - \frac{\varphi(c)(\varphi(c) + \eta(\varphi(d),\varphi(c)))}{\eta(\varphi(d),\varphi(c))} \int_{\varphi(c)}^{\varphi(c)+\eta(\varphi(d),\varphi(c))} \frac{\xi(z,.)}{z^2} dz \right|$$

$$= \left| \frac{\varphi(c)(\varphi(c) + \eta(\varphi(d),\varphi(c)))\eta(\varphi(d),\varphi(c))}{2} \right.$$

$$\left. \times \int_0^1 \frac{(1-2\delta)}{(\varphi(c) + \delta\eta(\varphi(d),\varphi(c)))^2} \xi'\left(\frac{\varphi(c)(\varphi(c) + \eta(\varphi(d),\varphi(c)))}{\varphi(c) + \delta\eta(\varphi(d),\varphi(c))},.\right) d\delta \right|$$

$$\leq \frac{\varphi(c)(\varphi(c) + \eta(\varphi(d),\varphi(c)))\eta(\varphi(d),\varphi(c))}{2}$$

$$\times \int_0^1 \left| \frac{(1-2\delta)}{(\varphi(c) + \delta\eta(\varphi(d),\varphi(c)))^2} \right| \left| \xi'\left(\frac{\varphi(c)(\varphi(c) + \eta(\varphi(d),\varphi(c)))}{\varphi(c) + \delta\eta(\varphi(d),\varphi(c))},.\right) \right| d\delta.$$

Applying Hölder's inequality and the definition of $Gh - HP_{\eta\varphi}SP$ in above inequality, we get

$$\left| \frac{\xi(\varphi(c),.) + \xi(\varphi(c) + \eta(\varphi(d),\varphi(c)),.)}{2} \right.$$

$$\left. - \frac{\varphi(c)(\varphi(c) + \eta(\varphi(d),\varphi(c)))}{\eta(\varphi(d),\varphi(c))} \int_{\varphi(c)}^{\varphi(c)+\eta(\varphi(d),\varphi(c))} \frac{\xi(z,.)}{z^2} dz \right|$$

$$\leq \frac{\varphi(c)(\varphi(c) + \eta(\varphi(d),\varphi(c)))\eta(\varphi(d),\varphi(c))}{2}$$

$$\times \left(\int_0^1 \frac{|1 - 2\delta|}{(\varphi(c) + \delta\eta(\varphi(d),\varphi(c)))^2} d\delta \right)^{\frac{q-1}{q}}$$

$$\times \left(\int_0^1 \frac{|1 - 2\delta|}{(\varphi(c) + \delta\eta(\varphi(d),\varphi(c)))^2} \left| \xi'\left(\frac{\varphi(c)(\varphi(c) + \eta(\varphi(d),\varphi(c)))}{\varphi(c) + \delta\eta(\varphi(d),\varphi(c))},. \right) \right|^q d\delta \right)^{\frac{1}{q}}$$

$$\leq \frac{\varphi(c)(\varphi(c) + \eta(\varphi(d),\varphi(c)))\eta(\varphi(d),\varphi(c))}{2}$$

$$\times \left(\int_0^1 \frac{|1 - 2\delta|}{(\varphi(c) + \delta\eta(\varphi(d),\varphi(c)))^2} d\delta \right)^{\frac{q-1}{q}}$$

$$\times \left(\int_0^1 \frac{|1 - 2\delta|}{(\varphi(c) + \delta\eta(\varphi(d),\varphi(c)))^2} [h(\delta)|\xi'(\varphi(c),.)|^q + h(1-\delta)|\xi'(\varphi(d),.)|^q] \right)^{\frac{1}{q}}$$

$$= \frac{\varphi(c)(\varphi(c) + \eta(\varphi(d),\varphi(c)))\eta(\varphi(d),\varphi(c))}{2}$$

$$\times \delta_1^{\frac{q-1}{q}} [\delta_2|\xi'(\varphi(c),.)|^q + \delta_3|\xi'(\varphi(d),.)|^q]^{\frac{1}{q}}.$$

This completes the proof.

Theorem 7.3.5. *Let $h : (0,1) \to \mathbb{R}$ and $\xi : K \times \Omega \subset \mathbb{R}\backslash\{0\} \times \Omega \to \mathbb{R}$ be a mean-square differentiable general stochastic process on K^0 with respect to η and φ with $\varphi(c) < \varphi(c) + \eta(\varphi(d),\varphi(c))$. If $|\xi'|^q$ is general $h-$harmonically preinvex on $[\varphi(c),\varphi(c) + \eta(\varphi(d),\varphi(c))]$ for $q > 1$, $\frac{1}{p} + \frac{1}{q} = 1$ then*

$$\left| \frac{\xi(\varphi(c),.) + \xi(\varphi(c) + \eta(\varphi(d),\varphi(c)),.)}{2} \right.$$

$$\left. - \frac{\varphi(c)(\varphi(c) + \eta(\varphi(d),\varphi(c)))}{\eta(\varphi(d),\varphi(c))} \int_{\varphi(c)}^{\varphi(c)+\eta(\varphi(d),\varphi(c))} \frac{\xi(z,.)}{z^2} dz \right|$$

$$\leq \frac{\varphi(c)(\varphi(c) + \eta(\varphi(d),\varphi(c)))\eta(\varphi(d),\varphi(c))}{2}$$

$$\times \left(\frac{1}{p+1} \right)^{\frac{1}{p}} [\delta_4|\xi'(\varphi(c),.)|^q + \delta_5|\xi'(\varphi(d),.)|^q]^{\frac{1}{q}},$$

where

$$\delta_4 = \int_0^1 \frac{h(\delta)}{(\varphi(c) + \delta\eta(\varphi(d), \varphi(c)))^{2q}} d\delta \text{ and}$$

$$\delta_5 = \int_0^1 \frac{h(1-\delta)}{(\varphi(c) + \delta\eta(\varphi(d), \varphi(c)))^{2q}} d\delta.$$

Proof From Lemma 7.3.2, we have

$$\left| \frac{\xi(\varphi(c), .) + \xi(\varphi(c) + \eta(\varphi(d), \varphi(c)), .)}{2} \right.$$

$$\left. - \frac{\varphi(c)(\varphi(c) + \eta(\varphi(d), \varphi(c)))}{\eta(\varphi(d), \varphi(c))} \int_{\varphi(c)}^{\varphi(c) + \eta(\varphi(d), \varphi(c))} \frac{\xi(z, .)}{z^2} dz \right|$$

$$\leq \frac{\varphi(c)(\varphi(c) + \eta(\varphi(d), \varphi(c)))\eta(\varphi(d), \varphi(c))}{2}$$

$$\times \int_0^1 \left| \frac{(1 - 2\delta)}{(\varphi(c) + \delta\eta(\varphi(d), \varphi(c)))^2} \right| \left| \xi' \left(\frac{\varphi(c)(\varphi(c) + \eta(\varphi(d), \varphi(c)))}{\varphi(c) + \delta\eta(\varphi(d), \varphi(c))}, . \right) \right| d\delta.$$

Applying Hölder's inequality and the definition of $Gh - HP_{\eta\varphi}SP$ in above inequality, we get

$$\left| \frac{\xi(\varphi(c), .) + \xi(\varphi(c) + \eta(\varphi(d), \varphi(c)), .)}{2} \right.$$

$$\left. - \frac{\varphi(c)(\varphi(c) + \eta(\varphi(d), \varphi(c)))}{\eta(\varphi(d), \varphi(c))} \int_{\varphi(c)}^{\varphi(c) + \eta(\varphi(d), \varphi(c))} \frac{\xi(z, .)}{z^2} dz \right|$$

$$\leq \frac{\varphi(c)(\varphi(c) + \eta(\varphi(d), \varphi(c)))\eta(\varphi(d), \varphi(c))}{2} \left(\int_0^1 |1 - 2\delta|^p d\delta \right)^{\frac{1}{p}}$$

$$\times \left(\int_0^1 \frac{1}{(\varphi(c) + \delta\eta(\varphi(d), \varphi(c)))^{2q}} \left| \xi' \left(\frac{\varphi(c)(\varphi(c) + \eta(\varphi(d), \varphi(c)))}{\varphi(c) + \delta\eta(\varphi(d), \varphi(c))}, . \right) \right|^q d\delta \right)^{\frac{1}{q}}$$

$$\leq \frac{\varphi(c)(\varphi(c) + \eta(\varphi(d), \varphi(c)))\eta(\varphi(d), \varphi(c))}{2} \left(\frac{1}{p+1} \right)^{\frac{1}{p}}$$

$$\times \left(\int_0^1 \frac{1}{(\varphi(c) + \delta\eta(\varphi(d), \varphi(c)))^{2q}} [h(\delta)|\xi'(\varphi(c), .)|^q + h(1 - \delta)|\xi'(\varphi(d), .)|^q] \right)^{\frac{1}{q}}$$

$$= \frac{\varphi(c)(\varphi(c) + \eta(\varphi(d), \varphi(c)))\eta(\varphi(d), \varphi(c))}{2}$$

$$\times \left(\frac{1}{p+1} \right)^{\frac{1}{p}} [\delta_4|\xi'(\varphi(c), .)|^q + \delta_5|\xi'(\varphi(d), .)|^q]^{\frac{1}{q}}.$$

This completes the proof.

7.4   Multidimensional General $h-$Harmonic Preinvex Stochastic Processes $(MGh - HP_\eta\varphi SP)$

In this section, we give the definition of multidimensional general h-harmonic preinvexity for stochastic processes with respect to η and φ on n coordinates and present Hermite–Hadamard type inequalities for $(MGh - HP_\eta\varphi SP)$ [155].

Definition 7.4.1. *Let $h : (0,1) \to \mathbb{R}$ be a nonnegative function. A stochastic process $\xi : K_{\eta\varphi}(\subset \mathbb{R}_n^+) \times \Omega \to \mathbb{R}$ is said to be $MGh - HP_\eta\varphi SP$ with respect to η and φ on $K_{\eta\varphi}$, if the following inequality holds almost everywhere*

$$\xi\left(\frac{\varphi(x_1)(\varphi(x_1) + \eta(\varphi(y_1), \varphi(x_1)))}{\varphi(x_1) + \delta\eta(\varphi(y_1), \varphi(x_1))}, ..., \frac{\varphi(x_n)(\varphi(x_n) + \eta(\varphi(y_n), \varphi(x_n)))}{\varphi(x_n) + \delta\eta(\varphi(y_n), \varphi(x_n))}, . \right)$$
$$\leq h(\delta)\xi(\varphi(x_1), ..., \varphi(x_n), .) + h(1 - \delta)\xi(\varphi(y_1), ..., \varphi(y_n), .),$$
for all $(\varphi(x_1), ..., \varphi(x_n)), (\varphi(y_1), ..., \varphi(y_n)) \in K_{\eta\varphi}$ and $\delta \in [0,1]$.

We now discuss some special cases of Definition 7.4.1:

1 If $h(\delta) = \delta$, then above definition reduces to the definition of multidimensional general harmonic preinvex stochastic process with respect to η and φ on $K_{\eta\varphi}$.

2 If $h(\delta) = \delta^s$, then above definition reduces to the definition of Breckner type of multidimensional general $s-$harmonic preinvex stochastic process with respect to η and φ on $K_{\eta\varphi}$.

3 If $h(\delta) = \delta^{-s}$, then above definition reduces to the definition of Godunova-Levin type of multidimensional general $s-$harmonic preinvex stochastic process with respect to η and φ on $K_{\eta\varphi}$.

4 If $h(\delta) = 1$, then above definition reduces to the definition of multidimensional general harmonic $P-$ preinvex stochastic process with respect to η and φ on $K_{\eta\varphi}$.

Definition 7.4.2. *Let $h : (0,1) \to (0,\infty)$. A stochastic process $\xi : K_{\eta\varphi}(\subset \mathbb{R}_n^+) \times \Omega \to \mathbb{R}$ is called $MGh - HP_\eta\varphi SP$ with respect to η and φ on $n-$coordinates if the following stochastic mappings $\xi^i_{\varphi(x_n)} : K \times \Omega \to \mathbb{R}$ are general $h-$harmonically preinvex with respect to η and φ on K almost everywhere for all $\varphi(x) \in K$:*

$$\xi^i_{\varphi(x_n)}(\varphi(x), .) = \xi\left((\wedge_{k=1}^{i-1}\varphi(x_k), \varphi(x), \wedge_{k=i+1}^n\varphi(x_k)), . \right).$$

Lemma 7.4.1. *Every $MGh - HP_{\eta\varphi}SP$ is $Gh - HP_{\eta\varphi}SP$ with respect to η and φ on $n-$ coordinates almost everywhere, but not conversely.*

Proof Suppose $\xi : K_{\eta\varphi} \times \Omega \to \mathbb{R}$ be $MGh - HP_{\eta\varphi}SP$ with respect to η and φ on $K_{\eta\varphi}$.

Define $\xi^i_{\varphi(x_n)} : K \times \Omega \to \mathbb{R}$ by

$\xi^i_{\varphi(x_n)}(\varphi(x),.) = \xi((\wedge^{i-1}_{k=1}\varphi(x_k), \varphi(x), \wedge^n_{k=i+1}\varphi(x_k)),.)$ almost everywhere for $\varphi(x) \in K$.

Then for $\varphi(x), \varphi(y) \in K$ and $\delta \in [0,1]$ almost everywhere

$$\xi^i_{\varphi(x_n)}\left(\frac{\varphi(x)(\varphi(x) + \eta(\varphi(y), \varphi(x)))}{\varphi(x) + (1-\delta)\eta(\varphi(y), \varphi(x))}, .\right)$$

$$= \xi\left(\left(\wedge^{i-1}_{k=1}\varphi(x_k), \frac{\varphi(x)(\varphi(x) + \eta(\varphi(y), \varphi(x)))}{\varphi(x) + (1-\delta)\eta(\varphi(y), \varphi(x))}, \wedge^n_{k=i+1}\varphi(x_k)\right), .\right)$$

$$\leq h(1-\delta)\xi((\wedge^{i-1}_{k=1}\varphi(x_k), \varphi(x), \wedge^n_{k=i+1}\varphi(x_k)), .)$$

$$+ h(\delta)\xi((\wedge^{i-1}_{k=1}\varphi(x_k), \varphi(y), \wedge^n_{k=i+1}\varphi(x_k)), .)$$

$$= h(1-\delta)\xi^i_{\varphi(x_n)}(\varphi(x), .) + h(\delta)\xi^i_{\varphi(x_n)}(\varphi(y), .).$$

For converse part, we give the following example.

Example 7.4.1. *Let $\xi : K_{\eta\varphi} = [1,2] \times [2,3] \times [3,4] \times \Omega \to \mathbb{R}$ be a stochastic process defined as*
$\xi(\varphi(x_1), \varphi(x_2), \varphi(x_3), .) = (\varphi(x_1) - 1)(\varphi(x_2) - 2)(\varphi(x_3) - 3)$ and $\eta : K_{\eta\varphi} \times K_{\eta\varphi} \to \mathbb{R}$, $\eta(\varphi(x), \varphi(y)) = \varphi(x) - \varphi(y)$, $\forall \varphi(x), \varphi(y) \in K_{\eta\varphi}$.
If we take $\varphi(x) = (1,3,4)$, $\varphi(y) = (2,2,4)$ and $h(\delta) = \delta$. Then,

$$\xi\left(\frac{\varphi(x_1)(\varphi(x_1) + \eta(\varphi(y_1), \varphi(x_1)))}{\varphi(x_1) + \delta\eta(\varphi(y_1), \varphi(x_1))}, \frac{\varphi(x_2)(\varphi(x_2) + \eta(\varphi(y_2), \varphi(x_2)))}{\varphi(x_2) + \delta\eta(\varphi(y_2), \varphi(x_2))}, \right.$$

$$\left. \frac{\varphi(x_3)(\varphi(x_3) + \eta(\varphi(y_3), \varphi(x_3)))}{\varphi(x_3) + \delta\eta(\varphi(y_3), \varphi(x_3))}, .\right)$$

$$= \xi\left(\frac{\varphi(x_1)\varphi(y_1)}{(1-\delta)\varphi(x_1) + \delta\varphi(y_1)}, \frac{\varphi(x_2)\varphi(y_2)}{(1-\delta)\varphi(x_2) + \delta\varphi(y_2)}, \right.$$

$$\left. \frac{\varphi(x_3)\varphi(y_3)}{(1-\delta)\varphi(x_3) + \delta\varphi(y_3)}, .\right)$$

$$= \left(\frac{2}{1+\delta} - 1\right)\left(\frac{6}{3-\delta} - 2\right) = \frac{2\delta(1-\delta)}{(1+\delta)(3-\delta)} > 0, \; (if \; \delta \neq 0) \quad (7.13)$$

and

$$h(\delta)\xi(\varphi(x_1), \varphi(x_2), \varphi(x_3), .) + h(1-\delta)\xi(\varphi(y_1), \varphi(y_2), \varphi(y_3), .)$$

$$= \delta.0 + (1-\delta).0 = 0. \quad (7.14)$$

From (7.13) and (7.14), we get

$$\xi\left(\frac{\varphi(x_1)(\varphi(x_1)+\eta(\varphi(y_1),\varphi(x_1)))}{\varphi(x_1)+\delta\eta(\varphi(y_1),\varphi(x_1))},\frac{\varphi(x_2)(\varphi(x_2)+\eta(\varphi(y_2),\varphi(x_2)))}{\varphi(x_2)+\delta\eta(\varphi(y_2),\varphi(x_2))},\right.$$
$$\left.\frac{\varphi(x_3)(\varphi(x_3)+\eta(\varphi(y_3),\varphi(x_3)))}{\varphi(x_3)+\delta\eta(\varphi(y_3),\varphi(x_3))},.\right)$$
$$> h(\delta)\xi(\varphi(x_1),\varphi(x_2),\varphi(x_3),.) + h(1-\delta)\xi(\varphi(y_1),\varphi(y_2),\varphi(y_3),.),$$

which is contradiction.
This completes the proof.

Consider $n-$dimensional interval $\diamondsuit^n = \Pi_{i=1}^n \triangle^n = \Pi_{i=1}^n[\varphi(w_i),\varphi(w_i)+\eta(\varphi(z_i),\varphi(w_i))] \subseteq \mathbb{R}_+^n$. For simplicity, let $\triangle_1^i = \varphi(w_i)$ and $\triangle_2^i = \varphi(w_i)+\eta(\varphi(z_i),\varphi(w_i))$ with $\eta(\varphi(z_i),\varphi(w_i) > 0$ for each $i = 1,2,...,n$.

Remark 7.4.1. *Let $h:(0,1)\to\mathbb{R}$ and $\xi:\diamondsuit^n\times\Omega\to\mathbb{R}_+$ is $MGh-HP_{\eta\varphi}SP$ with respect to η and φ on $\diamondsuit^n$. If the assumptions of Lemma 7.3.1 satisfy, then $\xi^i_{\varphi(x_n)}:[\varphi(w_i),\varphi(w_i)+\eta(\varphi(z_i),\varphi(w_i))]\times\Omega\to\mathbb{R}$ is $Gh-HP_{\eta\varphi}SP$ on $[\varphi(w_i),\varphi(w_i)+\eta(\varphi(z_i),\varphi(w_i))]$ with respect to η and φ for each $i=1,2,...,n$. Thus*

$$\frac{1}{2h(\frac{1}{2})}\xi^i_{\varphi(x_n)}\left(\frac{2\,\triangle_1^i\triangle_2^i}{\triangle_1^i+\triangle_2^i},.\right) \le \frac{\triangle_1^i\triangle_2^i}{\triangle_2^i-\triangle_1^i}\int_{\triangle_1^i}^{\triangle_2^i}\frac{\xi^i_{\varphi(x_n)}(\varphi(x_i),.)}{\varphi^2(x_i)}d\varphi(x_i)$$
$$\le [\xi^i_{\varphi(x_n)}(\varphi(w_i),.)+\xi^i_{\varphi(x_n)}(\varphi(z_i),.)]\int_0^1 h(\delta)d\delta,$$

almost everywhere for $i=1,2,...,n$.

Theorem 7.4.1. *Let $h:(0,1)\to\mathbb{R}$ and $\xi:\diamondsuit^n\times\Omega\to\mathbb{R}_+$ be $MGh-HP_{\eta\varphi}SP$ with respect to η and φ on $\diamondsuit^n$. If the assumptions of Lemma 7.3.1 satisfy, then almost everywhere*

$$\frac{1}{4h^2(\frac{1}{2})}\sum_{i=1}^{n-1}\xi\left(\left(\wedge_{k=1}^{i-1}\varphi(x_k),\frac{2\,\triangle_1^i\triangle_2^i}{\triangle_1^i+\triangle_2^i},\frac{2\,\triangle_1^{i+1}\triangle_2^{i+1}}{\triangle_1^{i+1}+\triangle_2^{i+1}},\wedge_{k=i+2}^n\varphi(x_k)\right),.\right)$$
$$\le \frac{1}{2h(\frac{1}{2})}\sum_{i=1}^{n-1}\frac{\triangle_1^i\triangle_2^i}{\triangle_2^i-\triangle_1^i}\int_{\triangle_1^i}^{\triangle_2^i}\frac{\xi^{i+1}_{\varphi(x_n)}\left(\frac{2\triangle_1^{i+1}\triangle_2^{i+1}}{\triangle_1^{i+1}+\triangle_2^{i+1}},.\right)}{\varphi^2(x_i)}d\varphi(x_i)$$
$$\le \sum_{i=1}^{n-1}\frac{\triangle_1^i\triangle_2^i}{\triangle_2^i-\triangle_1^i}\frac{\triangle_1^{i+1}\triangle_2^{i+1}}{\triangle_2^{i+1}-\triangle_1^{i+1}}\int_{\triangle_1^i}^{\triangle_2^i}\int_{\triangle_1^{i+1}}^{\triangle_2^{i+1}}\frac{\xi^{i+1}_{\varphi(x_n)}(\varphi(x_{i+1}),.)}{\varphi^2(x_{i+1})\varphi^2(x_i)}d\varphi(x_{i+1})d\varphi(x_i)$$
$$\le \left(\int_0^1 h(\delta)d\delta\right)$$
$$\times\sum_{i=1}^{n-1}\frac{\triangle_1^i\triangle_2^i}{\triangle_2^i-\triangle_1^i}\int_{\triangle_1^i}^{\triangle_2^i}\left[\frac{\xi^{i+1}_{\varphi(x_n)}(\varphi(w_{i+1}),.)+\xi^{i+1}_{\varphi(x_n)}(\varphi(z_{i+1}),.)}{\varphi^2(x_i)}\right]d\varphi(x_i)$$

$$\leq \left(\int_0^1 h(\delta)d\delta \right)^2 \sum_{i=1}^{n-1} \left[\xi \left(\left(\wedge_{k=1}^{i-1}\varphi(x_k), \varphi(w_i), \varphi(w_{i+1}), \wedge_{k=i+2}^n \varphi(x_k) \right), . \right) \right.$$

$$+ \xi \left(\left(\wedge_{k=1}^{i-1}\varphi(x_k), \varphi(z_i), \varphi(w_{i+1}), \wedge_{k=i+2}^n \varphi(x_k) \right), . \right)$$

$$+ \xi \left(\left(\wedge_{k=1}^{i-1}\varphi(x_k), \varphi(w_i), \varphi(z_{i+1}), \wedge_{k=i+2}^n \varphi(x_k) \right), . \right)$$

$$\left. + \xi \left(\left(\wedge_{k=1}^{i-1}\varphi(x_k), \varphi(z_i), \varphi(z_{i+1}), \wedge_{k=i+2}^n \varphi(x_k) \right), . \right) \right]. \tag{7.15}$$

Proof Recall Remark 7.4.1:

$$\frac{1}{2h(\frac{1}{2})} \xi_{\varphi(x_n)}^{i+1} \left(\frac{2 \, \Delta_1^{i+1} \Delta_2^{i+1}}{\Delta_1^{i+1} + \Delta_2^{i+1}}, . \right)$$

$$\leq \frac{\Delta_1^{i+1} \Delta_2^{i+1}}{\Delta_2^{i+1} - \Delta_1^{i+1}} \int_{\Delta_1^{i+1}}^{\Delta_2^{i+1}} \frac{\xi_{\varphi(x_n)}^{i+1}(\varphi(x_{i+1}), .)}{\varphi^2(x_{i+1})} d\varphi(x_{i+1})$$

$$\leq \left[\xi_{\varphi(x_n)}^{i+1}(\varphi(w_{i+1}), .) + \xi_{\varphi(x_n)}^{i+1}(\varphi(z_{i+1}), .) \right] \int_0^1 h(\delta)d\delta, \text{ almost everywhere.}$$

Integrating above inequality on Δ^i with respect to $\varphi(x_i)$, we have

$$\frac{1}{2h(\frac{1}{2})} \frac{\Delta_1^i \Delta_2^i}{\Delta_2^i - \Delta_1^i} \int_{\Delta_1^i}^{\Delta_2^i} \frac{\xi_{\varphi(x_n)}^{i+1} \left(\frac{2 \Delta_1^{i+1} \Delta_2^{i+1}}{\Delta_1^{i+1} + \Delta_2^{i+1}}, . \right)}{\varphi^2(x_i)} d\varphi(x_i)$$

$$\leq \frac{\Delta_1^i \Delta_2^i}{\Delta_2^i - \Delta_1^i} \frac{\Delta_1^{i+1} \Delta_2^{i+1}}{\Delta_2^{i+1} - \Delta_1^{i+1}} \int_{\Delta_1^i}^{\Delta_2^i} \int_{\Delta_1^{i+1}}^{\Delta_2^{i+1}} \frac{\xi_{\varphi(x_n)}^{i+1}(\varphi(x_{i+1}), .)}{\varphi^2(x_{i+1})\varphi^2(x_i)} d\varphi(x_{i+1})d\varphi(x_i)$$

$$\leq \left(\int_0^1 h(\delta)d\delta \right)$$

$$\times \frac{\Delta_1^i \Delta_2^i}{\Delta_2^i - \Delta_1^i} \int_{\Delta_1^i}^{\Delta_2^i} \left[\frac{\xi_{\varphi(x_n)}^{i+1}(\varphi(w_{i+1}), .) + \xi_{\varphi(x_n)}^{i+1}(\varphi(z_{i+1}), .)}{\varphi^2(x_i)} \right] d\varphi(x_i),$$

almost everywhere.

Applying Hermite–Hadamard integral inequality, we have

$$\frac{1}{4h^2(\frac{1}{2})} \xi \left(\left(\wedge_{k=1}^{i-1}\varphi(x_k), \frac{2 \, \Delta_1^i \Delta_2^i}{\Delta_1^i + \Delta_2^i}, \frac{2 \, \Delta_1^{i+1} \Delta_2^{i+1}}{\Delta_1^{i+1} + \Delta_2^{i+1}}, \wedge_{k=i+2}^n \varphi(x_k) \right), . \right)$$

$$\leq \frac{1}{2h(\frac{1}{2})} \frac{\Delta_1^i \Delta_2^i}{\Delta_2^i - \Delta_1^i} \int_{\Delta_1^i}^{\Delta_2^i} \frac{\xi_{\varphi(x_n)}^{i+1} \left(\frac{2 \Delta_1^{i+1} \Delta_2^{i+1}}{\Delta_1^{i+1} + \Delta_2^{i+1}}, . \right)}{\varphi^2(x_i)} d\varphi(x_i)$$

$$\leq \frac{\Delta_1^i \Delta_2^i}{\Delta_2^i - \Delta_1^i} \frac{\Delta_1^{i+1} \Delta_2^{i+1}}{\Delta_2^{i+1} - \Delta_1^{i+1}} \int_{\Delta_1^i}^{\Delta_2^i} \int_{\Delta_1^{i+1}}^{\Delta_2^{i+1}} \frac{\xi_{\varphi(x_n)}^{i+1}(\varphi(x_{i+1}), .)}{\varphi^2(x_{i+1})\varphi^2(x_i)} d\varphi(x_{i+1})d\varphi(x_i)$$

$$\leq \left(\int_0^1 h(\delta)d\delta \right) \frac{\Delta_1^i \Delta_2^i}{\Delta_2^i - \Delta_1^i} \int_{\Delta_1^i}^{\Delta_2^i} \left[\frac{\xi_{\varphi(x_n)}^{i+1}(\varphi(w_{i+1}), .) + \xi_{\varphi(x_n)}^{i+1}(\varphi(z_{i+1}), .)}{\varphi^2(x_i)} \right] d\varphi(x_i)$$

$$\leq \left(\int_0^1 h(\delta)d\delta \right)^2$$

$$\times \left[\xi \left(\left(\wedge_{k=1}^{i-1}\varphi(x_k), \varphi(w_i), \varphi(w_{i+1}), \wedge_{k=i+2}^{n}\varphi(x_k) \right), . \right) \right.$$

$$+ \xi \left(\left(\wedge_{k=1}^{i-1}\varphi(x_k), \varphi(z_i), \varphi(w_{i+1}), \wedge_{k=i+2}^{n}\varphi(x_k) \right), . \right)$$

$$+ \xi \left(\left(\wedge_{k=1}^{i-1}\varphi(x_k), \varphi(w_i), \varphi(z_{i+1}), \wedge_{k=i+2}^{n}\varphi(x_k) \right), . \right)$$

$$\left. + \xi \left(\left(\wedge_{k=1}^{i-1}\varphi(x_k), \varphi(z_i), \varphi(z_{i+1}), \wedge_{k=i+2}^{n}\varphi(x_k) \right), . \right) \right], \text{ almost everywhere,}$$

where $i = 1, 2, ..., n-1$.

Taking summation from 1 to $n-1$, we compute the result easily.

Remark 7.4.2. *If $h(\delta) = \delta$ and ξ is $MGP_{\eta\varphi}SP$, then above theorem reduces to Theorem 2.2 of [129], i.e.*

$$\sum_{i=1}^{n-1} \xi \left(\left(\wedge_{k=1}^{i-1}\varphi(x_k), \frac{\Delta_1^i + \Delta_2^i}{2}, \frac{\Delta_1^{i+1} + \Delta_2^{i+1}}{2}, \wedge_{k=i+2}^{n}\varphi(x_k) \right), . \right)$$

$$\leq \sum_{i=1}^{n-1} \frac{1}{\Delta_2^i - \Delta_1^i} \int_{\Delta_1^i}^{\Delta_2^i} \xi_{\varphi(x_n)}^{i+1} \left(\frac{\Delta_1^{i+1} + \Delta_2^{i+1}}{2}, . \right) d\varphi(x_i)$$

$$\leq \sum_{i=1}^{n-1} \frac{1}{(\Delta_2^i - \Delta_1^i)(\Delta_2^{i+1} - \Delta_1^{i+1})} \int_{\Delta_1^i}^{\Delta_2^i} \int_{\Delta_1^{i+1}}^{\Delta_2^{i+1}} \xi_{\varphi(x_n)}^{i+1}(\varphi(x_{i+1}), .)d\varphi(x_{i+1})d\varphi(x_i)$$

$$\leq \sum_{i=1}^{n-1} \frac{1}{2(\Delta_2^i - \Delta_1^i)} \int_{\Delta_1^i}^{\Delta_2^i} \left[\xi_{\varphi(x_n)}^{i+1}(\varphi(w_{i+1}), .) + \xi_{\varphi(x_n)}^{i+1}(\varphi(z_{i+1}), .) \right] d\varphi(x_i)$$

$$\leq \frac{1}{4} \sum_{i=1}^{n-1} \left[\xi \left(\left(\wedge_{k=1}^{i-1}\varphi(x_k), \varphi(w_i), \varphi(w_{i+1}), \wedge_{k=i+2}^{n}\varphi(x_k) \right), . \right) \right.$$

$$+ \xi \left(\left(\wedge_{k=1}^{i-1}\varphi(x_k), \varphi(z_i), \varphi(w_{i+1}), \wedge_{k=i+2}^{n}\varphi(x_k) \right), . \right)$$

$$+ \xi \left(\left(\wedge_{k=1}^{i-1}\varphi(x_k), \varphi(w_i), \varphi(z_{i+1}), \wedge_{k=i+2}^{n}\varphi(x_k) \right), . \right)$$

$$\left. + \xi \left(\left(\wedge_{k=1}^{i-1}\varphi(x_k), \varphi(z_i), \varphi(z_{i+1}), \wedge_{k=i+2}^{n}\varphi(x_k) \right), . \right) \right].$$

Theorem 7.4.2. *Let $h : (0,1) \to \mathbb{R}$ and $\xi : \Diamond^n \times \Omega \to \mathbb{R}_+$ be $MGh-HP_{\eta\varphi}SP$ with respect to η and φ on $\Diamond^n$. If the assumptions of Lemma 7.3.1 satisfy, then almost everywhere*

$$\sum_{i=1}^{n} \frac{\Delta_1^i \Delta_2^i}{\Delta_2^i - \Delta_1^i} \int_{\Delta_1^i}^{\Delta_2^i} \left[\frac{\xi_{\varphi(w_n)}^i(\varphi(x_i), .) + \xi_{\varphi(z_n)}^i(\varphi(x_i), .)}{\varphi^2(x_i)} \right] d\varphi(x_i)$$

$$\leq \left[n(\xi(\varphi(\boldsymbol{w}), .) + \xi(\varphi(\boldsymbol{z}), .)) + \sum_{i=1}^{n}(\xi_{\varphi(w_n)}^i(\varphi(z_i), .) \right.$$

$$\left. + \xi_{\varphi(z_n)}^i(\varphi(w_i), .)) \right] \int_0^1 h(\delta)d\delta. \tag{7.16}$$

Proof Applying Remark 7.4.1 for $\xi^i_{\varphi(w_n)}$ and $\xi^i_{\varphi(z_n)}$, $i = 1, 2, ..., n$, respectively, then almost everywhere

$$\frac{\Delta^i_1 \Delta^i_2}{\Delta^i_2 - \Delta^i_1} \int_{\Delta^i_1}^{\Delta^i_2} \frac{\xi^i_{\varphi(w_n)}(\varphi(x_i), .)}{\varphi^2(x_i)} d\varphi(x_i)$$

$$\leq [\xi^i_{\varphi(w_n)}(\varphi(w_i), .) + \xi^i_{\varphi(w_n)}(\varphi(z_i), .)] \int_0^1 h(\delta) d\delta$$

$$= [\xi(\varphi(\mathbf{w}), .) + \xi^i_{\varphi(w_n)}(\varphi(z_i), .)] \int_0^1 h(\delta) d\delta, \tag{7.17}$$

and

$$\frac{\Delta^i_1 \Delta^i_2}{\Delta^i_2 - \Delta^i_1} \int_{\Delta^i_1}^{\Delta^i_2} \frac{\xi^i_{\varphi(z_n)}(\varphi(x_i), .)}{\varphi^2(x_i)} d\varphi(x_i)$$

$$\leq [\xi^i_{\varphi(z_n)}(\varphi(w_i), .) + \xi^i_{\varphi(z_n)}(\varphi(z_i), .)] \int_0^1 h(\delta) d\delta$$

$$= [\xi^i_{\varphi(z_n)}(\varphi(w_i), .) + \xi(\varphi(\mathbf{z}), .)] \int_0^1 h(\delta) d\delta. \tag{7.18}$$

Adding (7.17) and (7.18), we obtain

$$\frac{\Delta^i_1 \Delta^i_2}{\Delta^i_2 - \Delta^i_1} \int_{\Delta^i_1}^{\Delta^i_2} \left[\frac{\xi^i_{\varphi(w_n)}(\varphi(x_i), .) + \xi^i_{\varphi(z_n)}(\varphi(x_i), .)}{\varphi^2(x_i)} \right] d\varphi(x_i)$$

$$\leq [\xi(\varphi(\mathbf{w}), .) + \xi^i_{\varphi(w_n)}(\varphi(z_i), .) + \xi^i_{\varphi(z_n)}(\varphi(w_i), .) + \xi(\varphi(\mathbf{z}), .)] \int_0^1 h(\delta) d\delta,$$

almost everywhere,

where $i = 1, 2, ..., n$.
Taking summation from 1 to n, we compute the result easily.

Remark 7.4.3. *If $h(\delta) = \delta$ and ξ is $MGP_{\eta\varphi}SP$, then above theorem reduces to Theorem 2.3 of [129], i.e.*

$$\sum_{i=1}^n \frac{1}{2(\Delta^i_2 - \Delta^i_1)} \int_{\Delta^i_1}^{\Delta^i_2} \left(\xi^i_{\varphi(w_n)}(\varphi(x_i), .) + \xi^i_{\varphi(z_n)}(\varphi(x_i), .) \right) d\varphi(x_i)$$

$$\leq \frac{n}{2} [\xi(\varphi(\boldsymbol{w}), .) + \xi(\varphi(\mathbf{z}), .)] + \frac{1}{2} \sum_{i=1}^n [\xi^i_{\varphi(w_n)}(\varphi(z_i), .)$$

$$+ v^i_{\varphi(z_n)}(\varphi(w_i), .)].$$

Theorem 7.4.3. *Let $h : (0, 1) \to \mathbb{R}$ and $\xi : \Diamond^n \times \Omega \to \mathbb{R}_+$ be $MGh-HP_{\eta\varphi}SP$ with respect to η and φ on $\Diamond^n$. If the assumptions of Lemma 7.3.1 satisfy, then* *almost everywhere*

$$\frac{1}{2^n h^n(\frac{1}{2})} \xi\left(\left(\frac{2\,\Delta_1^1\Delta_2^1}{\Delta_1^1+\Delta_2^1}, \frac{2\,\Delta_1^2\Delta_2^2}{\Delta_1^2+\Delta_2^2}, ..., \frac{2\,\Delta_1^n\Delta_2^n}{\Delta_1^n+\Delta_2^n}\right), \cdot\right)$$

$$\leq \prod_{i=1}^{n} \frac{\Delta_1^i\Delta_2^i}{\Delta_2^i-\Delta_1^i} \int_{\Delta_1^1}^{\Delta_2^1}\int_{\Delta_1^2}^{\Delta_2^2} ... \int_{\Delta_1^n}^{\Delta_2^n} \frac{\xi_{\varphi(x_n)}^n(\varphi(x_n),\cdot)}{\prod_{i=1}^{n}\varphi^2(x_i)}\,d\varphi(x_n)d\varphi(x_{n-1})...d\varphi(x_1)$$

$$\leq \left(\int_0^1 h(\delta)d\delta\right)^{n+1} \sum_{\rho\in m_i(n)} [\xi((h(\rho)\varphi(\boldsymbol{w})+h(1-\rho)\varphi(\boldsymbol{z})),\cdot)],$$

where $m_i(n) = \{\rho = (\rho_1,\rho_2,...,\rho_n) \in \mathbb{N}_0^n : \rho_i \leq 1; \ |\rho| = n+1-i; \ i = 1,2,...,n+1\}$;
$|\rho| : \rho_1 + \rho_2 + ... + \rho_n; \ \rho\varphi(\boldsymbol{w}) = (\rho_1\varphi(w_1),\rho_2\varphi(w_2),...,\rho_n\varphi(w_n)).$

Proof Recall Remark 7.4.1:

$$\frac{1}{2h(\frac{1}{2})} \xi_{\varphi(x_n)}^n\left(\frac{2\,\Delta_1^n\Delta_2^n}{\Delta_1^n+\Delta_2^n}, \cdot\right) \leq \frac{\Delta_1^n\Delta_2^n}{\Delta_2^n-\Delta_1^n} \int_{\Delta_1^n}^{\Delta_2^n} \frac{\xi_{\varphi(x_n)}^n(\varphi(x_n),\cdot)}{\varphi^2(x_n)}\,d\varphi(x_n)$$

$$\leq [\xi_{\varphi(x_n)}^n(\varphi(w_n),\cdot)+\xi_{\varphi(x_n)}^n(\varphi(z_n),\cdot)]\int_0^1 h(\delta)d\delta, \ \text{almost everywhere.}$$

Integrating above inequality on Δ^{n-1}, we obtain

$$\frac{1}{2h(\frac{1}{2})} \frac{\Delta_1^{n-1}\Delta_2^{n-1}}{\Delta_2^{n-1}-\Delta_1^{n-1}} \int_{\Delta_1^{n-1}}^{\Delta_2^{n-1}} \frac{\xi_{\varphi(x_n)}^n\left(\frac{2\Delta_1^n\Delta_2^n}{\Delta_1^n+\Delta_2^n}, \cdot\right)}{\varphi^2(x_{n-1})}\,d\varphi(x_{n-1})$$

$$\leq \frac{\Delta_1^{n-1}\Delta_2^{n-1}}{\Delta_2^{n-1}-\Delta_1^{n-1}} \frac{\Delta_1^n\Delta_2^n}{\Delta_2^n-\Delta_1^n}$$

$$\times \int_{\Delta_1^{n-1}}^{\Delta_2^{n-1}}\int_{\Delta_1^n}^{\Delta_2^n} \frac{\xi_{\varphi(x_n)}^n(\varphi(x_n),\cdot)}{\varphi^2(x_n)\varphi^2(x_{n-1})}\,d\varphi(x_n)d\varphi(x_{n-1})$$

$$\leq \left(\int_0^1 h(\delta)d\delta\right) \frac{\Delta_1^{n-1}\Delta_2^{n-1}}{\Delta_2^{n-1}-\Delta_1^{n-1}}$$

$$\times \int_{\Delta_1^{n-1}}^{\Delta_2^{n-1}} \frac{\xi_{\varphi(x_n)}^n(\varphi(w_n),\cdot)+\xi_{\varphi(x_n)}^n(\varphi(z_n),\cdot)}{\varphi^2(x_{n-1})}\,d\varphi(x_{n-1}),$$

almost everywhere.

Using Hermite–Hadamard integral inequality, we have

$$\frac{1}{4h^2(\frac{1}{2})} \xi\left(\left(\wedge_{k=1}^{n-2}\varphi(x_k), \frac{2\,\Delta_1^{n-1}\Delta_2^{n-1}}{\Delta_1^{n-1}+\Delta_2^{n-1}}, \frac{2\,\Delta_1^n\Delta_2^n}{\Delta_1^n+\Delta_2^n}\right), \cdot\right)$$

$$\leq \frac{1}{2h(\frac{1}{2})} \frac{\Delta_1^{n-1}\Delta_2^{n-1}}{\Delta_2^{n-1}-\Delta_1^{n-1}} \int_{\Delta_1^{n-1}}^{\Delta_2^{n-1}} \frac{\xi_{\varphi(x_n)}^n\left(\frac{2\Delta_1^n\Delta_2^n}{\Delta_1^n+\Delta_2^n}, \cdot\right)}{\varphi^2(x_{n-1})}\,d\varphi(x_{n-1})$$

$$\leq \frac{\Delta_1^{n-1}\Delta_2^{n-1}}{\Delta_2^{n-1} - \Delta_1^{n-1}} \frac{\Delta_1^n \Delta_2^n}{\Delta_2^n - \Delta_1^n}$$

$$\times \int_{\Delta_1^{n-1}}^{\Delta_2^{n-1}} \int_{\Delta_1^n}^{\Delta_2^n} \frac{\xi_{\varphi(x_n)}^n(\varphi(x_n),.)}{\varphi^2(x_n)\varphi^2(x_{n-1})} d\varphi(x_n) d\varphi(x_{n-1})$$

$$\leq \frac{\Delta_1^{n-1}\Delta_2^{n-1}}{\Delta_2^{n-1} - \Delta_1^{n-1}} \left(\int_0^1 h(\delta) d\delta \right)$$

$$\times \int_{\Delta_1^{n-1}}^{\Delta_2^{n-1}} \frac{\xi_{\varphi(x_n)}^n(\varphi(w_n),.) + \xi_{\varphi(x_n)}^n(\varphi(z_n),.)}{\varphi^2(x_{n-1})} d\varphi(x_{n-1})$$

$$\leq \left(\int_0^1 h(\delta) d\delta \right)^2 \left[\xi\left(\left(\wedge_{k=1}^{n-2}\varphi(x_k), \varphi(w_{n-1}), \varphi(w_n) \right), . \right) \right.$$

$$+ \xi\left(\left(\wedge_{k=1}^{n-2}\varphi(x_k), \varphi(z_{n-1}), \varphi(w_n) \right), . \right)$$

$$+ \xi\left(\left(\wedge_{k=1}^{n-2}\varphi(x_k), \varphi(w_{n-1}), \varphi(z_n) \right), . \right)$$

$$\left. + \xi\left(\left(\wedge_{k=1}^{n-2}\varphi(x_k), \varphi(z_{n-1}), \varphi(z_n) \right), . \right) \right], \text{ almost everywhere.}$$

Integrating above inequality on Δ^{n-2}, we get

$$\frac{1}{4h^2(\frac{1}{2})} \frac{\Delta_1^{n-2}\Delta_2^{n-2}}{\Delta_2^{n-2} - \Delta_1^{n-2}}$$

$$\times \int_{\Delta_1^{n-2}}^{\Delta_2^{n-2}} \frac{\xi\left(\left(\wedge_{k=1}^{n-2}\varphi(x_k), \frac{2\Delta_1^{n-1}\Delta_2^{n-1}}{\Delta_1^{n-1}+\Delta_2^{n-1}}, \frac{2\Delta_1^n\Delta_2^n}{\Delta_1^n+\Delta_2^n} \right), . \right)}{\varphi^2(x_{n-2})} d\varphi(x_{n-2})$$

$$\leq \frac{\Delta_1^{n-2}\Delta_2^{n-2}}{\Delta_2^{n-2} - \Delta_1^{n-2}} \frac{\Delta_1^{n-1}\Delta_2^{n-1}}{\Delta_2^{n-1} - \Delta_1^{n-1}} \frac{\Delta_1^n \Delta_2^n}{\Delta_2^n - \Delta_1^n}$$

$$\times \int_{\Delta_1^{n-2}}^{\Delta_2^{n-2}} \int_{\Delta_1^{n-1}}^{\Delta_2^{n-1}} \int_{\Delta_1^n}^{\Delta_2^n} \frac{\xi_{\varphi(x_n)}^n(\varphi(x_n),.)}{\varphi^2(x_n)\varphi^2(x_{n-1})\varphi^2(x_{n-2})} d\varphi(x_n) d\varphi(x_{n-1}) d\varphi(x_{n-2})$$

$$\leq \left(\int_0^1 h(\delta) d\delta \right)^2 \frac{\Delta_1^{n-2}\Delta_2^{n-2}}{\Delta_2^{n-2} - \Delta_1^{n-2}}$$

$$\times \int_{\Delta_1^{n-2}}^{\Delta_2^{n-2}} \frac{1}{\varphi^2(x_{n-2})} \left[\xi\left(\left(\wedge_{k=1}^{n-2}\varphi(x_k), \varphi(w_{n-1}), \varphi(w_n) \right), . \right) \right.$$

$$+ \xi\left(\left(\wedge_{k=1}^{n-2}\varphi(x_k), \varphi(z_{n-1}), \varphi(w_n) \right), . \right)$$

$$+ \xi\left(\left(\wedge_{k=1}^{n-2}\varphi(x_k), \varphi(w_{n-1}), \varphi(z_n) \right), . \right)$$

$$\left. + \xi\left(\left(\wedge_{k=1}^{n-2}\varphi(x_k), \varphi(z_{n-1}), \varphi(z_n) \right), . \right) \right] d\varphi(x_{n-2}), \text{ almost everywhere.}$$

Using Hermite–Hadamard integral inequality in above inequality, we obtain

$$\frac{1}{8h^3(\frac{1}{2})}\xi\left(\left(\wedge_{k=1}^{n-3}\varphi(x_k),\frac{2\,\Delta_1^{n-2}\Delta_2^{n-2}}{\Delta_1^{n-2}+\Delta_2^{n-2}},\frac{2\,\Delta_1^{n-1}\Delta_2^{n-1}}{\Delta_1^{n-1}+\Delta_2^{n-1}},\frac{2\,\Delta_1^{n}\Delta_2^{n}}{\Delta_1^{n}+\Delta_2^{n}}\right),.\right)$$

$$\leq\frac{1}{4h^2(\frac{1}{2})}\frac{\Delta_1^{n-2}\Delta_2^{n-2}}{\Delta_2^{n-2}-\Delta_1^{n-2}}$$

$$\times\int_{\Delta_1^{n-2}}^{\Delta_2^{n-2}}\frac{\xi\left(\left(\wedge_{k=1}^{n-2}\varphi(x_k),\frac{2\Delta_1^{n-1}\Delta_2^{n-1}}{\Delta_1^{n-1}+\Delta_2^{n-1}},\frac{2\Delta_1^{n}\Delta_2^{n}}{\Delta_1^{n}+\Delta_2^{n}}\right),.\right)}{\varphi^2(x_{n-2})}d\varphi(x_{n-2})$$

$$\leq\frac{\Delta_1^{n-2}\Delta_2^{n-2}}{\Delta_2^{n-2}-\Delta_1^{n-2}}\frac{\Delta_1^{n-1}\Delta_2^{n-1}}{\Delta_2^{n-1}-\Delta_1^{n-1}}\frac{\Delta_1^{n}\Delta_2^{n}}{\Delta_2^{n}-\Delta_1^{n}}$$

$$\times\int_{\Delta_1^{n-2}}^{\Delta_2^{n-2}}\int_{\Delta_1^{n-1}}^{\Delta_2^{n-1}}\int_{\Delta_1^{n}}^{\Delta_2^{n}}\frac{\xi_{\varphi(x_n)}^n(\varphi(x_n),.)}{\varphi^2(x_n)\varphi^2(x_{n-1})\varphi^2(x_{n-2})}d\varphi(x_n)d\varphi(x_{n-1})d\varphi(x_{n-2})$$

$$\leq\left(\int_0^1 h(\delta)d\delta\right)^2\frac{\Delta_1^{n-2}\Delta_2^{n-2}}{\Delta_2^{n-2}-\Delta_1^{n-2}}$$

$$\times\int_{\Delta_1^{n-2}}^{\Delta_2^{n-2}}\frac{1}{\varphi^2(x_{n-2})}\left[\xi\left(\left(\wedge_{k=1}^{n-2}\varphi(x_k),\varphi(w_{n-1}),\varphi(w_n)\right),.\right)\right.$$

$$+\xi\left(\left(\wedge_{k=1}^{n-2}\varphi(x_k),\varphi(z_{n-1}),\varphi(w_n)\right),.\right)$$

$$+\xi\left(\left(\wedge_{k=1}^{n-2}\varphi(x_k),\varphi(w_{n-1}),\varphi(z_n)\right),.\right)$$

$$+\xi\left(\left(\wedge_{k=1}^{n-2}\varphi(x_k),\varphi(z_{n-1}),\varphi(z_n)\right),.\right)\Big]d\varphi(x_{n-2})$$

$$\leq\left(\int_0^1 h(\delta)d\delta\right)^3\left[\xi\left(\left(\wedge_{k=1}^{n-3}\varphi(x_k),\varphi(w_{n-2}),\varphi(w_{n-1}),\varphi(w_n)\right),.\right)\right.$$

$$+\xi\left(\left(\wedge_{k=1}^{n-3}\varphi(x_k),\varphi(z_{n-2}),\varphi(w_{n-1}),\varphi(w_n)\right),.\right)$$

$$+\xi\left(\left(\wedge_{k=1}^{n-3}\varphi(x_k),\varphi(w_{n-2}),\varphi(z_{n-1}),\varphi(w_n)\right),.\right)$$

$$+\xi\left(\left(\wedge_{k=1}^{n-3}\varphi(x_k),\varphi(z_{n-2}),\varphi(z_{n-1}),\varphi(w_n)\right),.\right)$$

$$+\xi\left(\left(\wedge_{k=1}^{n-3}\varphi(x_k),\varphi(w_{n-2}),\varphi(w_{n-1}),\varphi(z_n)\right),.\right)$$

$$+\xi\left(\left(\wedge_{k=1}^{n-3}\varphi(x_k),\varphi(z_{n-2}),\varphi(w_{n-1}),\varphi(z_n)\right),.\right)$$

$$+\xi\left(\left(\wedge_{k=1}^{n-3}\varphi(x_k),\varphi(w_{n-2}),\varphi(z_{n-1}),\varphi(z_n)\right),.\right)$$

$$+\xi\left(\left(\wedge_{k=1}^{n-3}\varphi(x_k),\varphi(z_{n-2}),\varphi(z_{n-1}),\varphi(z_n)\right),.\right)\Big],\text{ almost everywhere.}$$

Doing this procedure successively, we obtain

$$
\frac{1}{2^n h^n(\frac{1}{2})} \xi \left(\left(\frac{2\,\Delta_1^1 \Delta_2^1}{\Delta_1^1 + \Delta_2^1}, \frac{2\,\Delta_1^2 \Delta_2^2}{\Delta_1^2 + \Delta_2^2}, \dots, \frac{2\,\Delta_1^n \Delta_2^n}{\Delta_1^n + \Delta_2^n} \right), . \right)
$$

$$
\leq \prod_{i=1}^{n} \frac{\Delta_1^i \Delta_2^i}{\Delta_2^i - \Delta_1^i} \int_{\Delta_1^1}^{\Delta_2^1} \int_{\Delta_1^2}^{\Delta_2^2} \cdots \int_{\Delta_1^n}^{\Delta_2^n} \frac{\xi_{\varphi(x_n)}^n(\varphi(x_n), .)}{\prod_{i=1}^{n} \varphi^2(x_i)} d\varphi(x_n) d\varphi(x_{n-1})...d\varphi(x_1)
$$

$$
\leq \left(\int_0^1 h(\delta) d\delta \right)^{n+1} \sum_{\rho \in m_i(n)} [\xi((h(\rho)\varphi(\mathbf{w}) + h(1-\rho)\varphi(\mathbf{z})), .)].
$$

This completes the proof.

Remark 7.4.4. *If $h(\delta) = \delta$ and ξ is $MGP_{\eta\varphi}SP$, then above theorem reduces to Theorem 2.4 of [129], i.e.*

$$
\xi \left(\left(\frac{\Delta_1^1 + \Delta_2^1}{2}, \frac{\Delta_1^2 + \Delta_2^2}{2}, \dots, \frac{\Delta_1^n + \Delta_2^n}{2} \right), . \right)
$$

$$
\leq \frac{1}{\prod_{i=1}^{n}(\Delta_2^i - \Delta_1^i)}
$$

$$
\times \int_{\Delta_1^1}^{\Delta_2^1} \int_{\Delta_1^2}^{\Delta_2^2} \cdots \int_{\Delta_1^n}^{\Delta_2^n} \xi((\varphi(x_1), \dots, \varphi(x_n)).)d\varphi(x_n)d\varphi(x_{n-1})...d\varphi(x_1)
$$

$$
\leq \frac{1}{2^n} \sum_{\rho \in m_i(n)} [\xi((\rho\varphi(\boldsymbol{w}) + (1-\rho)\varphi(\boldsymbol{z})), .)].
$$

Example 7.4.2. *Let $\xi : \lozenge^2 \times \Omega \to \mathbb{R}_+$ be two dimensional $Gh-HP_{\eta\varphi}SP$ with respect to η and φ on $\lozenge^2$. Then from Theorem 7.4.3 for $n = 2$ and $h(\delta) = \delta$, we get*

$$
\xi \left(\left(\frac{2\,\Delta_1^1 \Delta_2^1}{\Delta_1^1 + \Delta_2^1}, \frac{2\,\Delta_1^2 \Delta_2^2}{\Delta_1^2 + \Delta_2^2} \right), . \right)
$$

$$
\leq \frac{\Delta_1^1 \Delta_2^1}{\Delta_2^1 - \Delta_1^1} \frac{\Delta_1^2 \Delta_2^2}{\Delta_2^2 - \Delta_1^2} \int_{\Delta_1^1}^{\Delta_2^1} \int_{\Delta_1^2}^{\Delta_2^2} \frac{\xi_{\varphi(x_2)}^2(\varphi(x_2), .)}{\varphi^2(x_1)\varphi^2(x_2)} d\varphi(x_2)d\varphi(x_1)
$$

$$
\leq \frac{1}{8} \sum_{\rho \in m_i(2)} [\xi((\rho\varphi(\boldsymbol{w}) + (1-\rho)\varphi(\boldsymbol{z})), .)],
$$

where $m_i(2) = \{\rho = (\rho_1, \rho_2) \in \mathbb{N}_0^2 : \rho_i \leq 1;\ |\rho| = n + 1 - i;\ i = 1, 2, 3\};\ |\rho| :$ $\rho_1 + \rho_2;\ \rho\varphi(\boldsymbol{w}) = (\rho_1\varphi(w_1), \rho_2\varphi(w_2))$.
Therefore, $m_1(2) = \{(1, 1)\}, m_2(2) = \{(0, 1), (1, 0)\}$ and $m_3(2) = \{(0, 0)\}$.

Now using these values in above inequality, we compute

$$\xi\left(\left(\frac{2\,\Delta_1^1\Delta_2^1}{\Delta_1^1+\Delta_2^1},\frac{2\,\Delta_1^2\Delta_2^2}{\Delta_1^2+\Delta_2^2}\right),.\right)$$

$$\leq \frac{\Delta_1^1\Delta_2^1}{\Delta_2^1-\Delta_1^1}\frac{\Delta_1^2\Delta_2^2}{\Delta_2^2-\Delta_1^2}\int_{\Delta_1^1}^{\Delta_2^1}\int_{\Delta_1^2}^{\Delta_2^2}\frac{\xi_{\varphi(x_2)}^2(\varphi(x_2),.)}{\varphi^2(x_1)\varphi^2(x_2)}d\varphi(x_2)d\varphi(x_1)$$

$$\leq \frac{1}{8}[\xi((\varphi(w_1),\varphi(w_2)),.)+\xi((\varphi(z_1),\varphi(w_2)),.)+\xi((\varphi(w_1),\varphi(z_2)),.)$$

$$+\,\xi((\varphi(z_1),\varphi(z_2)),.)].$$

7.5 Strongly Generalized Convex Stochastic Processes

In this section, we introduce the concept of strongly η-convex stochastic processes [156].

Definition 7.5.1. *Let* (Ω,A,P) *be a probability space and* $K\subseteq\mathbb{R}$ *be an interval. Let* $\mu(.)$ *denote a positive random variable, then* $\xi:K\times\Omega\to\mathbb{R}$ *is said to be strongly* η-*convex stochastic process with respect to* $\eta:\xi(K)\times\xi(K)\to\mathbb{R}$ *and modulus* $\mu(.)>0$ *if*

$$\xi(\delta x+(1-\delta)y,.)\leq\xi(y,.)+\delta\eta(\xi(x,.),\xi(y,.))-\mu(.)\delta(1-\delta)(d-c)^2\ (a.e.)$$
$$\forall x,y\in K\ and\ \delta\in[0,1].$$

Remark 7.5.1. *If* $\eta(\xi(x,.),\xi(y,.))=\xi(x,.)-\xi(y,.)$, *then the definition of strongly* η-*convex stochastic process reduces to the definition of strongly convex stochastic process proposed by Kotrys [88]. When* $\mu(.)=0$, *then above definition reduces to the definition of* η-*convex stochastic process [69].*

Example 7.5.1. *Let* $\xi:(0,\infty)\times\Omega\to\mathbb{R}$ *be a stochastic process defined as* $\xi(u,.)=u$, *and* $\eta:\xi((0,\infty))\times\xi((0,\infty))\to\mathbb{R},\eta(\xi(u,.),\xi(v,.))=(\xi(u,.)-\xi(v,.))^2+\xi(u,.)+\xi(v,.)$. *Then* ξ *is strongly* η-*convex stochastic processes with modulus 1.*

Theorem 7.5.1. *A random variable* $\xi:K\times\Omega\to\mathbb{R}$ *is an strongly* η-*convex stochastic process with modulus* $\mu(.)>0$ *if and only if for any* $\kappa_1,\kappa_2,\kappa_3\in K$ *with* $\kappa_1\leq\kappa_2\leq\kappa_3$, *we have*

$$\begin{vmatrix} 0 & 1 & 1 \\ (\kappa_3-\kappa_2) & \xi(\kappa_2,.)-\xi(\kappa_3,.) & 0 \\ (\kappa_3-\kappa_1) & \eta(\xi(\kappa_1,.),\xi(\kappa_3,.)) & \mu(.)(\kappa_2-\kappa_1)(\kappa_3-\kappa_1) \end{vmatrix}\geq 0.$$

Proof Suppose that ξ is an strongly η-convex stochastic process and $\kappa_1, \kappa_2, \kappa_3 \in K$ with $\kappa_1 \leq \kappa_2 \leq \kappa_3$. Then there exist $\delta \in (0,1)$, such that $\kappa_2 = \delta\kappa_1 + (1 - \delta)\kappa_3$.

$$\xi(\kappa_2,.) = \xi(\delta\kappa_1 + (1 - \delta)\kappa_3,.) \leq \xi(\kappa_3,.) + \left(\frac{\kappa_2 - \kappa_3}{\kappa_1 - \kappa_3}\right)\eta(\xi(\kappa_1,.),\xi(\kappa_3,.))$$

$$- \mu(.)\left(\frac{\kappa_2 - \kappa_3}{\kappa_1 - \kappa_3}\right)\left(\frac{\kappa_1 - \kappa_2}{\kappa_1 - \kappa_3}\right)(\kappa_3 - \kappa_1)^2.$$

This implies,

$$(\xi(\kappa_3,.) - \xi(\kappa_2,.))(\kappa_3 - \kappa_1) + (\kappa_3 - \kappa_2)\eta(\xi(\kappa_1,.),\xi(\kappa_3,.))$$
$$- \mu(.)(\kappa_3 - \kappa_2)(\kappa_2 - \kappa_1)(\kappa_3 - \kappa_1) \geq 0,$$

or
$$\begin{vmatrix} 0 & 1 & 1 \\ (\kappa_3 - \kappa_2) & \xi(\kappa_2,.) - \xi(\kappa_3,.) & 0 \\ (\kappa_3 - \kappa_1) & \eta(\xi(\kappa_1,.),\xi(\kappa_3,.)) & \mu(.)(\kappa_2 - \kappa_1)(\kappa_3 - \kappa_1) \end{vmatrix} \geq 0.$$

For the converse part, take $x_1, x_2 \in I$ with $x_1 \leq x_2$. Choose any $\delta \in (0,1)$, then we have $x_1 \leq \delta x_1 + (1 - \delta)x_2 \leq x_2$.
The above determinant is

$$\begin{vmatrix} 0 & 1 & 1 \\ (x_2 - (\delta x_1 + (1 - \delta)x_2)) & \xi(\delta x_1 + (1 - \delta)x_2,.) - \xi(x_2,.) & 0 \\ (x_2 - x_1) & \eta(\xi(x_1,.),\xi(x_2,.)) & \mu(.)(\delta x_1 + (1 - \delta)x_2 - x_1)(x_2 - x_1) \end{vmatrix}$$

≥ 0.
This implies

$$\xi(\delta x_1 + (1 - \delta)x_2,.) \leq \xi(x_2,.) + \delta\eta(\xi(x_1,.),\xi(x_2,.)) - \mu(.)\delta(1 - \delta)(x_2 - x_1)^2.$$

Remark 7.5.2. *When $\mu(.) = 0$, then above theorem reduces to Theorem 1.10 of [69], i.e.*

$$\begin{vmatrix} (\kappa_3 - \kappa_2) & \xi(\kappa_2,.) - \xi(\kappa_3,.) \\ (\kappa_3 - \kappa_1) & \eta(\xi(\kappa_1,.),\xi(\kappa_3,.)) \end{vmatrix} \geq 0.$$

Now, we present Hermite–Hadamard inequality, Ostrowski inequality and some other interesting inequalities for strongly η-convex stochastic processes [156].

Theorem 7.5.2. *Suppose that $\xi : [c,d] \times \Omega \to \mathbb{R}$ is an strongly $\eta-$convex stochastic process with modulus $\mu(.) > 0$, such that η is bounded above on $\xi[c,d] \times \xi[c,d]$, then the following inequalities hold almost everywhere:*

$$\xi\left(\frac{c+d}{2},.\right) - \frac{M_\eta}{2} + \frac{\mu(.)}{12}(d - c)^2 \leq \frac{1}{d - c}\int_c^d \xi(x,.)dx$$

$$\leq \frac{\xi(c,.) + \xi(d,.)}{2} + \frac{1}{4}(\eta(\xi(c,.),\xi(d,.)) + \eta(\xi(d,.),\xi(c,.))) - \frac{\mu(.)}{6}(d - c)^2$$

$$\leq \frac{\xi(c,.) + \xi(d,.)}{2} + M_\eta - \frac{\mu(.)}{6}(d - c)^2,$$

where M_η is an upper bound of η.

Proof Since ξ is strongly η-convex stochastic process, therefore

$$\xi\left(\frac{c+d}{2},.\right) = \xi\left(\frac{1}{2}\left(\frac{c+d-\delta(d-c)}{2}\right) + \frac{1}{2}\left(\frac{c+d+\delta(d-c)}{2}\right),.\right)$$

$$\leq \xi\left(\frac{c+d+\delta(d-c)}{2},.\right)$$

$$+ \frac{1}{2}\eta\left(\xi\left(\frac{c+d-\delta(d-c)}{2},.\right),\xi\left(\frac{c+d+\delta(d-c)}{2},.\right)\right) - \frac{\mu(.)}{4}\delta^2(d-c)^2$$

$$\leq \xi\left(\frac{c+d+\delta(d-c)}{2},.\right) + \frac{M_\eta}{2} - \frac{\mu(.)}{4}\delta^2(d-c)^2.$$

This implies,

$$\xi\left(\frac{c+d}{2},.\right) - \frac{M_\eta}{2} + \frac{\mu(.)}{4}\delta^2(d-c)^2 \leq \xi\left(\frac{c+d+\delta(d-c)}{2},.\right).$$

Similarly,

$$\xi\left(\frac{c+d}{2},.\right) - \frac{M_\eta}{2} + \frac{\mu(.)}{4}\delta^2(d-c)^2 \leq \xi\left(\frac{c+d-\delta(d-c)}{2},.\right).$$

By using the change of variable technique, we have

$$\frac{1}{d-c}\int_c^d \xi(x,.)dx = \frac{1}{d-c}\left[\int_c^{(c+d)/2}\xi(x,.)dx + \int_{(c+d)/2}^d \xi(x,.)dx\right]$$

$$= \frac{1}{2}\int_0^1 \xi\left(\frac{c+d-\delta(d-c)}{2},.\right)d\delta + \frac{1}{2}\int_0^1 \xi\left(\frac{c+d+\delta(d-c)}{2},.\right)d\delta$$

$$\geq \int_0^1 \left[\xi\left(\frac{c+d}{2},.\right) - \frac{M_\eta}{2} + \frac{\mu(.)}{4}\delta^2(d-c)^2\right]d\delta$$

$$= \xi\left(\frac{c+d}{2},.\right) - \frac{M_\eta}{2} + \frac{\mu(.)}{12}(d-c)^2. \tag{7.19}$$

We now prove the right hand side of the theorem. Since ξ is strongly η-convex stochastic process with modulus $\mu(.) > 0$, we get

$$\xi(\delta c + (1-\delta)d,.) \leq \xi(d,.) + \delta\eta(\xi(c,.),\xi(d,.)) - \mu(.)\delta(1-\delta)(d-c)^2.$$

Integrating above inequality with respect to δ on both sides from 0 to 1, we have

$$\int_0^1 \xi(\delta c + (1-\delta)d,.)d\delta$$

$$\leq \int_0^1 (\xi(d,.) + \delta\eta(\xi(c,.),\xi(d,.)) - \mu(.)\delta(1-\delta)(d-c)^2)d\delta.$$

This implies,

$$\frac{1}{d-c}\int_c^d \xi(x,.)dx \leq \xi(d,.) + \frac{1}{2}\eta(\xi(c,.),\xi(d,.)) - \frac{\mu(.)}{6}(d-c)^2 = P.$$

Similarly,

$$\frac{1}{d-c}\int_c^d \xi(x,.)dx \leq \xi(c,.) + \frac{1}{2}\eta(\xi(d,.),\xi(c,.)) - \frac{\mu(.)}{6}(d-c)^2 = Q.$$

Therefore,

$$\frac{1}{d-c}\int_c^d \xi(x,.)dx \leq Min\{P,Q\}$$

$$\leq \frac{\xi(c,.)+\xi(d,.)}{2} + \frac{1}{4}(\eta(\xi(c,.),\xi(d,.)) + \eta(\xi(d,.),\xi(c,.)))$$

$$- \frac{\mu(.)}{6}(d-c)^2$$

$$\leq \frac{\xi(c,.)+\xi(d,.)}{2} + M_\eta - \frac{\mu(.)}{6}(d-c)^2. \tag{7.20}$$

From (7.19) and (7.20), we have

$$\xi\left(\frac{c+d}{2},..\right) - \frac{M_\eta}{2} + \frac{\mu(.)}{12}(d-c)^2 \leq \frac{1}{d-c}\int_c^d \xi(x,.)dx$$

$$\leq \frac{\xi(c,.)+\xi(d,.)}{2} + \frac{1}{4}(\eta(\xi(c,.),\xi(d,.)) + \eta(\xi(d,.),\xi(c,.))) - \frac{\mu(.)}{6}(d-c)^2$$

$$\leq \frac{\xi(c,.)+\xi(d,.)}{2} + M_\eta - \frac{\mu(.)}{6}(d-c)^2. \tag{7.21}$$

Remark 7.5.3. *When $\mu(.) = 0$, then above theorem reduces to Theorem 7.2.7. If we consider $\eta(\xi(x,.),\xi(y,.)) = \xi(x,.) - \xi(y,.)$ and $\mu(.) = 0$, then above theorem reduces to the classical Hermite–Hadamard inequality for convex stochastic process [87].*

Theorem 7.5.3. *If a stochastic process $\xi : K \times \Omega \to \mathbb{R}$ be an strongly η-convex with modulus $\mu(.) > 0$ and integrable on $K \times \Omega$, we have*

$$\frac{1}{d-c}\int_c^d \xi(x,.)\xi(c+d-x,.)dx \leq \xi(c,.)\xi(d,.) + \frac{1}{2}(\xi(c,.)\eta(\xi(c,.),\xi(d,.))$$

$$+ \xi(d,.)\eta(\xi(d,.),\xi(c,.))) + \frac{1}{3}\eta(\xi(c,.),\xi(d,.))\eta(\xi(d,.),\xi(c,.))$$

$$- \frac{\mu(.)}{6}(d-c)^2(\xi(c,.)+\xi(d,.)) - \frac{\mu(.)}{12}(d-c)^2(\eta(\xi(c,.),\xi(d,.))$$

$$+ \eta(\xi(d,.),\xi(c,.))) + \frac{\mu^2(.)}{30}(d-c)^4 \quad (a.e.).$$

Proof Since ξ is strongly η-convex stochastic process, therefore

$$\xi(\delta c + (1-\delta)d,.) \leq \xi(d,.) + \delta\eta(\xi(c,.),\xi(d,.)) - \mu(.)\delta(1-\delta)(d-c)^2 \quad (7.22)$$

and

$$\xi(\delta d + (1-\delta)c,.) \leq \xi(c,.) + \delta\eta(\xi(d,.),\xi(c,.)) - \mu(.)\delta(1-\delta)(d-c)^2. \quad (7.23)$$

From (7.22) and (7.23), we obtain

$$\xi(\delta c + (1-\delta)d,.)\xi(\delta d + (1-\delta)c,.) \leq \xi(c,.)\xi(d,.) + \delta(\xi(c,.)\eta(\xi(c,.),\xi(d,.))$$

$$+ \xi(d,.)\eta(\xi(d,.),\xi(c,.))) + \delta^2\eta(\xi(c,.),\xi(d,.))\eta(\xi(d,.),\xi(c,.))$$

$$- \mu(.)\delta(1-\delta)(d-c)^2(\xi(c,.) + \xi(d,.))$$

$$- \mu(.)\delta^2(1-\delta)(d-c)^2(\eta(\xi(c,.),\xi(d,.))$$

$$+ \eta(\xi(d,.),\xi(c,.))) + \mu^2(.)\delta^2(1-\delta)^2(d-c)^4.$$

Integrating above inequality from 0 to 1 on both sides with respect to δ, we have

$$\int_0^1 \xi(\delta c + (1-\delta)d,.)\xi(\delta d + (1-\delta)c,.)d\delta \leq \xi(c,.)\xi(d,.)$$

$$+ \frac{1}{2}(\xi(c,.)\eta(\xi(c,.),\xi(d,.)) + \xi(d,.)\eta(\xi(d,.),\xi(c,.)))$$

$$+ \frac{1}{3}\eta(\xi(c,.),\xi(d,.))\eta(\xi(d,.),\xi(c,.)) - \frac{\mu(.)}{6}(d-c)^2(\xi(c,.)$$

$$+ \xi(d,.)) - \frac{\mu(.)}{12}(d-c)^2(\eta(\xi(c,.),\xi(d,.)) + \eta(\xi(d,.),\xi(c,.))) + \frac{\mu^2(.)}{30}(d-c)^4.$$

This implies,

$$\frac{1}{d-c}\int_c^d \xi(x,.)\xi(c+d-x,.)dx \leq \xi(c,.)\xi(d,.) + \frac{1}{2}(\xi(c,.)\eta(\xi(c,.),\xi(d,.))$$

$$+ \xi(d,.)\eta(\xi(d,.),\xi(c,.))) + \frac{1}{3}\eta(\xi(c,.),\xi(d,.))\eta(\xi(d,.),\xi(c,.))$$

$$- \frac{\mu(.)}{6}(d-c)^2(\xi(c,.) + \xi(d,.)) - \frac{\mu(.)}{12}(d-c)^2(\eta(\xi(c,.),\xi(d,.))$$

$$+ \eta(\xi(d,.),\xi(c,.))) + \frac{\mu^2(.)}{30}(d-c)^4.$$

Remark 7.5.4. *When ξ is strongly log-convex stochastic process, then above theorem reduces to Theorem 3 of [166], i.e.*

$$\frac{1}{d-c}\int_c^d \xi(x,.)\xi(c+d-x,.)dx \le \xi(c,.)\xi(d,.) + \frac{\mu^2(.)}{30}(d-c)^4$$

$$-\frac{4\mu(.)(d-c)^2}{\ln[\xi(c,.)-\xi(d,.)]^2}[A(\xi(c,.),\xi(d,.)) + L(\xi(c,.),\xi(d,.))]$$

$$\le \frac{2[A(\xi(c,.),\xi(d,.))]^2 + [G(\xi(c,.),\xi(d,.))]^2}{3}$$

$$-\frac{\mu(.)A(\xi(c,.),\xi(d,.))(d-c)^2}{3} + \frac{\mu^2(.)}{30}(d-c)^4.$$

Theorem 7.5.4. *Let $\xi : K \times \Omega \to \mathbb{R}$ be a mean square stochastic process such that ξ' is mean square integrable on $[c,d]$, where $c,d \in K$ with $c < d$. If $|\xi'|$ is an strongly η-convex stochastic process with modulus $\mu(.) > 0$ on K and $|\xi'(t,.)| \le M$ for every t, then*

$$\left| \xi(t,.) - \frac{1}{d-c}\int_c^d \xi(x,.)dx \right|$$

$$\le \frac{M}{2(d-c)}((t-c)^2 + (d-t)^2) + \frac{1}{3(d-c)}[(t-c)^2\eta(|\xi'(t,.)|,|\xi'(c,.)|)$$

$$+ (d-t)^2\eta(|\xi'(t,.)|,|\xi'(d,.)|)] - \frac{\mu(.)}{12(d-c)}((t-c)^4 + (d-t)^4) \quad (a.e.).$$

Proof Recall Lemma 7.2.1:

$$\xi(t,.) - \frac{1}{d-c}\int_c^d \xi(x,.)dx = \frac{(t-c)^2}{d-c}\int_0^1 y\xi'(yt + (1-y)c,.)dy$$

$$-\frac{(d-t)^2}{d-c}\int_0^1 y\xi'(yt + (1-y)d,.)dy, \quad (a.e), \text{ for each } t \in [c,d].$$

Since $|\xi'|$ is strongly η-convex stochastic process, therefore

$$\left| \xi(t,.) - \frac{1}{d-c}\int_c^d \xi(x,.)dx \right|$$

$$\le \frac{(t-c)^2}{d-c}\int_0^1 y[|\xi'(c,.)| + y\eta(|\xi'(t,.)|,|\xi'(c,.)|) - \mu(.)y(1-y)(c-t)^2]dy$$

$$+ \frac{(d-t)^2}{d-c}\int_0^1 y[|\xi'(d,.)| + y\eta(|\xi'(t,.)|,|\xi'(d,.)|) - \mu(.)y(1-y)(d-t)^2]dy$$

$$\le \frac{(t-c)^2}{d-c}\left[\frac{M}{2} + \frac{\eta(|\xi'(t,.)|,|\xi'(c,.)|)}{3} - \frac{\mu(.)}{12}(c-t)^2\right]$$

$$+ \frac{(d-t)^2}{d-c}\left[\frac{M}{2} + \frac{\eta(|\xi'(t,.)|,|\xi'(d,.)|)}{3} - \frac{\mu(.)}{12}(d-t)^2\right]$$

$$= \frac{M}{2(d-c)}((t-c)^2 + (d-t)^2) + \frac{1}{3(d-c)}[(t-c)^2\eta(|\xi'(t,.)|, |\xi'(c,.)|)$$

$$+ (d-t)^2\eta(|\xi'(t,.)|, |\xi'(d,.)|)] - \frac{\mu(.)}{12(d-c)}((t-c)^4 + (d-t)^4).$$

Remark 7.5.5. When $\mu(.) = 0$, then above theorem reduces to Theorem 4.2 of [69], i.e.

$$\left| \xi(t,.) - \frac{1}{d-c} \int_c^d \xi(x,.)dx \right|$$

$$\leq \frac{M}{2(d-c)}((t-c)^2 + (d-t)^2) + \frac{1}{3(d-c)}[(t-c)^2\eta(|\xi'(t,.)|, |\xi'(c,.)|)$$

$$+ (d-t)^2\eta(|\xi'(t,.)|, |\xi'(d,.)|)] \quad (a.e.).$$

Theorem 7.5.5. *Let* $\xi : K \times \Omega \to \mathbb{R}$ *be a mean square differentiable stochastic process on* K^0 *and* ξ' *be a mean square integrable on* $[c,d]$, *where* $c, d \in K, c < d$. *If* $|\xi'|$ *is an strongly* η-*convex stochastic process on* $[c,d]$, *then we have almost everywhere:*

$$\left| \frac{\xi(c,.) + \xi(d,.)}{2} - \frac{1}{d-c} \int_c^d \xi(x,.)dx \right|$$

$$\leq \frac{d-c}{4}\left(|\xi'(d,.)| + \frac{1}{2}\eta(|\xi'(c,.)|, |\xi'(d,.)|) - \frac{\mu(.)}{8}(d-c)^2 \right).$$

Proof From Lemma 7.2.2, we have

$$\left| \frac{\xi(c,.) + \xi(d,.)}{2} - \frac{1}{d-c} \int_c^d \xi(x,.)dx \right| \leq \frac{d-c}{2} \int_0^1 |1 - 2\delta||\xi'(\delta c + (1-\delta)d,.)|d\delta.$$

Using the definition of strong η-convex stochastic process in above inequality, we have

$$\left| \frac{\xi(c,.) + \xi(d,.)}{2} - \frac{1}{d-c} \int_c^d \xi(x,.)dx \right|$$

$$\leq \frac{d-c}{2} \int_0^1 |1 - 2\delta|(|\xi'(d,.)| + \delta\eta(|\xi'(c,.)|, |\xi'(d,.)|)$$

$$- \mu(.)\delta(1-\delta)(d-c)^2)d\delta$$

$$= \frac{d-c}{2} \left(\int_0^1 |1 - 2\delta||\xi'(d,.)|d\delta + \int_0^1 \delta|1 - 2\delta|\eta(|\xi'(c,.)|, |\xi'(d,.)|)d\delta \right.$$

$$\left. - \mu(.)(d-c)^2 \int_0^1 \delta(1-\delta)|1 - 2\delta|d\delta \right)$$

$$= \frac{d-c}{2}\left(\frac{1}{2}|\xi'(d,.)| + \frac{1}{4}\eta(|\xi'(c,.)|,|\xi'(d,.)|) - \frac{\mu(.)}{16}(d-c)^2\right)$$

$$= \frac{d-c}{4}\left(|\xi'(d,.)| + \frac{1}{2}\eta(|\xi'(c,.)|,|\xi'(d,.)|) - \frac{\mu(.)}{8}(d-c)^2\right).$$

Corollary 7.5.1. *When $\mu(.) = 0$ in above theorem, then we obtain the following inequality for η-convex stochastic processes:*

$$\left|\frac{\xi(c,.) + \xi(d,.)}{2} - \frac{1}{d-c}\int_c^d \xi(x,.)dx\right|$$

$$\leq \frac{d-c}{4}\left(|\xi'(d,.)| + \frac{1}{2}\eta(|\xi'(c,.)|,|\xi'(d,.)|)\right) \quad (a.e.).$$

Remark 7.5.6. *When $\mu(.) = 0$ and $\eta(|\xi'(c,.)|,|\xi'(d,.)|) = |\xi'(c,.)| - |\xi'(d,.)|$, then above theorem reduces to Corollary 5.3 of Fu et al. [51], i.e.*

$$\left|\frac{\xi(c,.) + \xi(d,.)}{2} - \frac{1}{d-c}\int_c^d \xi(x,.)dx\right| \leq \frac{d-c}{4}A(|\xi'(c,.)|,|\xi'(d,.)|) \quad (a.e.).$$

Theorem 7.5.6. *Let $\xi : K \times \Omega \to \mathbb{R}$ be a mean square differentiable stochastic process on K^0 with $q > 1$, $\frac{1}{p} + \frac{1}{q} = 1$ and assume that ξ' be a mean square integrable on $[c,d]$, where $c,d \in K, c < d$. If $|\xi'|^q$ is an strongly η-convex stochastic process on $[c,d]$, then we have almost everywhere:*

$$\left|\frac{\xi(c,.) + \xi(d,.)}{2} - \frac{1}{d-c}\int_c^d \xi(x,.)dx\right| = \frac{d-c}{2}\left(\frac{1}{p+1}\right)^{1/p}$$

$$\times \left(|\xi'(d,.)|^q + \frac{1}{2}\eta(|\xi'(c,.)|^q,|\xi'(d,.)|^q) - \frac{\mu(.)}{6}(d-c)^2\right)^{1/q}.$$

Proof From Lemma 7.2.2, we have

$$\left|\frac{\xi(c,.) + \xi(d,.)}{2} - \frac{1}{d-c}\int_c^d \xi(x,.)dx\right|$$

$$\leq \frac{d-c}{2}\int_0^1 |1 - 2\delta||\xi'(\delta c + (1-\delta)d,.)|d\delta.$$

Using Hölder's inequality and the definition of strong η-convex stochastic process in above inequality, we obtain

$$\left| \frac{\xi(c,.) + \xi(d,.)}{2} - \frac{1}{d-c} \int_c^d \xi(u,.)du \right|$$

$$\leq \frac{d-c}{2} \left(\int_0^1 |1 - 2\delta|^p d\delta \right)^{1/p} \left(\int_0^1 |\xi'(\delta c + (1-\delta)d,.)|^q d\delta \right)^{1/q}$$

$$\leq \frac{d-c}{2} \left(\frac{1}{p+1} \right)^{1/p} \left(\int_0^1 (|\xi'(d,.)|^q + \delta\eta(|\xi'(c,.)|^q, |\xi'(d,.)|^q) \right.$$

$$\left. - \mu(.)\delta(1-\delta)(d-c)^2)d\delta \right)^{1/q}$$

$$= \frac{d-c}{2} \left(\frac{1}{p+1} \right)^{1/p}$$

$$\times \left(|\xi'(d,.)|^q + \frac{1}{2}\eta(|\xi'(c,.)|^q, |\xi'(d,.)|^q) - \frac{\mu(.)}{6}(d-c)^2 \right)^{1/q}.$$

Corollary 7.5.2. *When $\mu(.) = 0$ in above theorem, then we obtain the following inequality for η-convex stochastic processes:*

$$\left| \frac{\xi(c,.) + \xi(d,.)}{2} - \frac{1}{d-c} \int_c^d \xi(x,.)dx \right|$$

$$\leq \frac{d-c}{2} \left(\frac{1}{p+1} \right)^{1/p} \left(|\xi'(d,.)|^q + \frac{1}{2}\eta(|\xi'(c,.)|^q, |\xi'(d,.)|^q) \right)^{1/q} \quad (a.e.).$$

Remark 7.5.7. *When $\mu(.) = 0$ and $\eta(|\xi'(c,.)|^q, |\xi'(d,.)|^q) = |\xi'(c,.)|^q - |\xi'(d,.)|^q$, then above theorem reduces to Corollary 5.5 of Fu et al. [51], i.e.*

$$\left| \frac{\xi(c,.) + \xi(d,.)}{2} - \frac{1}{d-c} \int_c^d \xi(x,.)dx \right|$$

$$\leq \frac{d-c}{2} \left(\frac{1}{p+1} \right)^{1/p} A^{\frac{1}{q}}(|\xi'(c,.)|^q, |\xi'(d,.)|^q) \quad (a.e.).$$

Theorem 7.5.7. *Let $\xi : K \times \Omega \to \mathbb{R}$ be a mean square differentiable stochastic process on K^0 with $q \geq 1$, and assume that ξ' be a mean square integrable on $[c,d]$, where $c, d \in K, c < d$. If $|\xi'|^q$ is an strongly η-convex stochastic process on $[c,d]$, then we have almost everywhere:*

$$\left| \frac{\xi(c,.) + \xi(d,.)}{2} - \frac{1}{d-c} \int_c^d \xi(x,.)dx \right|$$

$$\leq \frac{d-c}{4} \left(|\xi'(d,.)|^q + \frac{1}{2}\eta(|\xi'(c,.)|^q, |\xi'(d,.)|^q) - \frac{\mu(.)}{8}(d-c)^2 \right)^{1/q}.$$

Proof For $q = 1$, we use the estimates from the proof of Theorem 7.5.5. Now we prove result for $q > 1$. From Lemma 7.2.2, we have

$$\left| \frac{\xi(c,.) + \xi(d,.)}{2} - \frac{1}{d-c} \int_c^d \xi(x,.)dx \right|$$

$$\leq \frac{d-c}{2} \int_0^1 |1 - 2\delta| |\xi'(\delta c + (1-\delta)d, .)| d\delta.$$

Using Hölder's inequality and the definition of strong η-convex stochastic process in above inequality, we obtain

$$\left| \frac{\xi(c,.) + \xi(d,.)}{2} - \frac{1}{d-c} \int_c^d \xi(x,.)dx \right|$$

$$\leq \frac{d-c}{2} \left(\int_0^1 |1 - 2\delta| d\delta \right)^{1 - \frac{1}{q}} \left(\int_0^1 |1 - 2\delta| |\xi'(\delta c + (1-\delta)d, .)|^q d\delta \right)^{1/q}$$

$$\leq \frac{d-c}{2} \left(\frac{1}{2} \right)^{1 - \frac{1}{q}} \left(\int_0^1 |1 - 2\delta| (|\xi'(d,.)|^q + \delta \eta(|\xi'(c,.)|^q, |\xi'(d,.)|^q) \right.$$

$$\left. - \mu(.)\delta(1-\delta)(d-c)^2) d\delta \right)^{1/q}$$

$$= \frac{d-c}{2} \left(\frac{1}{2} \right)^{1 - \frac{1}{q}} \left(\frac{1}{2} |\xi'(d,.)|^q + \frac{1}{4} \eta(|\xi'(c,.)|^q, |\xi'(d,.)|^q) - \frac{\mu(.)}{16}(d-c)^2 \right)^{1/q}$$

$$= \frac{d-c}{4} \left(|\xi'(d,.)|^q + \frac{1}{2} \eta(|\xi'(c,.)|^q, |\xi'(d,.)|^q) - \frac{\mu(.)}{8}(d-c)^2 \right)^{1/q}.$$

Corollary 7.5.3. *When $\mu(.) = 0$ in above theorem, then we obtain the following inequality for η-convex stochastic processes:*

$$\left| \frac{\xi(c,.) + \xi(d,.)}{2} - \frac{1}{d-c} \int_c^d \xi(x,.)dx \right|$$

$$\leq \frac{d-c}{4} \left(|\xi'(d,.)|^q + \frac{1}{2} \eta(|\xi'(c,.)|^q, |\xi'(d,.)|^q) \right)^{1/q} \quad (a.e.).$$

Remark 7.5.8. *When $\mu(.) = 0$ and $\eta(|\xi'(c,.)|^q, |\xi'(d,.)|^q) = |\xi'(c,.)|^q - |\xi'(d,.)|^q$, then above theorem reduces to Corollary 5.8 of Fu et al. [51], i.e.*

$$\left| \frac{\xi(c,.) + \xi(d,.)}{2} - \frac{1}{d-c} \int_c^d \xi(x,.)dx \right|$$

$$\leq \frac{d-c}{4} A^{\frac{1}{q}}(|\xi'(c,.)|^q, |\xi'(d,.)|^q) \quad (a.e.).$$

Theorem 7.5.8. *Let $\xi : K \times \Omega \to \mathbb{R}$ be a mean square differentiable stochastic process on K^0 with $q > 1$, $\frac{1}{p} + \frac{1}{q} = 1$ and assume that ξ' be a mean square* *integrable on $[c, d]$, where $c, d \in K, c < d$. If $|\xi'|^q$ is an strongly η-convex stochastic process on $[c, d]$, then we have almost everywhere:*

$$\left| \frac{\xi(c, .) + \xi(d, .)}{2} - \frac{1}{d-c} \int_c^d \xi(x, .)dx \right|$$

$$\leq \frac{d-c}{2} \left(\frac{1}{2(p+1)} \right)^{\frac{1}{p}}$$

$$\times \left[\left(\frac{1}{2}|\xi'(d, .)|^q + \frac{1}{6}\eta(|\xi'(c, .)|^q, |\xi'(d, .)|^q) - \frac{\mu(.)}{12}(d-c)^2 \right)^{1/q} \right.$$

$$\left. + \left(\frac{1}{2}(|\xi'(d, .)|^q + \frac{1}{3}\eta(|\xi'(c, .)|^q, |\xi'(d, .)|^q) - \frac{\mu(.)}{12}(d-c)^2 \right)^{1/q} \right].$$

Proof From Lemma 7.2.2, we have

$$\left| \frac{\xi(c, .) + \xi(d, .)}{2} - \frac{1}{d-c} \int_c^d \xi(x, .)dx \right|$$

$$\leq \frac{d-c}{2} \int_0^1 |1 - 2\delta||\xi'(\delta c + (1-\delta)d, .)|d\delta.$$

Using Hölder's inequality and the definition of strong η-convex stochastic process in above inequality, we get

$$\left| \frac{\xi(c, .) + \xi(d, .)}{2} - \frac{1}{d-c} \int_c^d \xi(x, .)dx \right|$$

$$\leq \frac{d-c}{2} \left(\int_0^1 (1-\delta)|1 - 2\delta|^p d\delta \right)^{\frac{1}{p}} \left(\int_0^1 (1-\delta)|\xi'(\delta c + (1-\delta)d, .)|^q d\delta \right)^{1/q}$$

$$+ \frac{d-c}{2} \left(\int_0^1 \delta|1 - 2\delta|^p d\delta \right)^{\frac{1}{p}} \left(\int_0^1 \delta|\xi'(\delta c + (1-\delta)d, .)|^q d\delta \right)^{1/q}$$

$$\leq \frac{d-c}{2} \left(\frac{1}{2(p+1)} \right)^{\frac{1}{p}}$$

$$\times \left[\left(\int_0^1 (1-\delta)(|\xi'(d, .)|^q + \delta\eta(|\xi'(c, .)|^q, |\xi'(d, .)|^q) - \mu(.)\delta(1-\delta)(d-c)^2)d\delta \right)^{1/q} \right.$$

$$\left. + \left(\int_0^1 \delta(|\xi'(d, .)|^q + \delta\eta(|\xi'(c, .)|^q, |\xi'(d, .)|^q) - \mu(.)\delta(1-\delta)(d-c)^2)d\delta \right)^{1/q} \right]$$

$$= \frac{d-c}{2} \left(\frac{1}{2(p+1)} \right)^{\frac{1}{p}} \left[\left(\frac{1}{2}|\xi'(d, .)|^q + \frac{1}{6}\eta(|\xi'(c, .)|^q, |\xi'(d, .)|^q) - \frac{\mu(.)}{12}(d-c)^2 \right)^{1/q} \right.$$

$$\left. + \left(\frac{1}{2}(|\xi'(d, .)|^q + \frac{1}{3}\eta(|\xi'(c, .)|^q, |\xi'(d, .)|^q) - \frac{\mu(.)}{12}(d-c)^2 \right)^{1/q} \right].$$

Corollary 7.5.4. *When $\mu(.) = 0$ in above theorem, then we obtain the following inequality for η-convex stochastic processes:*

$$
\left| \frac{\xi(c,.) + \xi(d,.)}{2} - \frac{1}{d-c} \int_c^d \xi(x,.)dx \right|
$$

$$
\leq \frac{d-c}{2} \left(\frac{1}{2(p+1)} \right)^{\frac{1}{p}} \left[\left(\frac{1}{2}|\xi'(d,.)|^q + \frac{1}{6}\eta(|\xi'(c,.)|^q, |\xi'(d,.)|^q) \right)^{1/q} \right.
$$

$$
\left. + \left(\frac{1}{2}(|\xi'(d,.)|^q + \frac{1}{3}\eta(|\xi'(c,.)|^q, |\xi'(d,.)|^q) \right)^{1/q} \right] \quad (a.e.).
$$

Remark 7.5.9. *When $\mu(.) = 0$ and $\eta(|\xi'(c,.)|^q, |\xi'(d,.)|^q) = |\xi'(c,.)|^q - |\xi'(d,.)|^q$, then above theorem reduces to Corollary 5.10 of Fu et al. [51], i.e.*

$$
\left| \frac{\xi(c,.) + \xi(d,.)}{2} - \frac{1}{d-c} \int_c^d \xi(x,.)dx \right| \leq \frac{d-c}{4} \left(\frac{1}{(p+1)} \right)^{\frac{1}{p}}
$$

$$
\times \left[\left(\frac{|\xi'(c,.)|^q + 2|\xi'(d,.)|^q}{3} \right)^{1/q} + \left(\frac{2|\xi'(c,.)|^q + |\xi'(d,.)|^q}{3} \right)^{1/q} \right] \quad (a.e.).
$$

Theorem 7.5.9. *Let $\xi : K \times \Omega \to \mathbb{R}$ be a mean square differentiable stochastic process on K^0 with $q \geq 1$, and assume that ξ' be a mean square integrable on $[c,d]$, where $c,d \in K, c < d$. If $|\xi'|^q$ is an strongly η-convex stochastic process on $[c,d]$, then we have almost everywhere:*

$$
\left| \frac{\xi(c,.) + \xi(d,.)}{2} - \frac{1}{d-c} \int_c^d \xi(x,.)dx \right|
$$

$$
\leq \frac{d-c}{8} \left[\left(|\xi'(d,.)|^q + \frac{1}{4}\eta(|\xi'(c,.)|^q, |\xi'(d,.)|^q) - \frac{\mu(.)}{8}(d-c)^2 \right)^{1/q} \right.
$$

$$
\left. \times \left(|\xi'(d,.)|^q + \frac{3}{4}\eta(|\xi'(c,.)|^q, |\xi'(d,.)|^q) - \frac{\mu(.)}{8}(d-c)^2 \right)^{1/q} \right].
$$

Proof For $q = 1$, we use the estimates from the proof of Theorem 7.5.5. Now we prove result for $q > 1$. From Lemma 7.2.2, we have

$$
\left| \frac{\xi(c,.) + \xi(d,.)}{2} - \frac{1}{d-c} \int_c^d \xi(x,.)dx \right|
$$

$$
\leq \frac{d-c}{2} \int_0^1 |1 - 2\delta||\xi'(\delta c + (1-\delta)d,.)|d\delta.
$$

Using improved power-mean integral inequality and the definition of strong η-convex stochastic process in above inequality, we get

$$\left| \frac{\xi(c,.) + \xi(d,.)}{2} - \frac{1}{d-c} \int_c^d \xi(x,.)dx \right|$$

$$\leq \frac{d-c}{2} \left(\int_0^1 (1-\delta)|1-2\delta|d\delta \right)^{1-\frac{1}{q}} \left(\int_0^1 (1-\delta)|1-2\delta||\xi'(\delta c + (1-\delta)d,.)|^q d\delta \right)^{1/q}$$

$$+ \frac{d-c}{2} \left(\int_0^1 \delta|1-2\delta|d\delta \right)^{1-\frac{1}{q}} \left(\int_0^1 \delta|1-2\delta||\xi'(\delta c + (1-\delta)d,.)|^q d\delta \right)^{1/q}$$

$$\leq \frac{d-c}{2} \left(\frac{1}{4} \right)^{1-\frac{1}{q}} \left[\left(\int_0^1 (1-\delta)|1-2\delta|(|\xi'(d,.)|^q + \delta\eta(|\xi'(c,.)|^q, |\xi'(d,.)|^q) \right. \right.$$

$$\left. - \mu(.)\delta(1-\delta)(d-c)^2)d\delta \right)^{1/q} + \left(\int_0^1 \delta|1-2\delta|(|\xi'(d,.)|^q + \delta\eta(|\xi'(c,.)|^q, |\xi'(d,.)|^q) \right.$$

$$\left. \left. - \mu(.)\delta(1-\delta)(d-c)^2)d\delta \right)^{1/q} \right]$$

$$= \frac{d-c}{2} \left(\frac{1}{4} \right)^{1-\frac{1}{q}} \left[\left(\frac{1}{4}|\xi'(d,.)|^q + \frac{1}{16}\eta(|\xi'(c,.)|^q, |\xi'(d,.)|^q) - \frac{\mu(.)}{32}(d-c)^2 \right)^{1/q} \right.$$

$$\left. + \left(\frac{1}{4}|\xi'(d,.)|^q + \frac{3}{16}\eta(|\xi'(c,.)|^q, |\xi'(d,.)|^q) - \frac{\mu(.)}{32}(d-c)^2 \right)^{1/q} \right]$$

$$= \frac{d-c}{8} \left[\left(|\xi'(d,.)|^q + \frac{1}{4}\eta(|\xi'(c,.)|^q, |\xi'(d,.)|^q) - \frac{\mu(.)}{8}(d-c)^2 \right)^{1/q} \right.$$

$$\left. + \left(|\xi'(d,.)|^q + \frac{3}{4}\eta(|\xi'(c,.)|^q, |\xi'(d,.)|^q) - \frac{\mu(.)}{8}(d-c)^2 \right)^{1/q} \right].$$

Corollary 7.5.5. *When $\mu(.) = 0$ in above theorem, then we obtain the following inequality for η-convex stochastic processes:*

$$\left| \frac{\xi(c,.) + \xi(d,.)}{2} - \frac{1}{d-c} \int_c^d \xi(x,.)dx \right|$$

$$\leq \frac{d-c}{8} \left[\left(|\xi'(d,.)|^q + \frac{1}{4}\eta(|\xi'(c,.)|^q, |\xi'(d,.)|^q) \right)^{1/q} \right.$$

$$\left. + \left(|\xi'(d,.)|^q + \frac{3}{4}\eta(|\xi'(c,.)|^q, |\xi'(d,.)|^q) \right)^{1/q} \right] \quad (a.e.).$$

Remark 7.5.10. *When $\mu(.) = 0$ and $\eta(|\xi'(c,.)|^q, |\xi'(d,.)|^q) = |\xi'(c,.)|^q - |\xi'(d,.)|^q$, then above theorem reduces to Corollary 5.13 of Fu et al. [51], i.e.*

$$\left| \frac{\xi(c,.) + \xi(d,.)}{2} - \frac{1}{d-c} \int_c^d \xi(x,.)dx \right|$$

$$\leq \frac{d-c}{8} \left[\left(\frac{|\xi'(c,.)|^q}{4} + \frac{3|\xi'(d,.)|^q}{4} \right)^{1/q} + \left(\frac{3|\xi'(c,.)|^q}{4} + \frac{|\xi'(d,.)|^q}{4} \right)^{1/q} \right]$$

$(a.e.)$.

Chapter 8

Applications

8.1 Hermite–Hadamard Inequality

The Hermite–Hadamard inequality [56] is the first fundamental result for convex functions defined on a interval of real numbers with a natural geometrical interpretation. Hermite–Hadamard inequality is now itself a special domain of the theory of inequalities with many powerful results and a large number of applications in numerical integration, probability theory and statistics, information theory and integral operator theory. The following examples show the importance of Hermite–Hadamard inequality in Calculus.

Example 8.1.1. For $\xi(x) = \frac{1}{1+x}, x \geqslant 0$ and $c = 0, d = x$ in Theorem 1.3.1, Hermite [59] noticed that $x - \frac{x^2}{x+2} < \log(1 + x) < x - \frac{x^2}{2(1+x)}$. In particular,

$$\frac{1}{n + \frac{1}{2}} < \log(n + 1) - \log n < \frac{1}{2}\left(\frac{1}{n} + \frac{1}{n + 1}\right)$$

for every $n \in \mathbb{N}$, and this fact is instrumental in deriving Stirling's formula,

$$n! \sim \sqrt{2\pi}.n^{n + \frac{1}{2}}e^{-n},$$

see [8].

Example 8.1.2. If $\xi(x) = e^{2x}$, then from Hermite–Hadamard inequality, we have

$$e^{\frac{2(c+d)}{2}} < \frac{e^{2d} - e^{2c}}{2d - 2c} < \frac{e^{2c} + e^{2d}}{2} \text{ for } c \neq d \text{ in } \mathbb{R};$$

i.e.,

$$\sqrt{uv} < \frac{u - v}{\log u - \log v} < \frac{u + v}{2} \text{ for } u \neq v \text{ in } (0, \infty),$$

which represents the Geometric, Logarithmic and Arithmetic mean inequality.

Example 8.1.3. For $\xi(x) = \cos x, x \in [0, \pi/2]$, we have

$$\frac{\cos c + \cos d}{2} < \frac{\sin d - \sin c}{d - c} < \cos\left(\frac{c + d}{2}\right);$$

DOI: 10.1201/9781003408284-8

$$or \quad \frac{2\cos\left(\frac{c+d}{2}\right)\cos\left(\frac{c-d}{2}\right)}{2} < \frac{2\cos\left(\frac{d+c}{2}\right)\sin\left(\frac{d-c}{2}\right)}{d-c} < \cos\left(\frac{c+d}{2}\right)$$

$$or \quad \cos\left(\frac{d-c}{2}\right) < \frac{2\sin\left(\frac{d-c}{2}\right)}{d-c} < 1$$

and this implies that $x\cos x < \sin x < x$ *for* $x \in [0, \frac{\pi}{2}]$, *see Figure (8.1).*

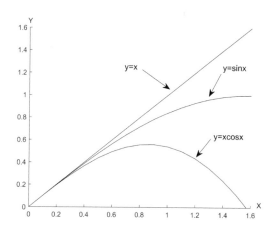

FIGURE 8.1: Graphical representation of inequality.

8.1.1   Higher dimensional Hermite–Hadamard inequality

Chen and Cheung [29] gave the following higher dimensional Hermite–Hadamard inequality for a function $\xi : \prod_{i=1}^{n}[c_i, d_i] \subset \mathbb{R}^n \to \mathbb{R}$, which is semiconvex of rate $(k_1, k_2, ..., k_n)$ on the coordinates, i.e.,

$$\xi(d_{\mu_1}, ..., d_{\mu_n})$$
$$\leq \frac{1}{n}\sum_{i=1}^{n}\int_{c_i}^{d_i}\xi(\hat{d}_{\mu_i}, x_i)d\mu_i(x_i) + \sum_{i=1}^{n}\frac{k_i}{2n}\int_{c_i}^{d_i}|x_i - d_{\mu i}|^2 d\mu_i(x_i) \qquad (8.1)$$
$$\leq \int_{\prod_{i=1,...n}[c_i,d_i]}\int\xi(x_1, ...x_n)d\mu_1(x_1) \otimes ... \otimes d\mu_n(x_n)$$
$$+ \sum_{i=1}^{n}\frac{k_i}{2}\int_{c_i}^{d_i}|x_i - d_{\mu i}|^2 d\mu_i(x_i) \qquad (8.2)$$

$$\leq \sum_{i=1}^{n} \int_{\Pi_{j\neq 1}[c_j,d_j]} \int \frac{d_i - d_{\mu_i}}{n(d_i - c_i)} \xi(\hat{x}_i, c_i) d\hat{\mu}_i(x_i)$$

$$+ \sum_{i=1}^{n} \int_{\Pi_{j\neq 1}[c_j,d_j]} \int \frac{d_{\mu_i} - c_i}{n(d_i - c_i)} \xi(\hat{x}_i, d_i) d\hat{\mu}_i(x_i)$$

$$+ \sum_{i=1}^{n} \frac{(n-1)k_i}{2n} \int_{c_i}^{d_i} |x_i - d_{\mu_i}|^2 d\mu_i(x_i) + \sum_{i=1}^{n} \frac{k_i}{2n}(d_{\mu_i} - c_i)(d_i - d_{\mu_i}) \quad (8.3)$$

$$\leq \sum_{x_i = c_i \text{ or } d_i, i=1,2,\dots n} \frac{\psi(x_1)\dots\psi(x_n)}{\prod_{i=1}^{n}(d_i - c_i)} \xi(x_1, ..., x_n) + \sum_{i=1}^{n} \frac{k_i}{2}(d_{\mu_i} - c_i)(d_i - d_{\mu_i}),$$

$$(8.4)$$

where

$$\psi(x_i) = \begin{cases} d_i - d_{\mu_i} & x_i = c_i, \\ d_{\mu_i} - c_i & x_i = d_i, i = 1, ..., n \end{cases}$$

and $\mu_i, i = 1, 2.., n$ are Borel probability measures on the intervals $[c_i, d_i], i = 1, ..., n$, respectively.

8.1.2 Mass transportation and higher dimensional Hermite–Hadamard inequality

Chen and Cheung [29] interpreted the meaning of the Hermite–Hadamard inequality **8.1.1** from the point of view of optimal mass transportation problems.

A typical optimal mass transport problem is the Kantorovich problem, which is formulated as follows:

$$\min_{y \in \Pi(v_1, v_2)} \int_{\mathbb{R}^n \times \mathbb{R}^n} C(x, y) d\gamma(x, y),$$

where $v_1, v_2 \in P(\mathbb{R}^n)$ with $P(\mathbb{R}^n)$ meaning the space of Borel probability measures on $\mathbb{R}^n$, $C(x, y) : \mathbb{R}^n \times \mathbb{R}^n \to [0, +\infty)$ is a cost function and

$$\prod(v_1, v_2) := \{\gamma \in P(\mathbb{R}^n \times \mathbb{R}^n) : (\pi_1)_\#(\gamma) = v_1, (\pi_2)_\#(\gamma) = v_2\}$$

is the set of transport plans between v_1 and v_2. Here, $\pi_1, \pi_2 : \mathbb{R}^n \times \mathbb{R}^n \to \mathbb{R}^n$ are the canonical projections on the first and second factors, respectively. For more details on optimal mass transportation theory, see [148, 170].

Let $\mu_i \in P([a_i, b_i]), i = 1, 2, ..., n$ and δ_x denotes the Dirac measure at the point $x \in \mathbb{R}$. The product measure $\delta_{d_{\mu_1}} \otimes ... \otimes \delta_{d_{\mu_n}} \in P(\mathbb{R}^n)$ of $\delta_{d_{\mu_i}}, i = 1, 2, .., n$ is given by

$$\delta_{d_{\mu_1}} \otimes ... \otimes \delta_{d_{\mu_n}}(A_1 \times ... \times A_n) = \begin{cases} 1, & d_{\mu_i} \in A_i, i = 1, 2, ..., n \\ 0, & \text{otherwise} \end{cases}$$

for any Borel measurable $A_i \subset [c_i, d_i], i = 1, 2, .., n$.

Mass transportation meaning of the Hermite–Hadamard inequality:
After adding the constant $\sum_{i=1}^{n} \frac{k_i}{2} d_{\mu_i}^2$ to each term in the higher dimensional Hermite–Hadamard inequality, Chen and Cheung [29] proved that each new term equals to the mass transport cost in the following series of transportation models with initial measure $v_1 = \delta_0 \otimes \dots \otimes \delta_0 \in P(\mathbb{R}^n)$ and cost function $C(x,y) : \mathbb{R}^n \times \mathbb{R}^n \to [0, +\infty)$ given by

$$C(x,y) = C(x_1, x_2, \dots x_n, y_1, y_2 \dots y_n) := \xi(y - x) + \sum_{i=1}^{n} \frac{k_i}{2}(x_i - y_i)^2,$$

where

$$\xi : \prod_{i=1}^{n}[c_i, d_i] \to [0, +\infty)$$

is continuous and semiconvex of rate $(k_1, k_2, \dots, k_n)$ on the coordinates.
The following examples are given by Chen and Cheung [29].

Example 8.1.4. *Take* $v_1 = \delta_0 \otimes \dots \otimes \delta_0, v_2 = \delta_{d_{\mu_1}} \otimes \dots \otimes \delta_{d_{\mu_n}} \in P(\mathbb{R}^n)$, *then* $\prod(v_1, v_2) = \{v_1 \otimes v_2\}$ *is a singleton, and the optimal transportation cost from* v_1 *to* v_2 *is the sum of expression (8.1) and* $\sum_{i=1}^{n} \frac{k_i}{2} d_{\mu_i}^2$.

Example 8.1.5. *Take* $v_1 = \delta_0 \otimes \dots \otimes \delta_0, v_2 = \frac{1}{n} \sum_{i=1}^{n} \delta_{d_{\mu_1}} \otimes \dots \otimes \delta_{d_{\mu_{i-1}}} \otimes \mu_i \otimes \delta_{d_{\mu_{i+1}}} \otimes \delta_{d_{\mu_n}} \in P(\mathbb{R}^n)$, *then* $\prod(v_1, v_2) = \{v_1 \otimes v_2\}$ *is a singleton, and the optimal transportation cost from* v_1 *to* v_2 *is the expression (8.2) and* $\sum_{i=1}^{n} \frac{k_i}{2} d_{\mu_i}^2$.

8.2 Jensen's Inequality

Jensen's inequality plays a significant role in almost all branches of mathematics as well as in other areas of science. For example, ecological physiologists and evolutionary biologists have applications of Jensen's inequality to model how plants and animals will respond to future changes in Earths climate, see [38, 40, 145, 169]. Biologists frequently deal with variation in physiological, environmental, and ecological processes by measuring how living systems perform under average conditions. However, performance at average conditions is rarely equal to average performance across a range of conditions. This fundamental property of nonlinear averaging known as 'Jensen's inequality' has huge consequences for all of biology.

Examples are the best way to understand the consequences of Jensen's inequality. First, we consider the simple case of a linear function. Oceanic phytoplankton are a major sink for the carbon dioxide released into the atmosphere by human activity, and the rate at which these unicellular organisms absorb CO_2 from seawater is directly proportional to the diameter of each cell, d (Figure 8.2) [17, 37]. In a hypothetical population, half the cells are small ($d = 8\ \mu m$) and half are large ($d = 16\mu m$). As shown in the graph 8.2, each of the small cells absorbs CO_2 at a rate of 320 $fmol\ s^{-1}$, and each of the large cells at a rate of 640 $fmol\ s^{-1}$. As a result, the average rate of absorption is 480 $fmol\ s^{-1}$ per cell. But consider the question in a different view. Average cell diameter in the population is $12\mu m$, and the response function of absorption versus diameter shows that a cell this size absorbs CO_2 at 480 $fmol\ s^{-1}$. In other words, for this linear function the average rate of absorption across cell sizes is equal to the rate of absorption by a cell of average size. In order to accurately measure the population's capacity to absorb society's waste CO_2, one may use average cell size.

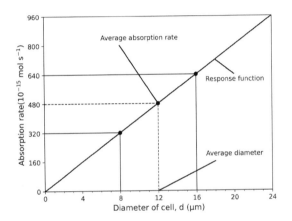

FIGURE 8.2: Graphical representation for the absorption rate of CO_2.

Now, we consider a case when the function is nonlinear. For instance, the relationship between metabolic rate and body temperature in ectotherms, see (Figure 8.3). In this hypothetical example, metabolic rate at any given temperature is twice that at a temperature 12^0C lower (i.e. $Q_{12} = 2$). Suppose that body temperature is constant at 24^0C during the 12 h of daylight, and constant at 12^0C at night, such that the average temperature is 18^0C.

To find out, the average metabolic rate we graphically took the average of day and night rates by drawing a line between the two rates and locating its midpoint. We used the midpoint because equal time is spent at the two temperatures. If more time were spent at one or the other, we would find the average by sliding the point proportionally along the line L_1 (Figure 8.3). Here, we notice that the average metabolic rate is greater than the metabolic rate at average temperature. That is, animals with variable body temperature use more energy per time than do animals with a constant body temperature, even when both have the same average temperature, a disparity that could have consequences for growth rate, reproductive output and foraging requirements.

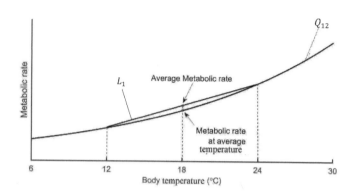

FIGURE 8.3: The graphical representation for the metabolic rate.

From the above graphical examples, we have two conclusions. First, we see why Jensen's inequality is true: only if a function is linear does the average response equal the response at average conditions. Second, the disparity between the average of a function and the function of the average depends on the functions shape. If the function is concave upward (e.g. metabolic rate as a function of temperature), the average of the function is greater than the function of the average.

Applications of Jensen's inequality in nature: Denny [36] discussed the importance of Jensen's inequality in explaining the properties of matter. For example, the London dispersion force is a result of nonlinear averaging at the atomic level and one important component of the suite of dipole interactions or van der waals forces that help keep materials together. He also discussed the significance role of Jensen's inequality in External fertilization, Species diversity, Predator-prey interactions and Enzyme kinetics.

8.3   Time Scales

The time scales calculus is the appropriate mathematics in which to construct the desired unification of discrete and continuous in a rigorous way. In principle, any system admitting discrete control injections within a traditionally continuous framework is amenable to a time scales approach. Poulsen *et al.* [137] discussed alternatives to real-time control that are made possible by the time scales calculus. Applications in finance, such as monetary policy, and in medicine, such as drug therapies, serve as illustrative examples. Researchers connected with pure mathematics have implemented time scales calculus in mathematical inequalities to unify discrete and continuous versions of inequalities. These integral inequalities are utilized in numerous areas for the boundedness, uniqueness, and so forth, of the solutions to integrodifferential equations.

The following example given by Rashid *et al.* [141], exhibits some practice related to the real world on time scale analysis.

Example 8.3.1. *Let $N(r)$ be the range of plant life of one specific type at time r in a certain vicinity. By experiments, Rashid et al. [141] recognized that N grows exponentially consistent with $N' = N$ during the months of May to August. At the beginning of September, all vegetation abruptly dies; however, the seeds remain in the ground and begin growing again at the start of May with N now being doubled. They modeled this case using the time scale*

$$T = \bigcup_{\lambda=0}^{\infty} [2\lambda, 2\lambda + 1],$$

where $r = 0$ is 1 May of the current year, $r = 1$ is 1 September of the current year, $r = 2$ is 1 May of the next year, $r = 3$ is 1 September of the subsequent year, and so on. We have

$$\delta(\lambda) = \begin{cases} 0, & 2\lambda \le r \le 2\lambda + 1 \\ 1, & r = 2\lambda + 1. \end{cases}$$

On $[2\lambda, 2\lambda+1]$ we have $N' = N$ that is, $N^{\Delta} = N$. Therefore, we have $N(2\lambda + 2) = 2N(2\lambda+1)$; that is, $\Delta N(2\lambda+1) = N(2\lambda+1)$ that is, $N^{\Delta} = N$ at $2\lambda+1$. As a result, N is a solution of the dynamic equation

$$N^{\Delta} = N.$$

Thus, if $N(0) = 1$ is given, N is exactly $e_1(.,0)$ on the time scale T.

8.4 Interval-Valued Functions

The importance of the study of interval-valued functions from a theoretical point of view as well as from their applications is well known. For instance, in topics as interval optimization, interval differential equations and random set, interval-valued functions and its properties play very important role. Particularly, integrable and differentiable interval-valued functions are tools essential in these topics. Since 1938, when Ostrowski [27] presented his famous inequality known as Ostrowski's inequality, many researchers have been working about and around it, in many different directions and with a lot of applications, in particular to numerical integration [44].

Cano *et al.* [26] derived the following Ostrowski type inequality for gH-differentiable (generalized Hukuhara differentiable) interval-valued functions.

Theorem 8.4.1. *Let $\xi : [a,b] \to K_c$ be a continuously gH-differentiable interval-valued function on $[a,b]$ with a finite number of switching points at precisely $a = t_0 < t_1 < ..t_{n-1} < t_n = b$ of $I = [a,b]$. Then, for $x \in [a,b]$ and $h \in [0,1]$, we have*

$$H\left(\int_a^b \xi(y)dy, \left(\xi(x)(1-\alpha) + \frac{\xi(a)+\xi(b)}{2}\alpha \right)(b-a) \right)$$
$$\leq \|\xi'\|_\infty \left[\left(\frac{(x-a)^2+(b-x)^2}{2} \right)(1-\alpha) + \frac{(b-a)^2}{2}\alpha \right], \qquad (8.5)$$

where $0 \leq x \leq 1$.

Using the inequality (8.5), Cano *et al.* [26] obtained an error estimation to quadrature rules for gH-differentiable interval-valued functions as follows:

8.4.1   Error estimation to quadrature rules Using Ostrowski type inequality for interval-valued functions

Consider the following class of interval-valued functions $\xi(x) = c\psi(x)$ where $\psi : T = [a,b] \to \mathbb{R}$ is a real function and $c = [\underline{c},\overline{c}] \in K_c$ is a fixed closed interval. Then,

$$\underline{\xi}(t) = \begin{cases} \underline{c}\psi(t), & \psi(t) \geq 0, \\ \overline{c}\psi(t), & \psi(t) < 0, \end{cases} \quad \overline{\xi}(t) = \begin{cases} \overline{c}\psi(t), & \psi(t) \geq 0, \\ \underline{c}\psi(t), & \psi(t) < 0. \end{cases} \qquad (8.6)$$

An important point is that the properties of the functions $\underline{\xi}$ and $\overline{\xi}$ in (8.6) are not necessarily inherited from ψ. For instance, it can be easily seen that if the functions $\underline{\xi}$ and $\overline{\xi}$ in (8.6) are both differentiable at t_0, then ψ is differentiable at t_0. The converse of this assertion is not true. Also, we can see that if ξ is gH-differentiable, then $\underline{\xi}$ and $\overline{\xi}$ in (8.6) are not necessarily differentiable. Cano *et al.* [26] obtained $\int_a^b \xi(x)dx = \left[\int_a^b \underline{\xi}(x)dx, \int_a^b \overline{\xi}(x)dx \right]$. For this, Cano *et al.* [26] obtained an approximation of $\int_a^b \underline{\xi}(x)dx$ and $\int_a^b \overline{\xi}(x)dx$ using classical quadrature rules (Trapezoidal, Simpson). However, since $\underline{\xi}$ and $\overline{\xi}$ in (8.6) are not necessarily differentiable, we do not have an error estimation to those quadrature rules, or the classical results of error estimation which assume differentiability. However, given an interval-valued function ξ, we can extend the quadrature rules for real functions to obtain an approximation of $\int_a^b \xi(x)dx$. For instance, for any partition $I_h : a = t_0 < t_1 < ..t_{n-1} < t_n = b$ of $I = [a, b]$ and any intermediate point vector $\phi = (\phi_1, \phi_2, ..., \phi_n)$ satisfying $\phi_i \in [t_{i-1}, t_i], (i = 1, 2, ..., n)$, we have the quadrature rule of Riemann-type defined by

$$A_R(\xi, I_h, \phi) = \sum_{i=1}^{n} \xi(\phi_i)h_i. \tag{8.7}$$

Note that A_R involves elementary interval arithmetic which can be obtained using software packages, see for instance [109, 111]. To obtain an error estimation for quadrature rules, Cano *et al.* [26] gave the following results.

Proposition 8.4.1. *Under the assumptions of Theorem 8.4.1, for any partition $I_h : a = t_0 < t_1 < ... < t_{n-1} < t_n = b$ of $I = [a, b]$ and any intermediate point vector $\phi = (\phi_1, \phi_2, ..., \phi_n)$ satisfying $\phi_i \in [t_{i-1}, t_i], (i = 1, 2, ..., n)$, we have*

(A)

$$R_R(\xi, I_h, \phi) = H\left(\int_a^b \xi(y)dy, A_R(\xi, I_h, \phi)\right)$$

$$\leq \|\xi'\|_\infty \sum_{i=1}^{n} \left[\frac{h_i^2}{4} + \left(\xi_i - \frac{t_{i-1} + t_i}{2}\right)^2\right],$$

where $h_i = t_i - t_{i-1}$ and A_R denotes the quadrature rules of Riemann-type defined by

$$A_R(\xi, I_h, \phi)) = \sum_{i=1}^{n} \xi(\phi_i)h_i.$$

When $\phi_i = \frac{t_{i-1} + t_i}{2}$, we have

$$R_R(\xi, I_h, \phi) \leq \frac{1}{4}\|\xi'\|_\infty \sum_{i=1}^{n} [h_i^2].$$

(B)

$$R_S(\xi, I_h, \phi) = H\left(\int_a^b \xi(y)dy, A_S(\xi, I_h, \phi)\right)$$

$$\leq \frac{1}{6}\|\xi'\|_\infty \sum_{i=1}^{n} \left[2(\phi_i - t_{i-1})^2 + (t_i - \phi_i)^2 + h_i^2\right],$$

where $h_i = t_i - t_{i-1}$ and A_S denotes the quadrature rules of Simpson-type defined by

$$A_S(\xi, I_h, \phi)) = \sum_{i=1}^{n} \frac{1}{6}[\xi(t_{i-1}) + 4\xi(\phi_i) + \xi(t_i))]h_i.$$

When $\phi_i = \frac{t_{i-1}+t_i}{2}$, we have

$$R_S(\xi, I_h, \phi) \leq \frac{1}{3}\|\xi'\|_{\infty} \sum_{i=1}^{n}[h_i^2].$$

(C)

$$R_T(\xi, I_h, \phi) = H\left(\int_{a}^{b} \xi(y)dy, A_T(\xi, I_h, \psi)\right) \leq \frac{1}{2}\|\xi'\|_{\infty} \sum_{i=1}^{n}[h_i^2],$$

where $h_i = t_i - t_{i-1}$ and A_T denotes the quadrature rules of Trapezoidal-type defined by

$$A_T(\xi, I_h, \phi) = \sum_{i=1}^{n} \frac{h_i}{2}[\xi(t_{i-1}) + \xi(t_i)].$$

Bibliography

[1] M Adil Khan, G Ali Khan, T Ali, T Batbold, and A Kiliçman. Further refinements of Jensen's type inequalities for the function defined on the rectangle. In *Abstract and Applied Analysis*, volume 2013. Hindawi, 2013.

[2] Gholamreza Alirezaei and Rudolf Mathar. On exponentially concave functions and their impact in information theory. In *2018 Information Theory and Applications Workshop (ITA)*, pages 1–10. IEEE, 2018.

[3] Linda JS Allen. *An introduction to stochastic processes with applications to biology*. CRC Press, 2010.

[4] Yanrong An, Guoju Ye, Dafang Zhao, and Wei Liu. Hermite- Hadamard type inequalities for interval (h1, h2)-convex functions. *Mathematics*, 7(5):436, 2019.

[5] Glen Douglas Anderson, Mavina Krishna Vamanamurthy, and Matti Vuorinen. Generalized convexity and inequalities. *Journal of Mathematical Analysis and Applications*, 335(2):1294–1308, 2007.

[6] Tadeusz Antczak. (p, r)-invex sets and functions. *Journal of Mathematical Analysis and Applications*, 263(2):355–379, 2001.

[7] Tadeusz Antczak. Relationships between pre-invex concepts. *Nonlinear Analysis: Theory, Methods & Applications*, 60(2):349–367, 2005.

[8] Emil Artin. *The gamma function*. Courier Dover Publications, 2015.

[9] Ferhan M Atici, Daniel C Biles, and Alex Lebedinsky. An application of time scales to economics. *Mathematical and Computer Modelling*, 43(7-8):718–726, 2006.

[10] Muhammad Uzair Awan, Muhammad Aslam Noor, Vishnu Narayan Mishra, and Khalida Inayat Noor. Some characterizations of general preinvex functions. *International Journal of Analysis and Applications*, 15(1):46–56, 2017.

[11] Muhammad Uzair Awan, Muhammad Aslam Noor, and Khalida Inayat Noor. Hermite- Hadamard inequalities for exponentially convex functions. *Appl. Math. Inf. Sci*, 12(2):405–409, 2018.

[12] Muhammad Uzair Awan, Muhammad Aslam Noor, Khalida Inayat Noor, and Farhat Safdar. On strongly generalized convex functions. *Filomat*, 31(18):5783–5790, 2017.

[13] Antonio Azócar, Kazimierz Nikodem, and Gari Roa. Fejér-type inequalities for strongly convex functions. *Annales Mathematicae Silesianae*, 26:43–54, 2012.

[14] Dumitru Baleanu, José António Tenreiro Machado, and Albert CJ Luo. *Fractional dynamics and control*. Springer Science & Business Media, 2011.

[15] CR Bector, S Chandra, S Gupta, and SK Suneja. Univex sets, functions and univex nonlinear programming. In *Generalized convexity*, pages 3–18. Springer, 1994.

[16] Adi Ben-Israel and Bertram Mond. What is invexity? *The ANZIAM Journal*, 28(1):1–9, 1986.

[17] HC Berg. Random walks in Biology. Princeton Univ. Press, Princeton, NJ. *BergRandom Walks in Biology1983*, 1983.

[18] Ajay Kumar Bhurjee and Geetanjali Panda. Efficient solution of interval optimization problem. *Mathematical Methods of Operations Research*, 76(3):273–288, 2012.

[19] Ajay Kumar Bhurjee and Geetanjali Panda. Multi-objective interval fractional programming problems: an approach for obtaining efficient solutions. *Opsearch*, 52(1):156–167, 2015.

[20] Ajay Kumar Bhurjee and Geetanjali Panda. Sufficient optimality conditions and duality theory for interval optimization problem. *Annals of Operations Research*, 243(1):335–348, 2016.

[21] Martin Bohner and Thomas Matthews. Ostrowski inequalities on time scales. *Journal of Inequalities in Pure and Applied Mathematics*, 9(1):8, 2008.

[22] Martin Bohner and Allan Peterson. *Dynamic equations on time scales: An introduction with applications*. Springer Science & Business Media, 2001.

[23] Martin Bohner and Nick Wintz. The Kalman filter for linear systems on time scales. *Journal of Mathematical Analysis and Applications*, 406(2):419–436, 2013.

[24] Huseyin Budak, Hasan Kara, Muhammad Aamir Ali, Sundas Khan, and Yuming Chu. Fractional Hermite- Hadamard-type inequalities for interval-valued co-ordinated convex functions. *Open Mathematics*, 19(1):1081–1097, 2021.

[25] Hüseyin Budak, Tuba Tunç, and Mehmet Sarikaya. Fractional Hermite-Hadamard-type inequalities for interval-valued functions. *Proceedings of the American Mathematical Society*, 148(2):705–718, 2020.

[26] Yurilev Chalco-Cano, Weldon A Lodwick, and W Condori-Equice. Ostrowski type inequalities and applications in numerical integration for interval-valued functions. *Soft Computing*, 19(11):3293–3300, 2015.

[27] Yurilev Chalco-Cano, Weldon A Lodwick, and W Condori-Equice. Ostrowski type inequalities and applications in numerical integration for interval-valued functions. *Soft Computing*, 19(11):3293–3300, 2015.

[28] Peng Chen, Alfio Quarteroni, and Gianluigi Rozza. Stochastic optimal robin boundary control problems of advection-dominated elliptic equations. *SIAM Journal on Numerical Analysis*, 51(5):2700–2722, 2013.

[29] Ping Chen and Wing-Sum Cheung. Higher dimensional Hermite–Hadamard inequality for semiconvex functions of rate (k 1, k 2,, kn) on the coordinates and optimal mass transportation. *Mathematical Methods in the Applied Sciences*, 44(17):12613–12629, 2021.

[30] Yu-Ming Chu, M Adil Khan, T Ullah Khan, and Tahir Ali. Generalizations of Hermite- Hadamard type inequalities for M T-convex functions. *Journal of Nonlinear Sciences and Applications*, 9(5):4305–4316, 2016.

[31] Yu-Ming Chu, Muhammad Adil Khan, Tahir Ali, and Sever Silvestru Dragomir. Inequalities for α-fractional differentiable functions. *Journal of Inequalities and Applications*, 2017(1):1–12, 2017.

[32] Yu-Ming Chu, Hua Wang, and Tie-Hong Zhao. Sharp bounds for the neuman mean in terms of the quadratic and second seiffert means. *Journal of Inequalities and Applications*, 2014(1):1–14, 2014.

[33] BD Craven. Invex functions and constrained local minima. *Bulletin of the Australian Mathematical society*, 24(3):357–366, 1981.

[34] Bruce D Craven and Barney M Glover. Invex functions and duality. *Journal of the Australian Mathematical Society*, 39(1):1–20, 1985.

[35] Gabriela Cristescu, Muhammad Aslam Noor, and Muhammad Uzair Awan. Bounds of the second degree cumulative frontier gaps of functions with generalized convexity. *Carpathian Journal of Mathematics*, pages 173–180, 2015.

[36] Mark Denny. The fallacy of the average: on the ubiquity, utility and continuing novelty of Jensen's inequality. *Journal of Experimental Biology*, 220(2):139–146, 2017.

[37] Mark Denny and Steven Gaines. Chance in biology. In *Chance in Biology*. Princeton, New Jersey: Princeton University Press, 2011.

[38] Michael E Dillon, George Wang, and Raymond B Huey. Global metabolic impacts of recent climate warming. *Nature*, 467(7316):704–706, 2010.

[39] Cristian Dinu. Hermite-Hadamard inequality on time scales. *Journal of Inequalities and Applications*, 2008:1–24, 2008.

[40] W Wesley Dowd, Felicia A King, and Mark W Denny. Thermal variation, thermal extremes and the physiological performance of individuals. *The Journal of experimental biology*, 218(12):1956–1967, 2015.

[41] Sever S Dragomir. Some majorisation type discrete inequalities for convex functions. *Mathematical Inequalities and Applications*, 7:207–216, 2004.

[42] Sever S Dragomir, Muhammad Iqbal Bhatti, Muhammad Iqbal, and Muhammad Muddassar. Some new Hermite-Hadamard 's type fractional integral inequalities. *Journal of Computational Analysis & Applications*, 18(4), 2015.

[43] Sever S Dragomir and CEM Pearce. Selected topics on Hermite-Hadamard inequalities and applications. Victoria University, 2000.

[44] Sever Silvestru Dragomir, Themistocles M Rassias, and Irene Aleksanova. *Ostrowski type inequalities and applications in numerical integration*. Springer, 2002.

[45] Silvestru S Dragomir. On the Hadamard's inequality for convex functions on the co-ordinates in a rectangle from the plane. *Taiwanese Journal of Mathematics*, pages 775–788, 2001.

[46] Silvestru Sever Dragomir and Ian Gomm. Some Hermite-Hadamard type inequalities for functions whose exponentials are convex. *Studia Universitatis Babes-Bolyai Mathematica*, 60(4):527–534, 2015.

[47] SS Dragomir and RP0938 Agarwal. Two inequalities for differentiable mappings and applications to special means of real numbers and to trapezoidal formula. *Applied Mathematics Letters*, 11(5):91–95, 1998.

[48] Ting-Song Du, Jia-Gen Liao, and Yu-Jiao Li. Properties and integral inequalities of Hadamard- Simpson type for the generalized (s, m)-preinvex functions. *Journal of Nonlinear Sciences and Applications*, 9(5):3112–3126, 2016.

[49] J Favard. Sur les valeurs moyennes. *Bulletin des Sciences Mathematiques*, 57(2):54–64, 1933.

[50] L Fejér. Uber die Fourierreihen, II, Math. *Naturwiss. Anz Ungar. Akad. Wiss*, 24:369–390, 1906.

[51] Haoliang Fu, Muhammad Shoaib Saleem, Waqas Nazeer, Mamoona Ghafoor, and Peigen Li. On Hermite-Hadamard type inequalities for n-polynomial convex stochastic processes. *AIMS Mathematics*, 6(6):6322–6339, 2021.

[52] Ladislas Fuchs. A new proof of an inequality of Hardy-Littlewood-Pólya. *Mat. Tidsskr. B*, pages 53–54, 1947.

[53] L Gonzales, Jesús Materano, and MV Lopez. Ostrowski-type inequalities via hconvex stochastic processes. *JP Journal of Mathematical Sciences*, 16(2):15–29, 2016.

[54] M. E. Gordji, M. R. Delavar, and S. S. Dragomir. Some inequalities related to η-convex functions. *RGMIA*, 18, 2015.

[55] M Eshaghi Gordji, M Rostamian Delavar, and M De La Sen. On ϕ-convex functions. *Journal of Mathematical Inequalities*, 10(1):173–183, 2016.

[56] Jacques Hadamard. Étude sur les propriétés des fonctions entières et en particulier d'une fonction considérée par riemann. *Journal de mathématiques pures et appliquées*, pages 171–216, 1893.

[57] Morgan A Hanson. On sufficiency of the Kuhn- Tucker conditions. *Journal of Mathematical Analysis and Applications*, 80(2):545–550, 1981.

[58] Godfrey Harold Hardy, John Edensor Littlewood, George Pólya, György Pólya, et al. *Inequalities*. Cambridge, England: Cambridge University Press, 1952.

[59] Ch Hermite. Sur deux limites d'une intégrale définie. *Mathesis*, 3(1):1–82, 1883.

[60] Stefan Hilger. Ein makettenkalkl mit anwendung auf zentrumsmannigfaltigkeiten. *Wurzburg: Universtat Wurzburg*, 1988.

[61] Roger A Horn and Charles R Johnson. *Matrix analysis*. Cambridge, England: Cambridge University Press, 2012.

[62] Henryk Hudzik and Lech Maligranda. Some remarks on s-convex functions. *Aequationes Mathematicae*, 48(1):100–111, 1994.

[63] Alawiah Ibrahim. On strongly h-convex stochastic processes. *Journal of Quality Measurement and Analysis JQMA*, 16(2):243–251, 2020.

[64] Imdat Işcan. Hermite-Hadamard's inequalities for preinvex functions via fractional integrals and related fractional inequalities. *arXiv preprint arXiv:1204.0272*, 2012.

[65] Imdat Işcan. Hermite-Hadamard-Fejér type inequalities for convex functions via fractional integrals. *arXiv preprint arXiv:1404.7722*, 2014.

[66] İmdat İşcan. Hermite-Hadamard type inequalities for harmonically convex functions. *Hacettepe Journal of Mathematics and statistics*, 43(6):935–942, 2014.

[67] Mrinal Jana and Geetanjali Panda. Solution of nonlinear interval vector optimization problem. *Operational Research*, 14(1):71–85, 2014.

[68] Wei-Dong Jiang, Da-Wei Niu, Yun Hua, and Feng Qi. Generalizations of Hermite–Hadamard inequality to n-time differentiable functions which are s-convex in the second sense. Analysis, 32: 209–220, 2012.

[69] Chahn Yong Jung, Muhammad Shoaib Saleem, Shamas Bilal, Waqas Nazeer, and Mamoona Ghafoor. Some properties of η-convex stochastic processes [j]. *AIMS Mathematics*, 6(1):726–736, 2021.

[70] Hasan Kara, Muhammad Aamir Ali, and Hüseyin Budak. Hermite-Hadamard-type inequalities for interval-valued coordinated convex functions involving generalized fractional integrals. *Mathematical Methods in the Applied Sciences*, 44(1):104–123, 2021.

[71] Vildan Karahan and Nurgül Okur. Hermite-Hadamard type inequalities for convex stochastic processes on n-coordinates. *Turkish Journal of Mathematics and Computer Science*, 10:256–262, 2018.

[72] S Karamardian. The nonlinear complementarity problem with applications, part 2. *Journal of Optimization Theory and Applications*, 4(3):167–181, 1969.

[73] Jovan Karamata. Sur une inégalité relative aux fonctions convexes. *Publications de l'Institut mathematique*, 1(1):145–147, 1932.

[74] M Adil Khan, T Ali, Sever S Dragomir, and MZ Sarikaya. Hermite–Hadamard type inequalities for conformable fractional integrals. *Revista de la Real Academia de Ciencias Exactas, Físicas y Naturales. Serie A. Matemáticas*, 112(4):1033–1048, 2018.

[75] M Adil Khan, T Ali, and T Ullah Khan. Hermite-Hadamard type inequalities with applications. *Fasciculi Mathematici*, 59(1):57–74, 2017.

[76] M Adil Khan, T Ali, A Kılıçman, and Q Din. Refinements of Jensen's inequality for convex functions on the co-ordinates in a rectangle from the plane. *Filomat*, 30(3):803–814, 2016.

[77] M Adil Khan, Sadia Khalid, and J Pecaric. Refinements of some majorization type inequalities. *Journal of Mathematical Inequalities*, 7(1):73–92, 2013.

[78] M Adil Khan, G Ali Khan, T Ali, and Adem Kilicman. On the refinement of Jensen's inequality. *Applied Mathematics and Computation*, 262:128–135, 2015.

[79] M Adil Khan, Jamroz Khan, and Josip Pecaric. Generalization of Jensen's and Jensen- Steffensen's inequalities by generalized majorization theorem. *Journal of Mathematical Inequalities*, 11(4):1049–1074, 2017.

[80] M Adil Khan, Naveed Latif, and J Pecaric. Generalization of majorization theorem. *Journal of Mathematical Inequalities*, 9(3):847–872, 2015.

[81] M Adil Khan, Naveed Latif, J Pecaric, and I Peric. On majorization for matrices. *Mathematica Balkanica*, 2013.

[82] Muhammad Adil Khan, Fayaz Alam, and Syed Zaheer Ullah. Majorization type inequalities for strongly convex functions. Turkish Journal of Inequalities 3(2):62–78, 2019.

[83] Muhammad Adil Khan, Yousaf Khurshid, Tahir Ali, and Nasir Rehman. Inequalities for three times differentiable functions. *Punjab University Journal of Mathematics*, 48(2), 2020.

[84] Muhammad Adil Khan, Syed Zaheer Ullah, and Yu-Ming Chu. The concept of coordinate strongly convex functions and related inequalities. *Revista de la Real Academia de Ciencias Exactas, Físicas y Naturales. Serie A. Matemáticas*, 113(3):2235–2251, 2019.

[85] Anatoliĭ Aleksandrovich Kilbas, Hari M Srivastava, and Juan J Trujillo. *Theory and applications of fractional differential equations*, volume 204. Amsterdam, Netherlands: Elsevier, 2006.

[86] Uur S Kirmaci. Inequalities for differentiable mappings and applications to special means of real numbers and to midpoint formula. *Applied Mathematics and Computation*, 147(1):137–146, 2004.

[87] Dawid Kotrys. Hermite– Hadamard inequality for convex stochastic processes. *Aequationes Mathematicae*, 83(1-2):143–151, 2012.

[88] Dawid Kotrys. Remarks on strongly convex stochastic processes. *Aequationes Mathematicae*, 86(1-2):91–98, 2013.

[89] Daniel Kuhn. *Generalized bounds for convex multistage stochastic programs*, volume 548. Germany: Springer Science & Business Media, 2006.

[90] Young Chel Kwun, Muhammad Shoaib Saleem, Mamoona Ghafoor, Waqas Nazeer, and Shin Min Kang. Hermite– Hadamard-type inequalities for functions whose derivatives are η-convex via fractional integrals. *Journal of Inequalities and Applications*, 2019(1):1–16, 2019.

[91] Kin Keung Lai, Jaya Bisht, Nidhi Sharma, and Shashi Kant Mishra. Hermite-Hadamard-type fractional inclusions for interval-valued preinvex functions. *Mathematics*, 10(2):264, 2022.

[92] Kin Keung Lai, Jaya Bisht, Nidhi Sharma, and Shashi Kant Mishra. Hermite-Hadamard type integral inequalities for the class of strongly convex functions on time scales. *Journal of Mathematical Inequalities*, 16(3):975–991, 2022.

[93] Kin Keung Lai, Shashi Kant Mishra, Jaya Bisht, and Mohd Hassan. Hermite–Hadamard type inclusions for interval-valued coordinated preinvex functions. *Symmetry*, 14(4):771, 2022.

[94] Muhammad Amer Latif. New Hermite–Hadamard type integral inequalities for GA-convex functions with applications. *Analysis*, 34(4):379–389, 2014.

[95] N Latif, J Pečarić, and I Perić. On majorization, favard and berwald inequalities. *Annals of Functional Analysis*, 2(1):31–50, 2011.

[96] Lifeng Li, Sanyang Liu, and Jianke Zhang. On interval-valued invex mappings and optimality conditions for interval-valued optimization problems. *Journal of Inequalities and Applications*, 2015(1):1–19, 2015.

[97] Gui-Hua Lin and Masao Fukushima. Some exact penalty results for nonlinear programs and mathematical programs with equilibrium constraints. *Journal of Optimization Theory and Applications*, 118(1):67–80, 2003.

[98] Kui Liu, JinRong Wang, and Donal O'Regan. On the Hermite–Hadamard type inequality for ψ-Riemann–Liouville fractional integrals via convex functions. *Journal of Inequalities and Applications*, 2019(1):1–10, 2019.

[99] Xuelong Liu, Gouju Ye, Dafang Zhao, and Wei Liu. Fractional Hermite–Hadamard type inequalities for interval-valued functions. *Journal of Inequalities and Applications*, 2019(1):1–11, 2019.

[100] Vasile Lupulescu. Hukuhara differentiability of interval-valued functions and interval differential equations on time scales. *Information Sciences*, 248:50–67, 2013.

[101] Vasile Lupulescu. Fractional calculus for interval-valued functions. *Fuzzy Sets and Systems*, 265:63–85, 2015.

[102] Lech Maligranda, Josip E Pecaric, and Lars Erik Persson. Weighted favard and berwald inequalities. *Journal of Mathematical Analysis and Applications*, 190(1):248–262, 1995.

[103] Albert W Marshall, Ingram Olkin, and Barry C Arnold. *Inequalities: theory of majorization and its applications*, volume 143. Springer New York Dordrecht Heidelberg London: Springer, 1979.

[104] Marian Matłoka. On some Hadamard-type inequalities for (h 1, h 2)-preinvex functions on the co-ordinates. *Journal of Inequalities and Applications*, 2013(1):1–12, 2013.

[105] Nelson Merentes and Kazimierz Nikodem. Remarks on strongly convex functions. *Aequationes Mathematicae*, 80(1-2):193–199, 2010.

[106] SK Mishra and Nidhi Sharma. On strongly generalized convex functions of higher order. *Mathematical Inequalities & Applications*, 22(1):111–121, 2019.

[107] DS Mitrinović and IB Lacković. Hermite and convexity. *Aequationes Mathematicae*, 28(1):229–232, 1985.

[108] SR Mohan and SK Neogy. On invex sets and preinvex functions. *Journal of Mathematical Analysis and Applications*, 189(3):901–908, 1995.

[109] Ramon E Moore. *Methods and applications of interval analysis.* Philadelphia: SIAM, 1979.

[110] Ramon E Moore, R Baker Kearfott, and Michael J Cloud. *Introduction to interval analysis.* Philadelphia: SIAM, 2009.

[111] RE Moore. Computational functional analysis(book). *Chichester, West Sussex, England/New York, Ellis Horwood, Ltd./Halsted Press, 1985, 156 p*, 1985.

[112] S Mubeen and GM Habibullah. k-fractional integrals and application. *International Journal of Contemporary Mathematical Sciences*, 7(2):89–94, 2012.

[113] Constantin P. Niculescu. Convexity according to the geometric mean. *Mathematical Inequalities & Applications*, 3(2):155–167, 2000.

[114] Constantin P. Niculescu and Lars-Erik Persson. *Convex functions and their applications*, volume 23 of *CMS Books in Mathematics/Ouvrages de Mathématiques de la SMC*. Springer, New York, 2006. A contemporary approach.

[115] Marek Niezgoda and Josip Pečarić. Hardy– Littlewood– Pólya-type theorems for invex functions. *Computers & Mathematics with Applications*, 64(4):518–526, 2012.

[116] Kazimierz Nikodem. On convex stochastic processes. *Aequationes Mathematicae*, 20(1):184–197, 1980.

[117] Kazimierz Nikodem and Zsolt Pales. Characterizations of inner product spaces by strongly convex functions. *Banach Journal of Mathematical Analysis*, 5(1):83–87, 2011.

[118] M Aslam Noor. Hermite-Hadamard integral inequalities for log-preinvex functions. *Journal of Numerical Analysis and Approximation Theory*, 2(2):126–131, 2007.

[119] M Aslam Noor. On Hadamard integral inequalities involving two log-preinvex functions. *Journal of Inequalities in Pure and Applied Mathematics*, 8(3):1–14, 2007.

[120] M Aslam Noor. Hadamard integral inequalities for product of two preinvex functions. In *Hadamard integral inequalities for product of two preinvex functions*. 14:167–173, 2009.

[121] MA Noor, KI Noor, and S Iftikhar. Integral inequalities for differentiable relative harmonic preinvex functions (survey). *TWMS Journal of Pure and Applied Mathematics*, 7(1):3–19, 2016.

[122] Muhammad Aslam Noor and Khalida Inayat Noor. On exponentially convex functions. *Journal of Orissa Mathematical Society ISSN*, 975:2323, 2019.

[123] Muhammad Aslam Noor and Khalida Inayat Noor. Strongly exponentially convex functions. *UPB Scientific Bulletin Applied Mathematics, Series A*, 81(4):75–84, 2019.

[124] Muhammad Aslam Noor, Khalida Inayat Noor, and Muhammad Uzair Awan. Geometrically relative convex functions. *Applied Mathematics & Information Sciences*, 8(2):607–616, 2014.

[125] Muhammad Aslam Noor, Khalida Inayat Noor, and Muhammad Uzair Awan. Some inequalities for geometrically-arithmetically h-convex functions. *Creative Mathematics and Informatics*, 23(1):91–98, 2014.

[126] Muhammad Aslam Noor, Khalida Inayat Noor, and Sabah Iftikhar. Hermite-Hadamard inequalities for harmonic preinvex functions. *Saussurea*, 6(1):34–53, 2016.

[127] Muhammad Aslam Noor, Khalida Inayat Noor, and Farhat Safdar. Generalized geometrically convex functions and inequalities. *Journal of Inequalities and Applications*, pages Paper No. 202, 19, 2017.

[128] Sofian Obeidat and Muhammad Amer Latif. Weighted version of Hermite- Hadamard type inequalities for geometrically quasi-convex functions and their applications. *Journal of Inequalities and Applications*, pages Paper No. 307, 11, 2018.

[129] Nurgul Okur and Rovshan Aliyev. Some Hermite–Hadamard type integral inequalities for multidimensional general preinvex stochastic processes. *Communications in Statistics-Theory and Methods*, 50: 3338–3351, 2020.

[130] BG Pachpatte. On some inequalities for convex functions. *RGMIA Research Report Collection*, 6(1):1–9, 2003.

[131] Soumik Pal and Ting-Kam Leonard Wong. Exponentially concave functions and a new information geometry. *The Annals of Probability*, 46(2):1070–1113, 2018.

[132] Jaekeun Park. Inequalities of Hermite- Hadamard- Fejér type for convex functions via fractional integrals. *International Journal of Mathematical Analysis*, 8(59):2927–2937, 2014.

[133] Josip E Peajcariaac and Yung Liang Tong. *Convex functions, partial orderings, and statistical applications*. Cambridge, Massachusetts: Academic Press, 1992.

[134] Bożena Piatek. On the Riemann integral of set-valued functions. *Zeszyty Naukowe. Matematyka Stosowana/Politechnika Śląska*, 2012.

[135] Rita Pini. Invexity and generalized convexity. *Optimization*, 22(4):513–525, 1991.

[136] Boris Teodorovich Polyak. Existence theorems and convergence of minimizing sequences for extremal problems with constraints. In *Doklady Akademii Nauk*, volume 166, pages 287–290. Russia: Russian Academy of Sciences, 1966.

[137] Dylan Poulsen, Ian Gravagne, and John M Davis. Is deterministic real time control always necessary? A time scales perspective. In *Dynamic Systems and Control Conference*, volume 46209, page V003T41A002. New York: American Society of Mechanical Engineers, 2014.

[138] Feng Qi, Zong-Li Wei, and Qiao Yang. Generalizations and refinements of Hermite- Hadamard's inequality. *The Rocky Mountain Journal of Mathematics*, 35(1):235–251, 2005.

[139] Feng Qi and Bo-Yan Xi. Some Hermite- Hadamard type inequalities for geometrically quasi-convex functions. *Indian Academy of Sciences. Proceedings. Mathematical Sciences*, 124(3):333–342, 2014.

[140] Saima Rashid, Muhammad Aslam Noor, and Khalida Inayat Noor. Some generalize Riemann- Liouville fractional estimates involving functions having exponentially convexity property. *Journal of Mathematics*, 51(11):01–15, 2019.

[141] Saima Rashid, Muhammad Aslam Noor, Khalida Inayat Noor, Farhat Safdar, and Yu-Ming Chu. Hermite- Hadamard type inequalities for the class of convex functions on time scale. *Mathematics*, 7(10):956, 2019.

[142] Saima Rashid, Muhammad Aslam Noor, Ahmet Ocak Akdemir, and Khalida Inayat Noor. Some fractional estimates of upper bounds involving functions having exponential convexity property. *TWMS Journal of Applied and Engineering Mathematics*, 11(1): 20–33, 2021.

[143] M Rostamian Delavar and M De La Sen. Some Hermite- Hadamard-Fejér type integral inequalities for differentiable-convex functions with applications. *Journal of Mathematics*, 2017, 2017.

[144] Priyanka Roy and Geetanjali Panda. Expansion of generalized Hukuhara differentiable interval valued function. *New Mathematics and Natural Computation*, 15(03):553–570, 2019.

[145] Jonathan J Ruel and Matthew P Ayres. Jensen's inequality predicts effects of environmental variation. *Trends in Ecology & Evolution*, 14(9):361–366, 1999.

[146] M. A. Noor, S. Rashid and K. I. Noor. Trapezoid type inequalities for exponentially convex functions via generalized fractional operators. *preprint*, 2019.

[147] Elżbieta Sadowska. Hadamard inequality and a refinement of jensen inequality for setvalued functions. *Results in Mathematics*, 32:332–337, 1997.

[148] Filippo Santambrogio. Optimal transport for applied mathematicians. *Birkäuser, NY*, 55(58-63):94, 2015.

[149] Mehmet Zeki Sarikaya, Erhan Set, Hatice Yaldiz, and Nagihan Başak. Hermite- Hadamard's inequalities for fractional integrals and related fractional inequalities. *Mathematical and Computer Modelling*, 57(9-10):2403–2407, 2013.

[150] Issai Schur. Uber eine klasse von mittelbildungen mit anwendungen auf die determinantentheorie. *Sitzungsberichte der Berliner Mathematischen Gesellschaft*, 22(9-20):51, 1923.

[151] John Seiffertt. Adaptive resonance theory in the time scales calculus. *Neural Networks*, 120:32–39, 2019.

[152] John Seiffertt and Donald C Wunsch. A quantum calculus formulation of dynamic programming and ordered derivatives. In *2008 IEEE International Joint Conference on Neural Networks (IEEE World Congress on Computational Intelligence)*, pages 3690–3695. Hong Kong, China: IEEE, 2008.

[153] Nidhi Sharma, Jaya Bisht, and SK Mishra. Hermite–Hadamard type inequalities for functions whose derivatives are strongly η–convex via

fractional integrals. In *Optimization, Variational Analysis and Applications: IFSOVAA-2020*, Springer Proceedings in Mathematics and Statistics, 355:83–102, Springer, Singapore.

[154] Nidhi Sharma, Jaya Bisht, SK Mishra, and A Hamdi. Some majorization integral inequalities for functions defined on rectangles via strong convexity. *Journal of Inequalities and Special Functions*, 10:21–34, 2019.

[155] Nidhi Sharma, Rohan Mishra, and Abdelouahed Hamdi. Hermite-Hadamard type integral inequalities for multidimensional general h-harmonic preinvex stochastic processes. *Communications in Statistics-Theory and Methods*, 51(19):6719–6740, 2022.

[156] Nidhi Sharma, Rohan Mishra, and Abdelouahed Hamdi. On strongly generalized convex stochastic processes. *Communications in Statistics-Theory and Methods*, pages 1–16, 2022.

[157] Nidhi Sharma, S. K. Mishra, and R. N. Mohapatra. Extensions of different type parameterized inequalities for generalized (m; h)-preunivex mappings via k-fractional integrals. *Communications on Applied Nonlinear Analysis*, 27(4):65–100, 2020.

[158] Nidhi Sharma, Shashi Kant Mishra, and Abdelouahed Hamdi. Hermite-Hadamard type inequality for ψ-Riemann-Liouville fractional integrals via preinvex functions. *International Journal of Nonlinear Analysis and Applications*, 13(1):3333–3345, 2022.

[159] Nidhi Sharma, Shashi Kant Mishra, and A Hamdi. A weighted version of Hermite-Hadamard type inequalities for strongly GA-convex functions. *International Journal of Advances in Applied Sciences*, 7(3):113–118, 2020.

[160] Nidhi Sharma, Sanjeev Kumar Singh, Shashi Kant Mishra, and Abdelouahed Hamdi. Hermite–Hadamard-type inequalities for interval-valued preinvex functions via Riemann–Liouville fractional integrals. *Journal of Inequalities and Applications*, 2021(1):1–15, 2021.

[161] Qin Sheng, M Fadag, Johnny Henderson, and John M Davis. An exploration of combined dynamic derivatives on time scales and their applications. *Nonlinear Analysis: Real World Applications*, 7(3):395–413, 2006.

[162] Fangfang Shi, Guoju Ye, Dafang Zhao, and Wei Liu. Some fractional Hermite– Hadamard type inequalities for interval-valued functions. *Mathematics*, 8(4):534, 2020.

[163] Ye Shuang, Hong-Ping Yin, and Feng Qi. Hermite- Hadamard type integral inequalities for geometric-arithmetically s-convex functions. *Analysis. International Mathematical Journal of Analysis and its Applications*, 33(2):197–208, 2013.

[164] Kazimierz Sobczyk. *Stochastic differential equations with applications to physics and engineering*, volume 40. Germany: Springer Science & Business Media, 2013.

[165] Saeeda Fatima Tahir, Muhammad Mushtaq, and Muhammad Muddassar. A new interpretation of Hermite- Hadamards type integral inequalities by the way of time scales. *Journal of Computational Analysis and Applications*, 26:223–233, 2019.

[166] Muharrem Tomar, Erhan Set, and Nurgül Okur Bekar. Hermite-Hadamard type inequalities for strongly-log-convex stochastic processes. *Journal of Global Engineering Studies*, 1(53-61), 2014.

[167] Mevlüt Tunc, Esra Gov, and Ümmügülsüm Şanal. On tgs-convex function and their inequalities. *Facta Universitatis, Series: Mathematics and Informatics*, 30(5):679–691, 2015.

[168] Mevlut Tunc, Yusuf Subas, and Ibrahim Karabayir. On some Hadamard type inequalities for M T-convex functions. *International Journal of Open Problems in Computer Science and Mathematics*, 238(1393):1–24, 2013.

[169] David A Vasseur, John P DeLong, Benjamin Gilbert, Hamish S Greig, Christopher DG Harley, Kevin S McCann, Van Savage, Tyler D Tunney, and Mary I O'Connor. Increased temperature variation poses a greater risk to species than climate warming. *Proceedings of the Royal Society B: Biological Sciences*, 281(1779):20132612, 2014.

[170] Cédric Villani. *Optimal transport: old and new*, volume 338. Verlag Berlin Heidelberg: Springer, 2009.

[171] JinRong Wang and Michal Fečkan. *Fractional Hermite-Hadamard Inequalities*. de Gruyter, 2018.

[172] T Weir and V Jeyakumar. A class of nonconvex functions and mathematical programming. *Bulletin of the Australian Mathematical Society*, 38(2):177–189, 1988.

[173] Shanhe Wu, Muhammad Adil Khan, Abdul Basir, and Reza Saadati. Some majorization integral inequalities for functions defined on rectangles. *Journal of Inequalities and Applications*, 2018(1):1–13, 2018.

[174] Shanhe Wu, Muhammad Adil Khan, and Hidayat Ullah Haleemzai. Refinements of majorization inequality involving convex functions via Taylor's theorem with mean value form of the remainder. *Mathematics*, 7(8):663, 2019.

[175] Bo-Yan Xi and Feng Qi. Some integral inequalities of Hermite-Hadamard type for convex functions with applications to means. *Journal of Function Spaces and Applications*, 2012, 2012.

[176] Qiang Xiao, Tingwen Huang, and Zhigang Zeng. Stabilization of nonautonomous recurrent neural networks with bounded and unbounded delays on time scales. *IEEE Transactions on Cybernetics*, 50(10):4307–4317, 2019.

[177] Yingxia Yang, Muhammad Shoaib Saleem, Mamoona Ghafoor, and Muhammad Imran Qureshi. Fractional integral inequalities of Hermite–Hadamard type for differentiable generalized-convex functions. *Journal of Mathematics*, 2020, 2020.

[178] Zeyang Yin, Jianjun Luo, and Caisheng Wei. Novel adaptive saturated attitude tracking control of rigid spacecraft with guaranteed transient and steady-state performance. *Journal of Aerospace Engineering*, 31(5):04018062, 2018.

[179] Syed Zaheer Ullah, Muhammad Adil Khan, and Yu-Ming Chu. Majorization theorems for strongly convex functions. *Journal of Inequalities and Applications*, 2019(1):1–13, 2019.

[180] Syed Zaheer Ullah, Muhammad Adil Khan, Zareen Abdulhameed Khan, and Yu-Ming Chu. Integral majorization type inequalities for the functions in the sense of strong convexity. *Journal of Function Spaces*, 2019, 2019.

[181] Tian-Yu Zhang, Ai-Ping Ji, and Feng Qi. Some inequalities of Hermite-Hadamard type for G A-convex functions with applications to means. *Le Matematiche*, 68(1):229–239, 2013.

[182] Yao Zhang, Ting-Song Du, Hao Wang, Yan-Jun Shen, and Artion Kashuri. Extensions of different type parameterized inequalities for generalized (m, h)-preinvex mappings via k-fractional integrals. *Journal of Inequalities and Applications*, 2018(1):1–30, 2018.

[183] Dafang Zhao, Muhammad Aamir Ali, and Ghulam Murtaza. On the Hermite- Hadmard inequalities for interval-valued coordinated convex functions. *Advances in Difference Equations*, 2020.

[184] Dafang Zhao, Tianqing An, Guoju Ye, and Wei Liu. Chebyshev type inequalities for interval-valued functions. *Fuzzy Sets and Systems*, 396:82–101, 2020.

[185] Dafang Zhao, Tianqing An, Guoju Ye, and Delfim FM Torres. On Hermite- Hadamard type inequalities for harmonical h-convex interval-valued functions. *arXiv preprint arXiv:1911.06900*, 2019.

[186] Dafang Zhao, Guohui Zhao, Guoju Ye, Wei Liu, and Silvestru Sever Dragomir. On Hermite- Hadamard-type inequalities for coordinated h-convex interval-valued functions. *Mathematics*, 9(19):2352, 2021.

Index

9781032526324